Infrared and Raman Spectra of Crystals

Infrared and Raman Spectra of Crystals

GEORGE TURRELL

Professeur Associé,
Laboratoire de Spectroscopie Infrarouge,
Faculté des Sciences de Bordeaux,
France

1972

ACADEMIC PRESS · LONDON AND NEW YORK

ACADEMIC PRESS INC. (LONDON) LTD.
24/28 Oval Road,
London NW1

United States Edition published by
ACADEMIC PRESS INC.
111 Fifth Avenue
New York, New York 10003

Library of Congress Catalog Card Number: 72–175834
ISBN: 0–12–705050–7

PRINTED IN GREAT BRITAIN BY
ROYSTAN PRINTERS LIMITED
Spencer Court, 7 Chalcot Road
London NW1

PREFACE

The present work is the outgrowth of a series of lectures on the theory of vibrational spectroscopy which I presented at the University of Bordeaux in 1968–69 and 1969–70. In the search for suitable textbooks for these courses I found that, although a number of authors had considered the dynamics of isolated molecules and their spectroscopic implications, none had treated the vibrations of solids. In "Infrared and Raman Spectra of Crystals", I have attempted to fill this gap and, thus, to provide a guide to the interpretation of the vibrational spectra of crystalline solids.

The first two chapters review the dynamics and symmetry of molecular vibrations and introduce the matrix formulation which is used throughout the book. For readers who are familiar with the theory of molecular vibrations at the level of, say, the book by Wilson, Decius and Cross, the first two chapters merely serve to define the notation which they will encounter later. However, those who are studying the theory of vibrational spectroscopy for the first time should find in the first two chapters the basic material necessary for the understanding of the later chapters. Little mathematical background is needed beyond elementary differential equations and the fundamentals of matrix algebra.

The dynamics of crystal lattices and the theory of space groups, which are developed in Chapters 3 and 4, provide the crystallographic basis for the interpretation of the vibrational spectra of solids. The so-called factor-group analysis (sometimes attributed to Bhagavantam) is derived from first principles and illustrated using examples of both molecular and ionic crystals. The Hermann–Mauguin and Schönflies systems of group notation are used simultaneously to aid crystallographers and molecular spectroscopists to understand each other. The work continues with the introduction of classical electromagnetic theory to treat the interaction of crystal vibrations with electromagnetic radiation.

In the final three chapters, applications of vibrational spectroscopy to the determination of potential constants and structures and to the study of polymer systems and impure crystals are considered. The methods presented for the vibrational analysis of polymer chains and crystalline polymers should prove useful to workers in the fields of polymer science and biochemistry, where spectroscopy can provide a considerable amount of structural information. The final chapter emphasizes the spectroscopic similarity of "doped" crystals and matrix-isolation systems. In both cases the vibrational spectra are characterized by the effects of guest-host-lattice interactions and "localized" lattice vibrations. The addition of a number of tables as appendices should increase the practical use of this work in the interpretation of the vibrational spectra of crystalline solids.

The material presented in this book was drawn from many sources, including books on molecular spectroscopy, solid-state physics, and optics, as well as numerous articles in the scientific journals. Those references from which illustrations or diagrams were taken directly, and for which copyright releases have been obtained, are listed separately.

I should like to take this opportunity to thank my many colleagues and students who have aided me in the preparation of this book. I am particularly indebted to Professor M.-L. Josien of the University of Paris, who invited me to France, and to Professor J. Lascombe and his group in the "Laboratoire de Spectroscopie Infrarouge" at the University of Bordeaux, who provided the facilities, help, and encouragement which were essential to the completion of this work. My many friendly and fruitful collaborations with Professor P. Hagenmuller and the members of his "Service de Chimie Minérale Structurale" at Bordeaux are most gratefully acknowledged. The manuscript was painstakingly typed by Mlles N. Jessel, L. Labadie and F. Feru.

The entire text was read by Professor G. W. Wilkinson of King's College, London, who made many helpful comments on the presentation and pointed out a number of errors. The editorial staff of Academic Press Inc. (London) Ltd. provided competent and friendly advice all along the route from manuscript to printed volume.

Finally, and most important of all, is my profound gratitude to a special former student—my wife, Sylvia—who provided untiring aid in the preparation of the manuscript and the endless patience to see me through to the last page.

GEORGE TURRELL

March, 1972
Université Nationale du Zaïre,
Faculté des Sciences,
Kisangani,
République du Zaïre

Present address.
Département de Chimie,
Université de Montréal,
Montréal, P.Q.,
Canada.

ACKNOWLEDGEMENTS

I am indebted to the following publishers for permission to use copyrighted material: Institute of Physics and the Physical Society (*Proceedings*): Fig. 11, Chap. 3; Cambridge University Press: Fig. 7, Chap. 4; American Institute of Physics (*J. Chem. Phys.*): Fig. 22, Chap. 4, Figs. 4, 6, Chap. 6, Figs. 3, 4, 13, Chap. 7, Figs. 2–4, 15, Chap. 8, (*Phys. Rev.*): Figs. 12, 13, Chap. 5, Fig. 10, Chap. 8; Elsevier Publishing Co.: Fig. 4, Chap. 5, (*J. Mol. Structure*): Fig. 13, Chap. 6; Pergamon Press Ltd. (*Spectrochim. Acta*): Figs. 6, 10, Chap. 5, Figs. 11, 12, Chap. 7; Academic Press Inc. New York (*J. Mol. Spectry.*): Fig. 11, Chap. 6, Figs. 8–10, 15–17, 23, Chap. 7; Springer–Verlag New York Inc.: Fig. 8, Chap. 5; W. H. Freeman and Co.: Fig. 8, Chap. 6; John Wiley and Sons Inc. (*J. Polymer Sci.*): Figs. 19–21, Chap. 7; Academia of the Czechoslovak Academy of Sciences: Figs. 7, 8, Chap. 8; The Royal Society (*Proceedings*): Fig. 13, Chap. 8; Taylor and Francis Ltd. (*Adv. Physics*): Table 1, Appendix G; McGraw-Hill Book Co.: Fig. 1 and Table 1, Appendix A; Institute of Petroleum: Fig. 11, Chap. 8; Kynoch Press: Tables 3, 4, Chap. 4.

CONTENTS

CHAPTER 3
Lattice Dynamics

CHAPTER 4
Crystal Symmetry

CHAPTER 5
Optical Properties of Crystals

CHAPTER 6

Determination of Force Fields and Structures

CHAPTER 7

Infrared and Raman Spectra of Polymers

CHAPTER 8

Spectra of Impure Crystals

Appendices

Chapter 1

Molecular Vibrations

I. Introduction

On the basis of the Born–Oppenheimer approximation (**1, 2**)† the total energy of a molecule can be written as

$$E = E_{\text{elec}} + E_{\text{nucl}}, \tag{1}$$

where E_{elec} is the energy associated with the electronic configuration of the molecule and E_{nucl} is the energy of displacements of the nuclei. In general, this approximation is an excellent one, hence, interaction terms between E_{elec} and E_{nucl} can be neglected.‡

Within the framework of the Born–Oppenheimer approximation, E_{elec} is a function of the instantaneous nuclear configuration of the molecule. Thus, the "electronic problem" consists of finding solutions to the Schrödinger equation for the movement of the electrons for each fixed nuclear configuration of the N atoms in the molecule. The resulting energy, E_{elec}, then serves as the effective potential function which governs the movement of the N nuclei. The solution to this problem yields the various possible values of E_{nucl} for each electronic state of the molecule.

It is possible to decompose the energy further by writing

$$E_{\text{nucl}} = E_{\text{trans}} + E_{\text{rot}} + E_{\text{vib}} + \text{interactions}. \tag{2}$$

In this case the interaction terms are not always negligible. However, for a free molecule, as in the ideal gaseous state at ordinary temperatures, E_{trans} can be neglected, while the coupling between rotational and vibrational degrees of freedom is often quite small.

† Bold figures in parentheses denote references which are listed at the end of each chapter.

‡ It should be noted, however, that certain observable spectroscopic effects arise directly from such interaction terms, Λ-doubling, L-uncoupling, etc. [See, for example, reference (**3**), pp. 182, 188].

II. Kinetic Energy

In order to appreciate the origin, as well as the limitations of the approximation represented by Eqn (2), consider the total classical kinetic energy, T, of a polyatomic molecule. This quantity is a function of the velocities of the N atoms of the molecule with respect to a space-fixed origin, O of Fig. 1.

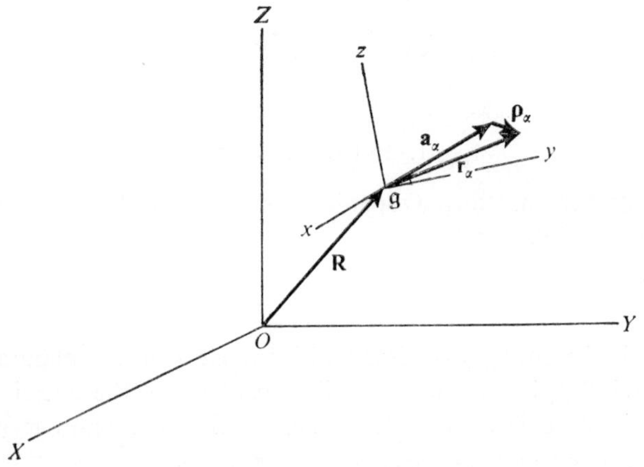

FIG. 1. Coordinate system used to indicate atomic positions in space.

Here, the center of gravity of the molecule, g, is located by the vector, $\mathbf{R}$, while the instantaneous position of each atom, α, with respect to the center of gravity is specified by $\mathbf{r}_\alpha$, and the corresponding equilibrium position by $\mathbf{a}_\alpha$. Thus, the vectors $\boldsymbol{\rho}_\alpha = \mathbf{r}_\alpha - \mathbf{a}_\alpha$ represent instantaneous displacements from equilibrium of the atoms $\alpha = 1, 2, \ldots, N$. The velocity of each atom with respect to point O is given by†

$$\mathbf{v}_\alpha = \dot{\mathbf{R}} + \dot{\mathbf{r}}_\alpha = \dot{\mathbf{R}} + \boldsymbol{v}_\alpha + \boldsymbol{\omega} \times \mathbf{r}_\alpha, \tag{3}$$

where $\boldsymbol{v}_\alpha$ is the velocity of atom α with respect to the center of gravity of the molecule and $\boldsymbol{\omega} \times \mathbf{r}_\alpha$ is the "velocity of following" which the atom must have to maintain its position in the molecule-fixed coordinate system x, y, z, which is rotating with angular velocity $\boldsymbol{\omega}$.

† A dot over a symbol is used to indicate its time derivative.

The total kinetic energy, T, resulting from displacements of the N atoms is then given by

$$\begin{aligned} 2T &= \sum_\alpha m_\alpha \mathbf{v}_\alpha^2 \\ &= \dot{\mathbf{R}}^2 \sum_\alpha m_\alpha + \sum_\alpha m_\alpha(\boldsymbol{\omega} \times \mathbf{r}_\alpha) \cdot (\boldsymbol{\omega} \times \mathbf{r}_\alpha) + \sum_\alpha m_\alpha \boldsymbol{v}_\alpha^2 \\ &\quad + 2\dot{\mathbf{R}} \cdot \boldsymbol{\omega} \times \sum_\alpha m_\alpha \mathbf{r}_\alpha + 2\dot{\mathbf{R}} \cdot \sum_\alpha m_\alpha \boldsymbol{v}_\alpha + 2\boldsymbol{\omega} \cdot \sum_\alpha (m_\alpha \mathbf{r}_\alpha \times \boldsymbol{v}_\alpha), \end{aligned} \tag{4}$$

where m_α is the mass of the αth atom.

For a free molecule, the Eckart conditions (**4, 5**) can be imposed, viz.

$$\sum_\alpha m_\alpha \mathbf{r}_\alpha = \sum_\alpha m_\alpha \boldsymbol{v}_\alpha = 0 \tag{5}$$

and

$$\sum_\alpha m_\alpha \mathbf{a}_\alpha \times \boldsymbol{v}_\alpha = 0. \tag{6}$$

Equation (5) is the condition for conservation of linear momentum which defines the position of the center of gravity of the molecule, while Eqn (6) is an approximate statement of the conservation of angular momentum of the system. Using the Eckart conditions of Eqns (5) and (6), the kinetic energy becomes

$$\begin{aligned} 2T &= \dot{\mathbf{R}}^2 \sum_\alpha m_\alpha + \sum_\alpha m_\alpha(\boldsymbol{\omega} \times \mathbf{r}_\alpha) \cdot (\boldsymbol{\omega} \times \mathbf{r}_\alpha) + \sum_\alpha m_\alpha \boldsymbol{v}_\alpha^2 \\ &\quad + 2\boldsymbol{\omega} \cdot \sum_\alpha m_\alpha(\boldsymbol{\rho}_\alpha \times \boldsymbol{v}_\alpha). \end{aligned} \tag{7}$$

The first term on the right-hand side of Eqn (7) results from the translation of the center of gravity of the molecule. For a gaseous molecule at ordinary temperatures, the translational energy levels form a quasi-continuum. Thus, this energy is not ordinarily involved in spectroscopic phenomena. The second term is the energy of overall molecular rotation about g, while the third term represents the total vibrational energy of the molecule resulting from relative displacements of the N atoms. The final term in Eqn (7) is the so-called Coriolis interaction, the principal part of the kinetic coupling between vibrational and rotational degrees of freedom.

It should be emphasized that the use of the Eckart conditions is justified only for a molecule in free space. In the case of a molecule trapped at a site in a solid or liquid, at which it is subject to external forces, not only will the translational energy, $\frac{1}{2}\dot{\mathbf{R}}^2 \Sigma_\alpha m_\alpha$ become important, but, furthermore, the coupling term between translation and rotation may play a significant role (See Chapter 8, Section IX).

For a free molecule, further simplification of the problem can be obtained by assuming that the displacements, $\boldsymbol{\rho}_\alpha$, of the atoms from their equilibrium positions are small. In this case the second term of Eqn (7) which represents the rotational energy of a rigid body with moment of inertia tensor, **I**, becomes†

$$2\,T_{\text{rot}} = \boldsymbol{\omega}^{\dagger}\,\mathbf{I}\,\boldsymbol{\omega}. \tag{8}$$

If principal axes of inertia are chosen, **I** is diagonal.

The vibrational contribution to the molecular kinetic energy [third term of Eqn (7)] can be written as a quadratic form in the time derivatives of Cartesian displacement coordinates $\boldsymbol{\xi}$ or in their conjugate momenta **p**. Thus,

$$2\,T_{\text{vib}} = \dot{\boldsymbol{\xi}}^{\dagger}\,\mathbf{M}\,\dot{\boldsymbol{\xi}} = \mathbf{p}^{\dagger}\,\mathbf{M}^{-1}\,\mathbf{p}, \tag{9}$$

where **M** is a diagonal matrix of atomic masses of rank $3N$.

The Cartesian displacement coordinates of all of the atoms form the elements of the column matrix $\boldsymbol{\xi}$. Thus, Δx_1, Δy_1 and Δz_1, the Cartesian displacements of the first atom from its equilibrium position form the first three elements, Δx_2, Δy_2, Δz_2 for the second atom form the next three, etc. The elements of the diagonal atomic-mass matrix **M** are arranged similarly, viz.

$$\mathbf{M} = \begin{pmatrix} m_1 & & & & & & & & & & & \\ & m_1 & & & & & & & & 0 & & \\ & & m_1 & & & & & & & & & \\ & & & m_2 & & & & & & & & \\ & & & & m_2 & & & & & & & \\ & & & & & m_2 & & & & & & \\ & & & & & & \cdot & & & & & \\ & & & & & & & \cdot & & & & \\ & & & & & & & & \cdot & & & \\ & & 0 & & & & & & & \cdot & & \\ & & & & & & & & & & m_N & \\ & & & & & & & & & & & m_N \\ & & & & & & & & & & & & m_N \end{pmatrix}. \tag{10}$$

† The moment of inertia is a symmetric tensor of the form

$$\mathbf{I} = \begin{pmatrix} I_{xx} & I_{xy} & I_{xz} \\ I_{yx} & I_{yy} & I_{yz} \\ I_{zx} & I_{zy} & I_{zz} \end{pmatrix},$$

which is symmetric in the sense that $I_{xy} = I_{yx}$, $I_{xz} = I_{zx}$, and $I_{yz} = I_{zy}$. The angular momentum vector can be represented by a column matrix $\boldsymbol{\omega}$ and its conjugate transpose by $\boldsymbol{\omega}^{\dagger}$. In this case, as the elements of $\boldsymbol{\omega}$ are real, $\boldsymbol{\omega}^{\dagger}$ is just it transpose; that is, $\boldsymbol{\omega}^{\dagger} = (\omega_x\, \omega_y\, \omega_z)$, and Eqn (8) is a quadratic form.

The Coriolis energy, the fourth term of Eqn (7), has been treated in detail by Meal and Polo (**6, 7**). This coupling term, although small, leads to perturbations of absorption lines in the rotation-vibration spectra of gaseous polyatomic molecules and to variations in the shapes of the envelopes of corresponding unresolved bands. Nevertheless, in the first approximation, Coriolis forces can be neglected and the kinetic energy of a rotating, vibrating molecule become

$$2\,T \approx 2\,T_{\mathrm{rot}} + 2\,T_{\mathrm{vib}} = \boldsymbol{\omega}^{\dagger}\,\mathbf{I}\,\boldsymbol{\omega} + \dot{\boldsymbol{\xi}}^{\dagger}\,\mathbf{M}\,\dot{\boldsymbol{\xi}}. \tag{11}$$

III. Internal Coordinates

In the treatment of molecular vibrations it is often convenient to introduce a set of internal coordinates, **S**. These coordinates may be, in the simplest case, a set of $3N - 6$ independent "valence coordinates" consisting of changes in the lengths of chemical bonds and angles between bonds.† However, it is often necessary to include other types of internal coordinates such as the change in angle between a bond and the plane defined by a pair of other bonds, or torsional angles between bonds. In the most general case, **S** may contain the six "external" degrees of freedom of translation and rotation, while symmetry considerations may require the inclusion of additional, redundant coordinates.

The linear transformation between Cartesian and internal coordinates can be written in the form

$$\mathbf{S} = \mathbf{B}\boldsymbol{\xi}, \tag{12}$$

and, for the corresponding velocities,

$$\dot{\mathbf{S}} = \mathbf{B}\,\dot{\boldsymbol{\xi}}. \tag{13}$$

The components of linear momenta are then given by

$$p_j \equiv \frac{\partial T}{\partial \dot{\xi}_j} = \sum_t \frac{\partial \dot{S}_t}{\partial \dot{\xi}_j} \cdot \frac{\partial T}{\partial \dot{S}_t} \,. \tag{14}$$

† In a molecule containing N atoms there are $3N$ degrees of freedom. Three of these degrees of freedom are associated with translation of the center of mass, while in the case of nonlinear molecules three arise from overall rotation. Hence, $3N - 6$ degrees of freedom are left for the internal or vibrational motion. In linear molecules there are $3N - 5$ degrees of internal fredeom, as only two coordinates are necessary to specify the molecular orientation. In general, the text refers to nonlinear molecules.

But, from Eqn (13), $\partial \dot{S}_t/\partial \dot{\xi}_j = B_{tj}$, and by definition, the momenta conjugate to the internal coordinates, $\mathbf{S}$, are $P_t \equiv \partial T/\partial \dot{S}_t$. Hence in matrix form Eqn (14) becomes

$$\mathbf{p} = \mathbf{B}^\dagger \mathbf{P}, \tag{15}$$

and the vibrational kinetic energy takes the form

$$2\,T_{\text{vib}} = \mathbf{P}^\dagger \mathbf{B}\,\mathbf{M}^{-1}\,\mathbf{B}^\dagger \mathbf{P} = \mathbf{P}^\dagger \mathbf{G}\,\mathbf{P}, \tag{16}$$

where $\mathbf{G} \equiv \mathbf{B}\,\mathbf{M}^{-1}\,\mathbf{B}^\dagger$.

An alternative expression of the vibrational kinetic energy can be obtained using one of Hamilton's equations, $\partial T/\partial P_t = \dot{S}_t$, which, when applied to Eqn (16), yields the relation

$$\dot{\mathbf{S}} = \mathbf{G}\,\mathbf{P}. \tag{17}$$

If $\mathbf{G}^{-1}$ exists,

$$\mathbf{P} = \mathbf{G}^{-1}\,\dot{\mathbf{S}} \tag{18}$$

and

$$\mathbf{P}^\dagger = \dot{\mathbf{S}}^\dagger(\mathbf{G}^{-1})^\dagger = \dot{\mathbf{S}}^\dagger\,\mathbf{G}^{-1}, \tag{19}$$

where the symmetric property of the inverse $\mathbf{G}$ matrix has been used.† Then, from Eqn (16),

$$2\,T_{\text{vib}} = \dot{\mathbf{S}}^\dagger\,\mathbf{G}^{-1}\,\mathbf{G}\,\mathbf{G}^{-1}\,\dot{\mathbf{S}} = \dot{\mathbf{S}}^\dagger\,\mathbf{G}^{-1}\,\dot{\mathbf{S}}. \tag{20}$$

The matrix $\mathbf{G}$, which, because of Eqn (20) is often referred to as the "inverse kinetic-energy matrix", is a function of the N atomic masses and the molecular geometry. One of the more important steps in the solution of the molecular vibration problem is the construction of the $\mathbf{G}$ matrix. The matrix $\mathbf{B}$ can be found directly by trigonometric considerations if the molecular geometry is known, thus allowing $\mathbf{G}$ to be calculated from the definition $\mathbf{G} \equiv \mathbf{B}\,\mathbf{M}^{-1}\,\mathbf{B}^\dagger$. However, this procedure can be greatly simplified using vector methods.

Consider the form of the $\mathbf{B}$ matrix. Each of the $3N$ columns is associated with a particular Cartesian displacement coordinate, Δx_1, Δy_1, Δz_1, for the first atom, Δx_2, Δy_2, Δz_2, for the second, etc. Each row is identified with an internal coordinate S_t. Thus, it is convenient to consider the three elements of $\mathbf{B}$ for a given atom α and an internal coordinate index t as components of a

† In general, $\mathbf{G}$ is square and symmetric. Hence, if the determinant $|\mathbf{G}| \neq 0$, $\mathbf{G}$ is said to be nonsingular and $\mathbf{G}^{-1}$ exists. Furthermore, since $\mathbf{G}$ is symmetric ($\mathbf{G} = \mathbf{G}^\dagger$), $\mathbf{G}^{-1}$ is also symmetric.

row vector $\mathbf{s}_{t\alpha}$. Then, for example, $\mathbf{s}_{11} = (B_{11}\, B_{12}\, B_{13})$, $\mathbf{s}_{12} = (B_{14}\, B_{15}\, B_{16})$,... and $\mathbf{s}_{21} = (B_{21}\, B_{22}\, B_{23})$, etc.

These vectors $\mathbf{s}_{t\alpha}$ allow any internal coordinate to be expressed as

$$S_t = \sum_{\alpha} \mathbf{s}_{t\alpha} \cdot \boldsymbol{\rho}_{\alpha}, \tag{21}$$

where the $\boldsymbol{\rho}_{\alpha}$'s are the atomic displacement vectors with components Δx_{α}, Δy_{α}, and Δz_{α}. In terms of the row matrix $\mathbf{s}$ whose elements are the vectors $\mathbf{s}_{t\alpha}$, the $\mathbf{G}$ matrix can be written in the form

$$\mathbf{G} = \mathbf{s}\,\mathbf{m}^{-1} \cdot \mathbf{s}^{\dagger}, \tag{22}$$

where $\mathbf{m}$ is a diagonal matrix of rank N of the atomic masses $m_1, m_2, \ldots, m_N$. Using Eqn (21), specific expressions have been developed for the $\mathbf{s}_{t\alpha}$'s for the types of internal coordinates mentioned above (**8, 9**). From these results, Decius (**10–12**) has derived tables of $\mathbf{G}$ elements which are very useful in setting up the vibrational problem.† It should be noted that the vector method outlined above forms the basis of computer programs which are now in current use for the numerical calculation of $\mathbf{G}$ from the coordinates of the equilibrium atomic positions (**13, 14**). See Appendix B.

IV. Potential Energy

In general, a molecule has one or more well-defined equilibrium configurations.‡ The criterion for the determination of this configuration is the minimization of the electronic energy, which is the potential energy for the vibrational problem. At ordinary temperatures the atoms undergo small oscillatory motions about their equilibrium positions. Hence, it is convenient to carry out a Taylor's expansion of the vibrational potential energy in atomic displacements from equilibrium. If, for example, it is assumed that the internal coordinates S_t are linearly independent and form a complete set, the potential energy is given by

$$V_{\text{vib}} = V_0 + \sum_{t=1}^{3N-6} \left(\frac{\partial V}{\partial S_t}\right)_0 S_t + \frac{1}{2!} \sum_{t,t'}^{3N-6} \left(\frac{\partial^2 V}{\partial S_t \partial S_{t'}}\right)_0 S_t S_{t'} + \ldots. \tag{23}$$

† A short table of $\mathbf{G}$ elements for stretching and bending coordinates is included in Appendix A.

‡ Note, however, that in molecules with essentially free internal rotation or inversion, there may be no well-defined equilibrium positions. In this case a more general approach, such as that of Longuet-Higgins, may be necessary. See references [(**15, 16**)].

The potential energy of the molecule in its equilibrium configuration can be arbitrarily set equal to zero; thus, $V_0 = 0$. Furthermore, the minimal condition on the potential energy requires that each quantity $(\partial V/\partial S_t)_0$ be set equal to zero. Thus, the first nonvanishing terms in Eqn (23) are the quadratic ones, which correspond to the potential for simple harmonic motion in classical mechanics. Although cubic, quartic, and higher terms in Eqn (23), will in general be present, they are usually assumed to be small and are neglected. Hence, in this, the harmonic approximation, the potential energy becomes

$$V_{\text{vib}} = \frac{1}{2}\sum_{t,t'}^{3N-6}\left(\frac{\partial^2 V}{\partial S_t \partial S_{t'}}\right)_0 S_t S_{t'} = \frac{1}{2}\sum_{t,t'}^{3N-6} f_{tt'}\, S_t\, S_{t'}, \tag{24}$$

where the force constants are defined by

$$f_{tt'} \equiv \left(\frac{\partial^2 V}{\partial S_t \partial S_{t'}}\right)_0.$$

In matrix form the potential energy is then given by the quadratic form

$$2\,V_{\text{vib}} = \mathbf{S}^\dagger\,\mathbf{F}\,\mathbf{S}, \tag{25}$$

where $\mathbf{F}$ is a symmetric matrix of the force constants.

As attempts at *a priori* calculations of force constants from molecular electronic wave functions have met with but limited success, these quantities have usually been treated as variable parameters in the vibrational problem. Their evaluation from experimental spectroscopic data is nevertheless of interest, both as a means of testing the results of electronic calculations and for their chemical significance.† For example, the knowledge of the value of a force constant for the stretching of a chemical bond provides some insight into the nature of the bond, including semi-quantitative measures of the length of the bond and its dissociation energy.‡

† It should be pointed out here that the harmonic approximation is usually used in calculating force constants from observed vibrational frequencies. Hence, the values obtained may be in error by as much as 5%, even though four-digit accuracy may be required to match the observed frequencies to within the limits of experimental error.

‡ Several empirical or semi-empirical relations have been proposed between the length of a bond and its force constant, The most famous of these expressions is Badger's rule, which can be written in the form $f = 1.86 \times 10^5/(r_e - d_{ij})$ where f is the stretching force constant in dyn/cm for the bond connecting atoms i and j, r_e is the bond length in Angstroms, and d_{ij} is a constant determined by the positions of the atoms i and j in the periodic table **(17, 18)**.

V. Normal Vibrations

In general, any complex vibration of a polyatomic molecule can be expanded in terms of a set of $3N - 6$ normal modes of motion. In a given normal mode, k, all atoms vibrate with the same frequency and pass through their equilibrium positions at the same instant. The corresponding coordinate Q_k, called the normal coordinate, is linearly related to the internal coordinates in the limit of small vibrations. Thus,

$$\mathbf{S} = \mathbf{L}\,\mathbf{Q} \tag{26}$$

and

$$\dot{\mathbf{S}} = \mathbf{L}\,\dot{\mathbf{Q}}. \tag{27}$$

Substitution of these expressions into Eqns (20) and (25) for the vibrational kinetic and potential energies, respectively, yields

$$2\,T_{\mathrm{vib}} = \dot{\mathbf{Q}}^{\dagger}\,\mathbf{L}^{\dagger}\,\mathbf{G}^{-1}\,\mathbf{L}\,\dot{\mathbf{Q}} = \dot{\mathbf{Q}}^{\dagger}\,\mathbf{E}\,\dot{\mathbf{Q}} \tag{28}$$

and

$$2\,V_{\mathrm{vib}} = \mathbf{Q}^{\dagger}\,\mathbf{L}^{\dagger}\,\mathbf{F}\,\mathbf{L}\,\mathbf{Q} = \mathbf{Q}^{\dagger}\,\mathbf{\Lambda}\,\mathbf{Q}, \tag{29}$$

where $\mathbf{E}$ is the unit matrix and $\mathbf{\Lambda}$ is a diagonal matrix with elements $\lambda_k = 4\pi^2\,\nu_k{}^2$, in which ν_k is the vibrational frequency of the kth normal mode and $k = 1, 2, \ldots 3N - 6$. The conditions represented by the right-hand parts of Eqns (28) and (29) form the mathematical definition of the normal modes, which requires that only "square terms" appear when the total energy is expressed as a function of normal coordinates and their derivatives. Equations (28) and (29) imply the relations

$$\mathbf{L}^{\dagger}\,\mathbf{G}^{-1}\,\mathbf{L} = \mathbf{E} \tag{30}$$

and

$$\mathbf{L}^{\dagger}\,\mathbf{F}\,\mathbf{L} = \mathbf{\Lambda}. \tag{31}$$

From Eqn (30),

$$\mathbf{L}^{\dagger} = \mathbf{L}^{-1}\,\mathbf{G}, \tag{32}$$

and

$$\mathbf{L}^{-1}\,\mathbf{G}\,\mathbf{F}\,\mathbf{L} = \mathbf{\Lambda}. \tag{33}$$

Hence, as is often stated, the determination of normal coordinates is equivalent to the successful search for a matrix $\mathbf{L}$ which diagonalizes the product $\mathbf{G}\,\mathbf{F}$ *via* a similarity transformation.

Alternatively, Eqn (33) can be written in the form

$$\mathbf{G\,F\,L} = \mathbf{L\,\Lambda}, \tag{34}$$

or

$$\sum_{t'} [(GF)_{tt'} - \delta_{tt'}\,\lambda_k]\,L_{t'k} = 0, \tag{35}$$

with the Kronecker delta defined by

$$\delta_{tt'} = \begin{cases} 1, & \text{if} \quad t = t' \\ 0, & \text{if} \quad t \neq t'. \end{cases}$$

Equation (35) is a set of simultaneous, linear, homogeneous equations for the unknown elements of **L**. A nontrivial solution to this problem exists only when the determinant of the coefficients vanishes, or

$$|\mathbf{G\,F} - \mathbf{E}\,\lambda_k| = 0. \tag{36}$$

This condition on the so-called secular determinant, is the basis of the vibrational problem. The roots of Eqn (36), the frequency parameters λ_k, are the eigenvalues of **G F**, while the columns of **L**, the eigenvectors, are related to the normal modes of vibration.†

Other forms of the secular determinant can be derived. For example, from Eqn (33),

$$\mathbf{L}^{-1}\,\mathbf{G\,F} = \mathbf{\Lambda\,L}^{-1}, \tag{37}$$

which becomes, when transposed,

$$\mathbf{F\,G}\,(\mathbf{L}^{-1})^{\dagger} = (\mathbf{L}^{-1})^{\dagger}\,\mathbf{\Lambda}, \tag{38}$$

leading to the secular determinant in "*F G* form",

$$|\mathbf{F\,G} - \mathbf{E}\,\lambda_k| = 0. \tag{39}$$

Clearly, the roots of Eqns (39) and (36) are the same. However, it is important to note that the eigenvectors are in this case different from those of Eqn (35). Furthermore, although both **F** and **G** are symmetric, neither of their products **G F** nor **F G** is, in general, symmetric. Thus, in order to calculate the

† As the Eqns (35) are homogeneous in $L_{t'k}$, these elements cannot be uniquely determined. However, their ratios can be found and can be resolved using the normalization condition,

$$\sum_{t'} L_{t'k}^2 = 1.$$

vibrational frequencies the eigenvalues of a nonsymmetric matrix must be determined.

In the event Cartesian atomic displacements are used as basis coordinates for the construction of the **G** matrix, it becomes equal to the diagonal matrix $\mathbf{M}^{-1}$. This result is evident from Eqn (11), where **M** plays the role of $\mathbf{G}^{-1}$ in the kinetic energy expression. In general, if **G** is diagonal, it can be written as

$$\mathbf{G} = \mathbf{G}^{\frac{1}{2}}\,\mathbf{G}^{\frac{1}{2}}, \tag{40}$$

where

$$\mathbf{G}^{\frac{1}{2}}\,\mathbf{G}^{-\frac{1}{2}} = \mathbf{E} \tag{41}$$

and $\mathbf{G}^{\frac{1}{2}}$ is a diagonal matrix of elements $m_\alpha^{-\frac{1}{2}}$ of rank $3N$. From Eqn (34)

$$\mathbf{G\,F\,L} = \mathbf{L}\,\mathbf{\Lambda} \tag{42}$$

and

$$\mathbf{G}^{\frac{1}{2}}\,\mathbf{F\,L} = \mathbf{G}^{\frac{1}{2}}\,\mathbf{F}\,\mathbf{G}^{\frac{1}{2}}\,\mathbf{G}^{-\frac{1}{2}}\,\mathbf{L} = \mathbf{G}^{-\frac{1}{2}}\,\mathbf{L}\,\mathbf{\Lambda}, \tag{43}$$

and the eigenvectors

$$\mathbf{L}' = \mathbf{G}^{-\frac{1}{2}}\,\mathbf{L} \tag{44}$$

can be defined. Thus, the secular equations become

$$\mathbf{G}^{\frac{1}{2}}\,\mathbf{F}\,\mathbf{G}^{\frac{1}{2}}\,\mathbf{L}' = \mathbf{L}'\,\mathbf{\Lambda} \tag{45}$$

and the corresponding determinant is

$$|\mathbf{G}^{\frac{1}{2}}\,\mathbf{F}\,\mathbf{G}^{\frac{1}{2}} - \mathbf{E}\,\lambda_k| = 0. \tag{46}$$

The above form of the secular determinant is particularly convenient because the matrix $\mathbf{G}^{\frac{1}{2}}\,\mathbf{F}\,\mathbf{G}^{\frac{1}{2}}$, whose eigenvalues are desired, is symmetric.† In Chapter 5 it will be found that this result is especially useful in the treatment of the vibrations of crystals. Even if an initial set of coordinates is chosen which does not lead to a diagonal **G** matrix, the matrix can be diagonalized by a suitable similarity transformation. This procedure is used in many computer programs (See Appendix B).

In order to find the atomic displacements for each normal mode of vibration, use is made of Eqns (12) and (26),

$$\mathbf{S} = \mathbf{B}\boldsymbol{\xi} = \mathbf{L\,Q}. \tag{47}$$

† If complex basis coordinates are used, this matrix is Hermitian.

The matrix of Cartesian displacements, ξ, could in principle be found by multiplying Eqn (47) by the inverse of **B**. However, **B** is not, in general, square and nonsingular, and, hence, cannot be inverted. Nevertheless, as Crawford and Fletcher have shown (**19**), from the definition $\mathbf{G} = \mathbf{B}\mathbf{M}^{-1}\mathbf{B}^{\dagger}$, it is evident that the product $\mathbf{M}^{-1}\mathbf{B}^{\dagger}\mathbf{G}^{-1}$ serves as the inverse of **B** in the sense that $\mathbf{B}(\mathbf{M}^{-1}\mathbf{B}^{\dagger}\mathbf{G}^{-1}) = (\mathbf{M}^{-1}\mathbf{B}^{\dagger}\mathbf{G}^{-1})\mathbf{B} = \mathbf{E}$. Hence, premultiplying Eqn (47) by $\mathbf{M}^{-1}\mathbf{B}^{\dagger}\mathbf{G}^{-1}$ yields the result

$$\xi = \mathbf{M}^{-1}\mathbf{B}^{\dagger}\mathbf{G}^{-1}\mathbf{S} = \mathbf{M}^{-1}\mathbf{B}^{\dagger}\mathbf{G}^{-1}\mathbf{L}\mathbf{Q}, \tag{48}$$

which is the desired transformation. Current computer programs for the vibrational problem usually include calculations of the normal modes in Cartesian displacement coordinates based on Eqn (48).

From the above discussion it is clear that a measurement of the frequencies of the normal vibrations should provide some structural information about a polyatomic molecule. If, for example, the molecular geometry is known, certain force constants, which characterize the electronic structure of the molecule, can, in principle, be calculated from the observed frequencies. However, in practice, since **F** contains as many as $\frac{1}{2}(3N-6)(3N-5)$ different elements corresponding to $3N-6$ frequencies, the problem is grossly underdetermined. This difficulty is at least partially alleviated by, (1) observing the vibrational frequencies of various isotopically substituted species, whose **F** matrices should be identical, (2) assuming special types of force fields which reduce the number of independent force constants (**20–22**), and (3) using other experimental data. Some other sources of experimental data relating to force constants include the analysis of band shapes in the infrared absorption spectra, which can be used to calculate Coriolis coupling constants (**23**), and electron diffraction (**24**), which provides some estimates of force constants in addition to vibrational amplitudes. Furthermore, as will be shown in Chapter 2, the symmetry of a molecule can be used to simplify the calculation of its vibrational frequencies.

As a simple illustration of the development of the secular determinant, consider the water molecule. A reasonable set of internal coordinates consists

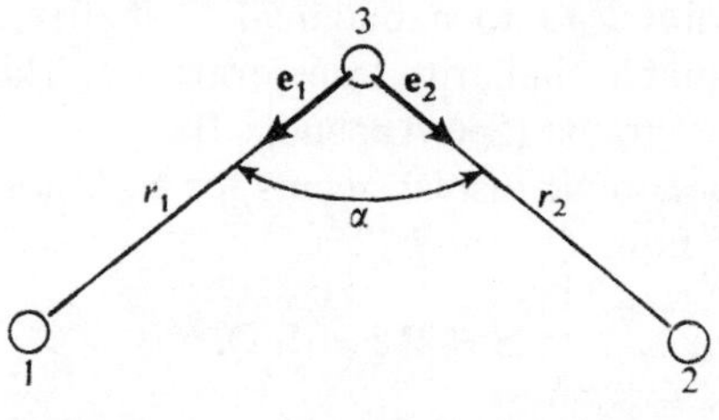

FIG. 2. Internal coordinate system for the water molecule.

of the changes in lengths of the two bonds and the variation in the bond angle. Thus, from Fig. 2, and Eqn (12),

$$\mathbf{S} = \begin{pmatrix} \Delta r_1 \\ \Delta r_2 \\ \Delta\alpha \end{pmatrix} = \mathbf{B}\boldsymbol{\xi}$$

$$= \left(\begin{array}{ccc|ccc|ccc} 0 & -s & -c & 0 & 0 & 0 & 0 & s & c \\ 0 & 0 & 0 & 0 & s & -c & 0 & -s & c \\ 0 & -c/r & s/r & 0 & c/r & s/r & 0 & 0 & -2s/r \end{array}\right) \begin{pmatrix} \Delta x_1 \\ \Delta y_1 \\ \Delta z_1 \\ \Delta x_2 \\ \Delta y_2 \\ \Delta z_2 \\ \Delta x_3 \\ \Delta y_3 \\ \Delta z_3 \end{pmatrix}, \quad (49)$$

where $s = \sin\alpha/2$, $c = \cos\alpha/2$, and r is the equilibrium O–H bond length. In terms of the atomic displacement vectors $\boldsymbol{\rho}_1$, $\boldsymbol{\rho}_2$, and $\boldsymbol{\rho}_3$, Eqn (49) can be rewritten in the form

$$\mathbf{S} = \mathbf{s}\cdot\boldsymbol{\rho}$$

$$= \begin{pmatrix} \mathbf{e}_1 & 0 & -\mathbf{e}_1 \\ 0 & \mathbf{e}_2 & -\mathbf{e}_2 \\ -\dfrac{\mathbf{e}_2 - \mathbf{e}_1\cos\alpha}{r\sin\alpha} & -\dfrac{\mathbf{e}_1 - \mathbf{e}_2\cos\alpha}{r\sin\alpha} & \dfrac{(\mathbf{e}_1 + \mathbf{e}_2)(1 - \cos\alpha)}{r\sin\alpha} \end{pmatrix} \cdot \begin{pmatrix} \boldsymbol{\rho}_1 \\ \boldsymbol{\rho}_2 \\ \boldsymbol{\rho}_3 \end{pmatrix}, \quad (50)$$

where the unit vectors $\mathbf{e}_1$ and $\mathbf{e}_2$ are directed toward hydrogen atoms 1 and 2, respectively, as shown in Fig. 2. For the water molecule, using **s** from Eqn (50) and

$$\mathbf{m} = \begin{pmatrix} m_H & 0 & 0 \\ 0 & m_H & 0 \\ 0 & 0 & m_O \end{pmatrix}, \quad (51)$$

the matrix **G**, which can be found directly from Eqn (22), is given by

$$\mathbf{G} = \begin{pmatrix} \mu_H + \mu_O & \mu_O \cos\alpha & -\dfrac{\mu_O \sin\alpha}{r} \\ & \mu_H + \mu_O & -\dfrac{\mu_O \sin\alpha}{r} \\ \text{symmetric} & & \dfrac{2\mu_H + 2\mu_O(1 - \cos\alpha)}{r^2} \end{pmatrix}. \quad (52)$$

Here, use has been made of the vector relations $\mathbf{e}_1 \cdot \mathbf{e}_1 = \mathbf{e}_2 \cdot \mathbf{e}_2 = 1$ and $\mathbf{e}_1 \cdot \mathbf{e}_2 = \cos\alpha$, and the definitions of the reciprocal masses $\mu_H = 1/m_H$ and $\mu_O = 1/m_O$. The **G** matrix given in Eqn (52) can be easily verified using the table of Appendix A.

The potential energy matrix for the water molecule, which can be written down directly in the above system of internal coordinates, becomes

$$\mathbf{F} = \begin{pmatrix} f_r & f_{rr} & f_{r\alpha} \\ & f_r & f_{r\alpha} \\ \text{sym} & & f_\alpha \end{pmatrix}, \quad (53)$$

where f_r is the coefficient of $(\Delta r_1)^2$ and $(\Delta r_2)^2$, and f_α is the coefficient of $(\Delta\alpha)^2$ in the potential energy expression. These constants are referred to as the principal or valence force constants. The off-diagonal constant f_{rr} represents an interaction constant which is the coefficient of $(\Delta r_1\, \Delta r_2)$ and $(\Delta r_2\, \Delta r_1)$ in the potential energy expression, while the interaction constant $f_{r\alpha}$ is the coefficient for all four terms of the type $(\Delta r_1\, \Delta\alpha)$, etc. By taking the product of the above **G** and **F** matrices, the secular equation in the form of Eqn (36) can be readily found.

The **F** matrix of Eqn (53) contains four independent force constants. Since the water molecule has but three vibrational frequencies, at least one interaction constant must be neglected or some other constraint introduced. If all off-diagonal elements of **F** are neglected, the two principal constants f_r and f_α constitute the *valence force field* for this molecule. However, in order to reproduce the three observed vibrational frequencies this force field must be modified by the inclusion of the interaction constant f_{rr}. Thus, as might be expected, three force constants are needed to fit the three frequencies. This example will be reconsidered in Chapter 2 after the question of molecular symmetry has been treated.

VI. Quantum-Mechanical Viewpoint

Molecular spectroscopists usually speak of the vibrational frequencies of a polyatomic molecule, as observed, for example, in an infrared spectrum. However, as the vibrations of a molecule are subject to the laws of quantum, rather than classical, mechanics, it should be realized that it is actually the energy of a transition between two different molecular energy levels that is being measured. Nevertheless, this energy difference is related by the Bohr frequency rule,

$$\Delta E = h\,\nu, \tag{54}$$

where h is Planck's constant, to ν, a quantity having the dimensions of a frequency.

A complete quantum-mechanical treatment of molecular vibrations requires the solution of the Schrödinger equation for the vibrational wave functions using a general potential energy function [Eqn (23)]. While this problem cannot be solved exactly, the harmonic approximation of Eqn (24) leads to exact wave functions which are essentially products of Hermite polynomials and to energy levels defined by

$$E_{k,v_k} = h\,\nu_k\,(v_k + \tfrac{1}{2}), \tag{55}$$

for each normal mode of vibration k. Here, $v_k = 0, 1, 2, \ldots$ is the vibrational quantum number, and ν_k is found to be identical to the classical frequency of vibration of the kth normal mode. A spectroscopic transition occurs when one or more of the v_k's changes, although all transitions are not, in general, allowed.

VII. Selection Rules

In absorption spectroscopy, the electric vector, $\mathscr{E}$, of the incident radiation couples with the electric dipole moment, $\mathfrak{m}$, of the molecular system. The resulting time-dependent perturbation,

$$\mathscr{H}' = \mathfrak{m}\cdot\mathscr{E}, \tag{56}$$

induces a transition in the molecule and a corresponding absorption of energy. Hence, at time t the Hamiltonian of the system can be written in the form

$$\mathscr{H} = \mathscr{H}^0 + \mathscr{H}', \tag{57}$$

where $\mathscr{H}^0$ is the Hamiltonian of the system at $t = 0$, at which time the wave function Ψ^0 is the solution of the Schrödinger equation,

$$\mathscr{H}^0\,\Psi^0 = -\frac{h}{2\pi i}\,\frac{\partial\Psi^0}{\partial t}\,. \tag{58}$$

Following the method of first-order, time-dependent perturbation theory, [See reference (**2**), p. 294], the wave function, Ψ, a function of the coordinates of the system and the time, is expanded in zero-order wave functions, namely,

$$\Psi(\xi, t) = \Sigma\, a_n(t)\, \Psi_n{}^0, \tag{59}$$

where the coefficients $a_n(t)$ are time-dependent. If at time $t = 0$ the system is in the state n, the quantity $a_m(t)^* a_m(t)$ represents the probability of the system being in state m at time t. The coefficient $a_m(t)$ can be found by solution of the equation

$$\frac{\partial a_m(t)}{\partial t} = -\frac{2\pi i}{h} \int \Psi_m{}^{0*}\, \mathscr{H}'\, \Psi_n{}^0\, \mathrm{d}\tau, \tag{60}$$

where τ represents the ensemble of spatial coordinates and the time. Substituting Eqn (56) into Eqn (60) and integrating with respect to time yields

$$a_m{}^*(t)\, a_m(t) = \frac{\pi^2}{3h^2} [\langle m|\mathfrak{m}_X|n\rangle^2 + \langle m|\mathfrak{m}_Y|n\rangle^2 + \langle m|\mathfrak{m}_Z|n\rangle^2]\, \bar{\mathscr{E}}^2 t$$

$$= \frac{8\pi^3}{3h^2} [\langle m|\mathfrak{m}_X|n\rangle^2 + \langle m|\mathfrak{m}_Y|n\rangle^2 + \langle m|\mathfrak{m}_Z|n\rangle^2]\, \rho t = B_{n\to m}\, \rho t, \tag{61}$$

where $\bar{\mathscr{E}}$ is the average magnitude of the electric field, $\rho = \bar{\mathscr{E}}^2/8\pi = \mathscr{I}/c$ is the radiant energy density of a light beam of intensity $\mathscr{I}$, and $B_{n\to m}$ is called the *Einstein coefficient* for absorption. Here the notation of Dirac,

$$\langle m|\mathfrak{m}_X|n\rangle \equiv \int \psi_m{}^*\, \mathfrak{m}_X\, \psi_n\, \mathrm{d}\tau, \text{ etc.} \tag{62}$$

has been used to represent the matrix elements of the dipole moment components with respect to the time-independent wave functions ψ_n and ψ_m for the initial and final states, respectively. Whether a given transition $n \to m$ is forbidden or allowed thus depends on the vanishing or nonvanishing of integrals such as that given in Eqn (62). These *selection rules* depend on the mathematical properties of the wave functions, the orientation of the dipole moment vector, and the direction and polarization of the incident radiation.

The total energy of nuclear displacements in the molecule was written as the sum of translational, rotational, and vibrational terms. In the same approximation, the wave function for a free molecule becomes

$$\psi \approx \psi\,(\text{trans}) \,.\, \psi\,(\text{rot}) \,.\, \psi\,(\text{vib}), \tag{63}$$

where for simplicity each factor is assumed to be separately normalized. Since the dipole moment is independent of translation of the molecule, the integrals appearing in Eqn (61) can be factored in the form

$$\langle m|\mathfrak{m}_X|n\rangle = \int \psi_m^* \,(\text{trans})\, \psi_n \,(\text{trans})\, d\tau_{\text{trans}}$$
$$\times \iint \psi_m^* \,(\text{vib})\, \psi_m^* \,(\text{rot})\, \mathfrak{m}_X\, \psi_n \,(\text{vib})\, \psi_n \,(\text{rot})\, d\tau_{\text{vib}}\, d\tau_{\text{rot}}. \quad (64)$$

Furthermore, the factor depending on the translational coordinates normalizes to unity. The remaining double integral is reduced by first transforming the molecular dipole moment to the space fixed coordinate system, X, Y, Z. Thus,

$$\mathbf{m}(X, Y, Z) = \mathbf{\Phi m}(x, y, z) \quad (65)$$

where $\mathbf{\Phi}$ is a three-by-three matrix whose elements are direction cosines connecting the various pairs of axes.† For the component $\mathfrak{m}_X$, Eqn (64) becomes

$$\langle m|\mathfrak{m}_X|n\rangle = \int \psi_m^* \,(\text{rot})\, \Phi_{Xx}\, \psi_n \,(\text{rot})\, d\tau_{\text{rot}} \cdot \int \psi_m^* \,(\text{vib})\, \mathfrak{m}_x\, \psi_n \,(\text{vib})\, d\tau_{\text{vib}}$$
$$+ \int \psi_m^* \,(\text{rot})\, \Phi_{Xy}\, \psi_n \,(\text{rot})\, d\tau_{\text{rot}} \cdot \int \psi_m^* \,(\text{vib})\, \mathfrak{m}_y\, \psi_n \,(\text{vib})\, d\tau_{\text{vib}} \quad (66)$$
$$+ \int \psi_m^* \,(\text{rot})\, \Phi_{Xz}\, \psi_n \,(\text{rot})\, d\tau_{\text{rot}} \cdot \int \psi_m^* \,(\text{vib})\, \mathfrak{m}_z\, \psi_n \,(\text{vib})\, d\tau_{\text{vib}}.$$

In order to separate completely the integrals over rotational and vibrational coordinates, each component of dipole moment is expanded about the equilibrium configuration in a Taylor's series in normal coordinates. That is,

$$\mathfrak{m}_x = \mathfrak{m}_x^0 + \sum_{k=1}^{3N-6} \left(\frac{\partial \mathfrak{m}_x}{\partial Q_k}\right)_0 Q_k + \text{higher terms}. \quad (67)$$

The higher terms, which represent electrical anharmonicity, are almost invariably neglected. Expressions of the form of Eqn (67) for the various components are substituted into the three equations for matrix elements, Eqn (66) being one of them. Then, for example, the first term of Eqn (66) becomes

$$\langle m|\mathfrak{m}_{Xx}|n\rangle = \mathfrak{m}_x^0 \int \psi_m^*(\text{rot})\, \Phi_{Xx}\, \psi_n(\text{rot})\, d\tau_{\text{rot}}$$
$$+ \sum_{k=1}^{3N-6} \left(\frac{\partial \mathfrak{m}_x}{\partial Q_k}\right)_0 \int \psi_m^*(\text{vib})\, Q_k\, \psi_n(\text{vib})\, d\tau_{\text{vib}} \cdot \int \psi_m^*(\text{rot})\, \Phi_{Xx}\, \psi_n(\text{rot})\, d\tau_{\text{rot}}, \quad (68)$$

† For linear molecules only two direction cosines are necessary to describe the orientation of the molecule.

where the integral over vibrational coordinates normalizes to unity in the first term. Evaluation of integrals such as the first term of Eqn (68) results in rotational selection rules which govern pure-rotational transitions in the microwave and far-infrared regions. These transitions involve no change in the vibrational state of the molecule.

The second term of Eqn (68) contains a product of two integrals, one over rotational coordinates and the other over vibrational coordinates. Therefore, the non-vanishing of this term places simultaneous restrictions on changes of rotational and vibrational states. This term is the origin of the selection rules for rotation-vibration absorption bands in the infrared spectral region. The form of the integrals over rotational coordinates depends on the type of rotor involved, being relatively straightforward for linear or symmetric-top rotors. [See reference (**25**), p. 44 ff]. It should be noted however, that in the case of solids, in which rotational motion is usually suppressed, the rotational integral of Eqn (68) becomes constant, and only the vibrational part of the problem need be considered.

The integral over vibrational coordinates in Eqn (68) can be easily evaluated in the harmonic approximation. In this case, the vibrational wave functions are of the form

$$\psi_n(\text{vib}) = \prod_{k} \psi_{v_k}, \qquad v_k = 0, 1, 2 \ldots, \tag{69}$$

where the ψ_{v_k}'s are essentially the Hermite polynomials of degree v_k in the argument Q_k. Using Eqn (69) the vibrational selection rules $\Delta v_k = \pm 1$ are readily obtained from Eqn (68) [See reference (**2**), p. 306]. However, this result depends on the harmonic approximation and, hence, will break down when anharmonicity is important. More general selection rules can be found by an analysis of the symmetry properties of the integrals of Eqn (68). This topic will be treated in some detail in the following chapter.

As indicated above, the selection rule $\Delta v_k = \pm 1$ is applicable in the harmonic approximation. As most molecules are in their vibrational ground states ($v_k = 0$ for all k) at ordinary temperatures, the most important absorption features are those corresponding to the transitions $v_k = 0 \rightarrow v_k = 1$ for a given k, that is, absorption of but a single quantum of vibrational energy. These absorption bands are called the *fundamentals*. From Eqn (55) it is seen that the frequency of a given fundamental absorption is given by

$$\frac{E_{k,1} - E_{k,0}}{h} = \nu_k(1 + \tfrac{1}{2}) - \nu_k(\tfrac{1}{2}) = \nu_k. \tag{70}$$

Thus the frequencies of the fundamentals can be identified with the frequencies of the normal modes of vibration of the classical system, as determined by the vanishing of the secular determinant [Eqn (36)].

If cubic and higher terms are considered in the expansion of Eqn (23), the molecular vibrations become anharmonic. The effect of anharmonicity is, (1) to modify the regular spacing of energy levels [Eqn (55)], and (2) to relax the selection rule $v_k = \pm 1$. Thus, in real molecules, absorption features are often observed at frequencies of approximately $2\nu_k$, $3\nu_k$, etc., the *overtones*, and at approximately $l\nu_k \pm l'\nu_{k'}$ $(l, l' = 1, 2, \ldots)$, the *combinations*. The detailed analysis of these effects requires the addition to Eqn (55) of terms in $(v_k + \frac{1}{2})^2$, $(v_k + \frac{1}{2})^3$, etc. whose coefficients are, in principle, related to cubic and higher terms in the expansion of Eqn (23). Such anharmonic effects have been neglected in treatments of all but the simplest polyatomic (triatomic) molecules.

VIII. Absolute Intensities

From the above discussion it is apparent that the possibility of absorption of radiation by a molecule depends on its dipole moment in the case of pure rotational spectra and on its dipole moment derivatives in the case of vibrational spectra. Hence, it is not surprising that the observed intensities of absorption lines or bands can be used as quantitative measures of the molecular dipole moment or its derivatives.

If a beam of light of intensity $\mathscr{I}$ falls on a layer of sample material of infinitesimal thickness, dl, the decrease in the intensity of the beam is given by

$$-\mathrm{d}\mathscr{I} = \mathfrak{a}\mathscr{I}\,\mathrm{d}l, \tag{71}$$

where $\mathfrak{a}$ is the absorption coefficient. The quantity $-\mathrm{d}\mathscr{I}$ can also be expressed in terms of the Einstein absorption coefficient introduced above,

$$-\mathrm{d}\mathscr{I} = h\nu_{n\to m} B_{n\to m}\,\rho(N_n - N_m)\,\mathrm{d}l, \tag{72}$$

where $\nu_{n\to m}$ is the frequency of the transition and N_n and N_m are the numbers of molecules per unit volume in the states n and m, respectively. Comparison of Eqns (71) and (72), and with the relation given by Eqn (61),

$$B_{n\to m} = \frac{8\pi^3}{3h^2}\left[\langle m|\mathfrak{m}_X|n\rangle^2 + \langle m|\mathfrak{m}_Y|n\rangle^2 + \langle m|\mathfrak{m}_Z|n\rangle^2\right], \tag{73}$$

leads to the expression

$$\mathfrak{a}_{n\to m} = \frac{8\pi^3}{3hc}\,\nu_{n\to m}(N_n - N_m)\left[\langle m|\mathfrak{m}_X|n\rangle^2 + \langle m|\mathfrak{m}_Y|n\rangle^2 + \langle m|\mathfrak{m}_Z|n\rangle^2\right] \tag{74}$$

for the absorption coefficient for the transition $n \to m$.

In the development of Eqn (74), it was assumed that the absorption due to the transition $n \to m$ gives rise to an infinitely sharp spectral line. However, as spectral lines are always somewhat broadened due to the uncertainty principle, the Doppler effect, and the influence of intermolecular forces, the absorption coefficient, $\mathfrak{a}_{n \to m}$, of Eqn (74) is usually replaced by an integral over the absorption line. Hence, the absolute or integrated intensity of a given line is represented by

$$\mathfrak{A}_{n \to m} \equiv \int_{\text{line}} \mathfrak{a}_{n \to m}\, d\nu = \frac{8\pi^3}{3hc} \nu_{n \to m}(N_n - N_m)\,[\langle m|\mathfrak{m}_X|n\rangle^2 + \langle m|\mathfrak{m}_Y|n\rangle^2 + \langle m|\mathfrak{m}_Z|n\rangle^2]. \qquad (75)$$

In the event the area of an entire rotation-vibration band is being measured, Eqn (75) must be summed over all of the rotational components present in order to determine the absolute intensity of the band.

Since the dipole moment matrix elements occurring in Eqn (75) contain terms of the form of Eqn (68), it is evident that the intensities of pure-rotational lines of gaseous molecules are determined by the permanent dipole moment, $\mathfrak{m}_0$. In fact, the absolute intensities of such lines are proportional to $\mathfrak{m}_0^2$, thus affording a direct method of determining dipole moments.†

In vibrational infrared spectroscopy, it is the second term of Eqn (68) which accounts for the observed band intensity. Thus, the intensity of each fundamental, $\mathfrak{A}_k$, is proportional to the square of the quantity $(\partial\mathfrak{m}/\partial Q_k)_0$. With the increasing reliability of absolute intensity measurements in recent years, the magnitudes of $(\partial\mathfrak{m}/\partial Q_k)_0$ have been determined for the various absorption bands of a large number of molecules. These results have stimulated further interest in normal-coordinate calculations, since the chemical significance of these data becomes more apparent when they are transformed to valence coordinates *via* relations involving **L**; for example,

$$\mathfrak{P}^\dagger = \mathbf{A}^\dagger \mathbf{L}^{-1}, \qquad (76)$$

where $\mathfrak{P}$ is a column matrix of elements $(\partial\mathfrak{m}/\partial S_t)_0$ and **A** is a column matrix of the $(\partial\mathfrak{m}/\partial Q_k)_0$'s. This procedure allows bond-moment derivatives to be determined, although in many cases it has been found necessary to include various interaction terms **(26–28)**.

† It should be pointed out, however, that much more accurate values of $\mathfrak{m}_0$ can be determined by a study of the rotational Stark effect in the microwave region. [See, for example, reference (3), p. 264].

REFERENCES

1. Born, M. and Oppenheimer, J. R. *Ann. Physik* **84,** 457 (1927).
2. Pauling, L. and Wilson, E. B. Jr. "Introduction to Quantum Mechanics", McGraw-Hill, New York, (1935).
3. Townes, C. H. and Schawlow, A. L. "Microwave Spectroscopy", McGraw-Hill, New York, (1955).
4. Eckart, C. *Phys. Rev.* **47,** 552 (1935).
5. Barchewitz, M.-P. "Spectroscopie infrarouge. I. Vibrations moléculaires", Gauthier-Villars, Paris, (1961).
6. Meal, J. H. and Polo, S. R. *J. Chem. Phys.* **24,** 1119, 1126 (1956).
7. Allen, H. C. Jr. and Cross, P. C. "Molecular Vib-Rotors", Wiley, New York, (1963).
8. Eliashevich, M. *Compt. Rend. Acad. Sci.* U.R.S.S. **28,** 605 (1940).
9. Wilson, E. B. Jr. *J. Chem. Phys.* **9,** 76 (1941).
10. Decius, J. C. *J. Chem. Phys.* **16,** 1025 (1948).
11. Ferigle, S. M. and Meister, A. G. *J. Chem. Phys.* **19,** 982 (1951).
12. Wilson, E. B. Jr., Decius, J. C. and Cross, P. C. "Molecular Vibrations", McGraw-Hill, New York, (1955).
13. Overend, J. and Scherer, J. R. *J. Chem. Phys.* **32,** 1289 (1960).
14. Schachtschneider, J. H. "Vibrational Analysis of Polyatomic Molecules V", Tech. Rep. No. 231–64, Shell Development Company, Emeryville, Calif., (1966).
15. Longuet-Higgins, H. C., *Mol. Phys.* **6,** 445 (1963).
16. Turrell, G. C. *J. Mol. Structure* **5,** 245 (1970).
17. Badger, R. M. *J. Chem. Phys.* **2,** 128 (1934); **3,** 710 (1935).
18. Herzberg, G. "Molecular Spectra and Molecular Structure I. Spectra of Diatomic Molecules", Second Edition, Van Nostrand, Princeton, (1950).
19. Crawford, B. L. and Fletcher, W. H. *J. Chem. Phys.* **19,** 141 (1951).
20. Urey, H. C. and Bradley, C. A. *Phys. Rev.* **38,** 1969 (1931).
21. Heath, D. F. and Linnett, J. W. *Trans. Faraday Soc.* **44,** 873 (1948).
22. Herzberg, G. "Molecular Spectra and Molecular Structure II. Infrared and Raman Spectra of Polyatomic Molecules", Van Nostrand, Princeton, (1945).
23. Edgell, W. F. and Moynihan, R. E. *J. Chem. Phys.* **27,** 155 (1957).
24. Cyvin, S. J. "Molecular Vibrations and Mean Square Amplitudes", Elsevier, Amsterdam, (1968).
25. Wollrab, J. E. "Rotational Spectra and Molecular Structure", Academic Press, New York, (1967).
26. Biarge, J. F., Herranz, J. and Morcillo, J. *Anales Real. Soc. Españ. Fis. y Quim. (Madrid)* **A55,** 267 (1959); **A57,** 81 (1961).
27. Sverdlov, L. M. *Optika i Spektroskopiya* **10,** 152 (1961).
28. Gribov, L. A. "Intensity Theory for Infrared Spectra of Polyatomic Molecules", Academy of Sciences, Moscow, 1963. English translation, Consultants' Bureau, New York, (1964).

Chapter 2

Molecular Symmetry

In Chapter 1 it was pointed out that, although a molecule is a complex collection of electrons and nuclei, a number of simplifications can be made in the development of its dynamics. In addition, the existence of a well-defined equilibrium nuclear configuration makes it possible to describe the structure of a molecule in geometrical terms. Thus, one speaks of the tetrahedral structure of methane or the linear structure of acetylene. These descriptions can be generalized by introducing the concept of symmetry operations, to which can then be applied the basic principles of the theory of groups.†

I. Symmetry Operations

Such molecules as water, boron trifluoride, benzene, and urea are said to be planar. That is, when the structural formulas of these molecules are drawn on a piece of paper, as in Fig. 1, the plane of the paper represents an *element of symmetry* in that, in a given structure, reflection of all points through the plane yields an equivalent (superimposable) structure. The actual process of carrying out the reflection is referred to as the *symmetry operation* of reflection.

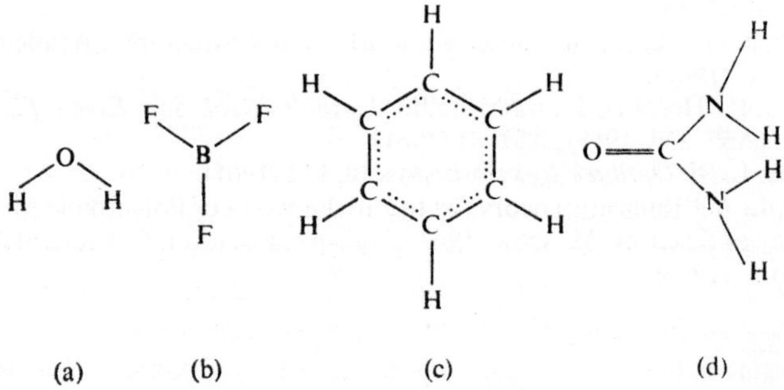

FIG. 1. Planar molecules: (a) water, (b) boron trifluoride, (c) benzene, (d) urea.

† For general discussions of the molecular application of group theory, see for example, references (**1–3**).

In general, a molecule may have a number of elements of symmetry. For example, the water molecule has, in addition to the plane of symmetry indicated above, a second plane of symmetry perpendicular to the first one (see Fig. 2). The latter plane (z, x) is a consequence of the geometrical equivalence of the two hydrogen atoms in this molecule. The symmetry operation of reflection through the plane z, x, usually represented by the symbol $\sigma(zx)$, results in the exchange of the two hydrogen atoms.

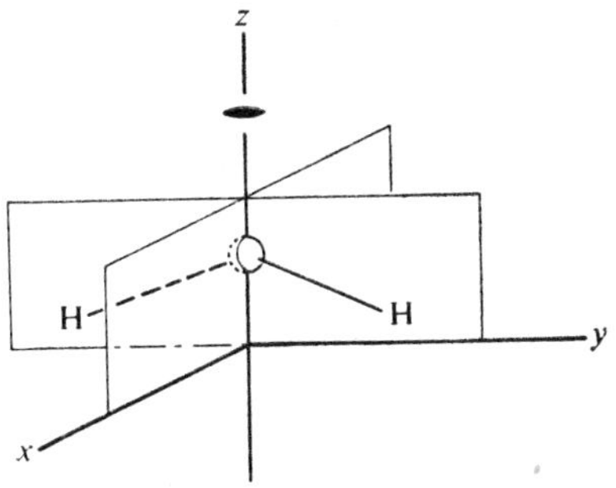

FIG. 2. Symmetry elements of the water molecule.

The presence of two perpendicular planes of symmetry implies the existence of another element of symmetry, the two-fold axis. From Fig. 2, it is seen that the intersection of the yz and zx planes is the bisector of the H–O–H angle. This line, the z axis, is referred to as a two-fold axis of symmetry because rotation of the molecule about it by 360°/2 in either sense results in a configuration which can be superimposed on the original one. As this operation must be carried out twice in succession in order to obtain the structure which is identical to the original one, it is called a two-fold rotation (designated by the symbol C_2) and the corresponding element is a two-fold or binary axis represented by ⬮. Molecules of higher symmetry may contain three-fold, four-fold, etc., axes, with corresponding operations C_3, C_4, etc. The symbol C_n thus represents a rotation of a molecule by $360°/n$ to give a configuration which is superimposable on the original one.

A further element of symmetry is the center of inversion. Of the molecules shown in Fig. 1, only benzene contains such an element (represented by ○). If every point in the molecule is carried through the center of inversion and an equal distance beyond, an equivalent structure is obtained. This operation, the inversion, is represented by the symbol i. Clearly, a given molecule can have but one center of inversion.

One more element is needed in order to complete the description of molecular symmetry. Molecular spectroscopists usually choose the rotation–

reflection or alternating axis, with which is associated an operation that is, in a sense, a combination of C_n and σ. This operation, sometimes called an improper rotation, is illustrated in Fig. 3 using the staggered configuration of ethane. The elements of symmetry of this molecule are: (1) a three-fold rotation axis (z), represented by the symbol ▲, (2) three two-fold axes perpendicular to z, (3) the center of inversion (○), and (4) an alternating six-fold axis coincident with z. The six-fold rotation–reflection operation, designated S_6, can be visualized by first carrying out a rotation about z by an angle of $360°/6 = 60°$, followed by reflection through the xy plane. Thus, although the xy plane is not in itself a symmetry element, its presence, along with the three-fold rotation axis perpendicular to it, results in an element of symmetry associated with the operation S_6.

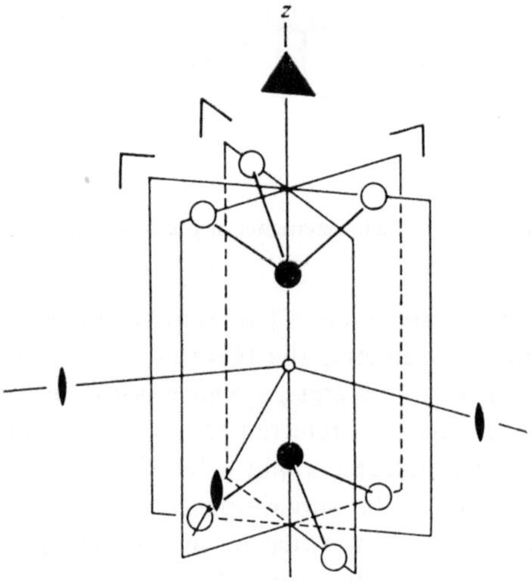

FIG. 3. Symmetry elements of the staggered configuration of ethane.

In some of the cases considered above, more than one operation can be associated with a given element of symmetry. For example, the BF_3 molecule shown in Fig. 1, when rotated in the counterclockwise sense by $360°/3 = 120°$ results in an equivalent configuration. However, rotation in the same sense by 240° produces a similar result. These rotations, both of which are symmetry operations, will be represented by the symbols $C_3^{\,1}$ and $C_3^{\,2}$, respectively. In general, a superscript will be used to indicate the number of successive counterclockwise rotations by $360°/n$. Thus, C_n^k becomes the complete symbol for any proper rotation. When $k = n$, the operation becomes a rotation of the molecule by 360° resulting in a configuration which is identical to (not merely equivalent to) the original one. This trivial operation, which then corresponds to a rotation by zero degrees is called the *identity*, and is usually given the symbol E (German: *Einheit*). Similar considerations apply to the rotation–reflection operation, represented in general by S_n^k. Thus, $S_4{}^4 \equiv E$ and $S_4{}^2 \equiv C_2{}^1$. However, while $S_2{}^1 \equiv i$ and $S_2{}^2 \equiv E$, note that $S_3{}^3 \equiv \sigma$, rather than E, and that $S_3{}^4 \equiv C_3$. As symmetry operations are not always easy to visualize, the reader is urged to construct molecular models of the structures being considered. The validity of this suggestion will become more apparent in Chapter 4, where crystal symmetry will be considered.

II. Algebra of Symmetry Operations

The relations written above between certain of the symmetry operations suggest that a complete algebra of such operations can be built up. Thus, a rotation by an angle of $2\pi k/n$ about an axis, followed by a rotation by $2\pi k'/n$ about the same axis, can be represented by

$$C_n^{k'}\, C_n^k = C_n^{k'+k}, \tag{1}$$

a relation which is consistent with the rules for combining exponents in ordinary algebra. As a further example, consider the ammonia molecule, NH_3, whose symmetry elements are indicated in Fig. 4. If the hydrogen

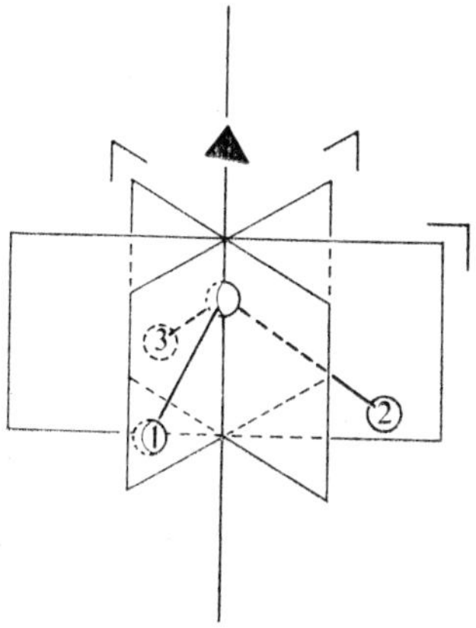

Fig. 4. Symmetry elements of the ammonia molecule.

atoms are arbitrarily distinguished by indices 1, 2, 3, the result of a given symmetry operation can be easily determined. Thus, the effect of the operation $C_3{}^1$, followed by σ_v is shown in Fig. 5. Here, σ_v is the operation of reflection through one of the vertical planes of symmetry. Reflections through the other two vertical planes are designated by σ_v' and σ_v''. The result of these successive operations is represented algebraically by

$$\sigma_v C_3^{\,1} = \sigma_v', \tag{2}$$

while, if the order of the two operations is reversed, the result

$$C_3^{\,1} \sigma_v = \sigma_v'' \tag{3}$$

is obtained, as is also shown in Fig. 5.

This example illustrates the fact that symmetry operations do not always commute. Hence, the order of operations must be preserved, and to be consistent with the above example, the convention that successive operations are written from right to left must be followed.

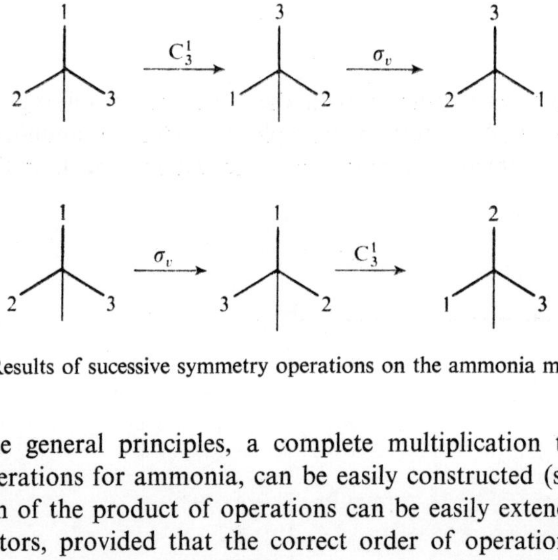

FIG. 5. Results of sucessive symmetry operations on the ammonia molecule.

Using these general principles, a complete multiplication table of the symmetry operations for ammonia, can be easily constructed (see Table 1). The definition of the product of operations can be easily extended to more than two factors, provided that the correct order of operations is always maintained. Furthermore, using Table 1, the associative property of these operations can be readily demonstrated.

An additional property of symmetry operations can be developed by inspection of Table 1. For any operation Q there can be found an operation P such that $PQ = E$. The effect of P is thus to "undo" the operation Q. Hence, P is called the inverse of Q, or $P = Q^{-1}$ and $Q^{-1}\,Q = E$. The inverse thus plays the role of division in ordinary algebra. From Table 1, it is seen that all reflection operations are their own inverses. The same is true for i, the inversion.

TABLE 1

↗	E	C_3^1	C_3^2	σ_v	σ_v'	σ_v''
E	E	C_3^1	C_3^2	σ_v	σ_v'	σ_v''
C_3^1	C_3^1	C_3^2	E	σ_v''	σ_v	σ_v'
C_3^2	C_3^2	E	C_3^1	σ_v'	σ_v''	σ_v
σ_v	σ_v	σ_v'	σ_v''	E	C_3^1	C_3^2
σ_v'	σ_v'	σ_v''	σ_v	C_3^2	E	C_3^1
σ_v''	σ_v''	σ_v	σ_v'	C_3^1	C_3^2	E

III. POINT GROUPS

Now consider the set of all the symmetry operations which can be carried out on a given molecule. This set of operations has the following four general properties:

(1) The product PQ of any two operations Q and P is an operation in the set.

(2) The product is associative, i.e. $R(PQ) = (RP)Q$, where R is also an operation in the set.

(3) The set contains the identity operation E having the property $RE = ER = R$, which clearly bears a certain analogy to the number one in ordinary algebra.

(4) Each operation R of the set has an inverse R^{-1}, which is also an operation of the set, and where $RR^{-1} = R^{-1}R = E$.

These four properties define a *group* in the mathematical sense whose members are the molecular symmetry operations. Thus, the symmetry of any molecule can be precisely described by its *symmetry group* or, *point group*.† The number of independent operations in a group is called the *order* of the group.

† The term point group is used because there is always some point in a molecule that is not affected by any of the operations of the molecular symmetry group. In Chapter 4, additional operations involving translations in space will be used to describe the symmetry of crystals. The groups which include such operations are known as *space groups*.

The various possible point groups are given in Table 2 along with their basic or defining operations. The notation used is that of Schönflies, as usually employed by molecular spectroscopists. However, an equivalent notation (Hermann–Mauguin), which is preferred by crystallographers, will be introduced in Chapter 4.

TABLE 2

Point group	Generating operation
$\mathscr{C}_n$	C_n
$\mathscr{D}_n$	$C_n, nC_2 \perp C_n$
$\mathscr{D}_{nh}$	$C_n, nC_2 \perp C_n, \sigma_h$
$\mathscr{S}_n$†	S_n
$\mathscr{C}_{nh}$†	$C_n, \sigma_h \perp C_n$
$\mathscr{C}_{nv}$	$C_n, n\sigma_v$
$\mathscr{D}_{nd}$	$C_n, nC_2 \perp C_n, n\sigma_a$
$\mathscr{T}, \mathscr{T}_d$, etc.	tetrahedron
$\mathscr{O}, \mathscr{O}_h$, etc.	octahedron or cube
$\mathscr{I}$ etc.	icosahedron

†Note, however, the particular notations

$$\mathscr{S}_2 \equiv \mathscr{C}_i \text{ and } \mathscr{C}_{1h} \equiv \mathscr{C}_s.$$

IV. Equivalent Atoms and Subgroups

An important concept in the study of molecular vibrations is that of *symmetrical equivalence.* Consider the staggered configuration of ethane represented in Fig. 3. The symmetry of this molecule is specified by the twelve operations E, C_3^1, C_3^2, C_2, C_2', C_2'', i, S_6^5, σ_v, σ_v', and σ_v'', corresponding to the point group $\mathscr{D}_{3d}$ (see Appendix C). All six of the hydrogen atoms are said to be equivalent in that by starting with a given hydrogen, all others can be generated by application of the group operations. These atoms form a set of symmetrically equivalent atoms. Similarly, the two carbon atoms of the $\mathscr{D}_{3d}$ structure of ethane also form an equivalent set.

From the above example it is evident that symmetrical atoms must be of the same species. However, this condition is not sufficient, as illustrated, for example, by the propane molecule (Fig. 6). Here, the two methylene hydrogens belong to one equivalent set, methyl hydrogens 1 and 2 to another, and the remaining hydrogen atoms to yet a third. It should be noted that the number of symmetrically equivalent atoms is always a divisor of the order of the molecular point group.

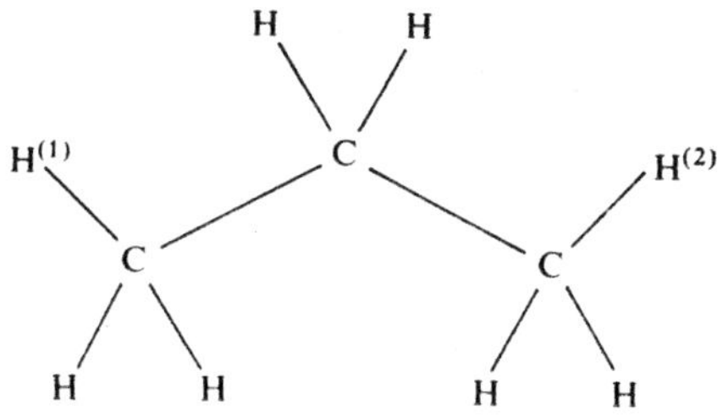

FIG. 6. Structure of the propane molecule.

A further classification of the symmetry operations can be made by considering the collection of operations which leave a given atom fixed. For the example of ethane considered above, the operations E, $C_3^{\,1}$, $C_3^{\,2}$, σ_v, σ_v', and σ_v'' do not modify the position of either carbon atom. This set of operations also forms a group, $\mathscr{C}_{3v}$, which is said to be a *subgroup* of the molecular point group $\mathscr{D}_{3d}$, Similarly, the operations which leave a given hydrogen atom in its original position are E and σ_v, corresponding to another subgroup of $\mathscr{D}_{3d}$, namely, $\mathscr{C}_{1h} \equiv \mathscr{C}_s$. The concept of subgroups is extremely important as a means of specifying the effect of isotopic substitution in a molecule (see Section XIV in this chapter), and, as will be shown in Chapter 4, as a method of relating internal vibrations of a molecule to those of a crystal lattice in which it is placed.

V. Coordinate Transformations

In the foregoing discussion, the symmetry operations have been described as "rotations", "reflections", etc. that is, as dynamical operations. However, strictly speaking, symmetry operations are really coordinate transformations which modify the mathematical statement of the atomic positions, rather than the positions themselves. Furthermore, these transformations are subject to the condition that they cannot modify the potential energy of the molecule. This property follows from the fact that any coordinate system is merely a bookkeeping scheme without physical significance.

To illustrate the coordinate transformation associated with a given symmetry operation, consider once again the operation C_3^1 on the ammonia molecule, as shown in Fig. 7. The initial displacement coordinates of each atom in the molecule are Δx_1, Δy_1, etc. which form the elements of the column matrix ξ, (See Chapter 1, p. 4). After the operation C_3^1, which is now simply a coordinate transformation, the corresponding displacements are $\Delta x_1'$, $\Delta y_1'$, etc., of ξ'. The specific transformation involved in this case is:

$$\begin{aligned}
\Delta x_1' &= -\tfrac{1}{2}\Delta x_2 + \frac{\sqrt{3}}{2}\Delta y_2 \\
\Delta y_1' &= -\frac{\sqrt{3}}{2}\Delta x_2 - \tfrac{1}{2}\Delta y_2 \\
\Delta z_1' &= \Delta z_2 \\
\Delta x_2' &= -\tfrac{1}{2}\Delta x_3 + \frac{\sqrt{3}}{2}\Delta y_3 \\
\Delta y_2' &= -\frac{\sqrt{3}}{2}\Delta x_3 - \tfrac{1}{2}\Delta y_3 \\
\Delta z_2' &= \Delta z_3 \\
\Delta x_3' &= -\tfrac{1}{2}\Delta x_1 + \frac{\sqrt{3}}{2}\Delta y_1 \\
\Delta y_3' &= -\frac{\sqrt{3}}{2}\Delta x_1 - \tfrac{1}{2}\Delta y_1 \\
\Delta z_3' &= \Delta z_1 \\
\Delta x_4' &= -\tfrac{1}{2}\Delta x_4 + \frac{\sqrt{3}}{2}\Delta y_4 \\
\Delta y_4' &= -\frac{\sqrt{3}}{2}\Delta x_4 - \tfrac{1}{2}\Delta y_4 \\
\Delta z_4' &= \Delta z_4
\end{aligned} \tag{4}$$

This transformation written in matrix form becomes $\xi' = \mathbf{R}\xi$, and $\mathbf{R}$ is a transformation matrix having the form shown in Fig. 8, where submatrices containing other than all zero elements have been shaded. Since any linear transformation is characterized by its trace, the sum of its diagonal elements, only those elements falling along the diagonal of the submatrix in the lower, righthand corner need be considered. This block, which is associated with atom 4, remains on the diagonal because the nitrogen atom is the only one not affected by the operation $C_3{}^1$. The specific form of this block is

$$\begin{pmatrix} \cos\phi & -\sin\phi & 0 \\ \sin\phi & \cos\phi & 0 \\ 0 & 0 & 1 \end{pmatrix} = \begin{pmatrix} -\frac{1}{2} & \frac{\sqrt{3}}{2} & 0 \\ -\frac{\sqrt{3}}{2} & -\frac{1}{2} & 0 \\ 0 & 0 & 1 \end{pmatrix}, \tag{5}$$

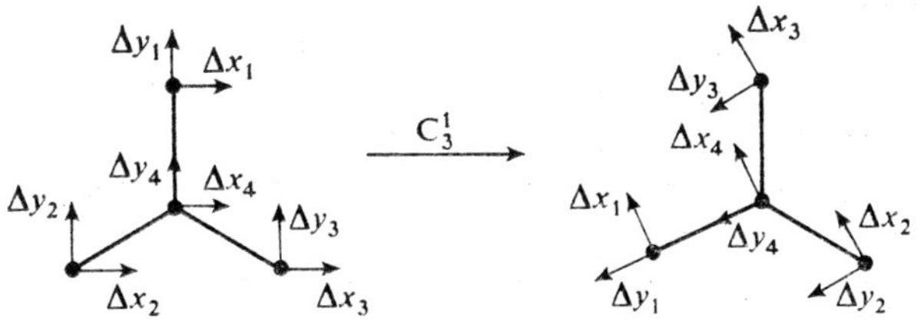

FIG. 7. Effect of the operation $C_3{}^1$ on displacement coordinates for ammonia.

R	Δx_1	Δy_1	Δz_1	Δx_2	Δy_2	Δz_2	Δx_3	Δy_3	Δz_3	Δx_4	Δy_4	Δz_4
$\Delta x_1'$	0	0	0	$-\frac{1}{2}$	$\frac{\sqrt{3}}{2}$	0	0	0	0	0	0	0
$\Delta y_1'$	0	0	0	$-\frac{\sqrt{3}}{2}$	$-\frac{1}{2}$	0	0	0	0	0	0	0
$\Delta z_1'$	0	0	0	0	0	1	0	0	0	0	0	0
$\Delta x_2'$	0	0	0	0	0	0	$-\frac{1}{2}$	$\frac{\sqrt{3}}{2}$	0	0	0	0
$\Delta y_2'$	0	0	0	0	0	0	$-\frac{\sqrt{3}}{2}$	$-\frac{1}{2}$	0	0	0	0
$\Delta z_2'$	0	0	0	0	0	0	0	0	1	0	0	0
$\Delta x_3'$	$-\frac{1}{2}$	$\frac{\sqrt{3}}{2}$	0	0	0	0	0	0	0	0	0	0
$\Delta y_3'$	$-\frac{\sqrt{3}}{2}$	$-\frac{1}{2}$	0	0	0	0	0	0	0	0	0	0
$\Delta z_3'$	0	0	1	0	0	0	0	0	0	0	0	0
$\Delta x_4'$	0	0	0	0	0	0	0	0	0	$-\frac{1}{2}$	$\frac{\sqrt{3}}{2}$	0
$\Delta y_4'$	0	0	0	0	0	0	0	0	0	$-\frac{\sqrt{3}}{2}$	$-\frac{1}{2}$	0
$\Delta z_4'$	0	0	0	0	0	0	0	0	0	0	0	1

FIG. 8. Form of the transformation matrix **R** based on Cartesian coordinates for the operation $C_3{}^1$ on ammonia.

where $\phi = 2\pi/n$ is the angle of rotation; thus, the trace (or spur) of **R** is given by

$$\chi_R = -\tfrac{1}{2} - \tfrac{1}{2} + 1 = 0. \tag{6}$$

This quantity is known as the *character* of the transformation. Only those atoms which remain "fixed" under a given symmetry operation can contribute to the character of the transformation matrix associated with the operation R.

The importance of the character of a linear transformation arises from the fact that it is independent of the choice of basis coordinates.† Thus, it is truly charactersitic of the transformation considered, and, hence, the corresponding symmetry operation. It can be readily shown that matrices such as that represented in Fig. 8 for the operation $C_3{}^1$, which can be constructed for any symmetry operation, possess all four of the group properties and, thus, form a group. Furthermore, as these matrices follow the same multiplication table as do the corresponding symmetry operations, the two groups are said to be isomorphic, and one group is a representation of the other. Since there is nothing unique in the choice of basis coordinates, there may be many different possible matrix representations of a given group. However, there exists in general for each point group a definite number of simplest representations, and it is one of the objectives of the group-theoretical method to express any representation as a function of these so-called *irreducible representations.* This problem will be developed in a later section.

VI. Conjugate Operations and Classes

Two group operations P and Q are said to be *conjugate* if $P = R^{-1}QR$, where R is a member of the group. This relation is symmetric in the sense that $Q = RPR^{-1}$, with R^{-1} also a member of the group. If P and Q are conjugate to the same operation W, they are mutually conjugate. If $P = S^{-1}WS$ and $Q = TWT^{-1}$, where S and T are both members of the group, it follows that $W = T^{-1}QT$ and $P = (S^{-1}T^{-1})Q(TS) = (TS)^{-1}Q(TS)$, where the product TS is also a member of the group.

For a given operation, W, a subset of operations can be chosen which are conjugate to W, and, hence, mutually conjugate, by considering all of the products of the form $R^{-1}WR$. Using this method a class of conjugate symmetry operations can be determined. If there is an operation which commutes with all other groups operation, this operation is in a class by

† For a proof of the invariance of the trace, see, for example, reference (**4**).

itself. If every class contains but one operation, the group is said to be *Abelian*.

Consider, for example, the operation $C_3^{\ 1}$ in point group $\mathscr{C}_{3v}$, the symmetry group of ammonia. Using the multiplication table (Table 1) the following products can be evaluated:

$$\begin{aligned}
&E^{-1}\, C_3^{\ 1}\, E = C_3^{\ 1} \\
&C_3^{\ -1}\, C_3^{\ 1}\, C_3^{\ 1} = C_3^{\ 1} \\
&C_3^{\ -2}\, C_3^{\ 1}\, C_3^{\ 2} = C_3^{\ 1} \\
&\sigma_v^{\ -1}\, C_3^{\ 1}\, \sigma_v = \sigma_v^{\ -1}\, \sigma_v'' = \sigma_v\, \sigma_v'' = C_3^{\ 2} \\
&\sigma_v'^{-1}\, C_3^{\ 1}\, \sigma_v' = \sigma_v'\, \sigma_v = C_3^{\ 2} \\
&\sigma_v''^{-1}\, C_3^{\ 1}\, \sigma_v'' = \sigma_v''\, \sigma_v' = C_3^{\ 2}.
\end{aligned} \tag{7}$$

Therefore, the operations $C_3^{\ 1}$ and $C_3^{\ 2}$ belong to the same class, designated $2C_3$. By replacing the operation $C_3^{\ 1}$ in Eqn (7) by each of the reflection operations in turn, it is easily found that σ_v, σ_v', and σ_v'' belong to the same class, $3\sigma_v$. The identity operation, E, is always in a class by itself, as it commutes with all group operations. Furthermore, all operations of the same class have the same characters. The latter general principle can be demonstrated for the case of ammonia by carrying out transformations similar to that of Eqn (4) for each of the operations of the point group $\mathscr{C}_{3v}$.

In the case of symmetry groups the classes of operations can be interpreted geometrically. The criterion for placing two operations, P and Q, in the same class is that there exists another group operation, R, which can be applied to the coordinate system so that the operation Q in the transformed coordinate system is equivalent to the operation P in the original coordinate system [(**3**), pp. 47–49]. It should be noted that all operations of a given class must be of the same type (reflections, two-fold rotations, three-fold rotations, etc.). However, all operations of the same type do not necessarily belong to the same class.

VII. Irreducible Representations

As indicated above in this chapter, under Section V, more than one equivalent matrix representation may exist for a given point group. The actual form of these representations depends on the particular choice of basis coordinates, although one quantity, the character, is independent of such a choice. Furthermore, for every molecule there exists at least one particular set of basis coordinates in terms of which the representation matrix is reduced to block–diagonal form, as shown symbolically in Fig. 9. This result can be expressed mathematically by the relation

$$\mathbf{\Gamma} = \sum_{\gamma} n^{(\gamma)} \mathbf{\Gamma}^{(\gamma)}, \tag{8}$$

where $\mathbf{\Gamma}$ is the representation being considered, which is now written as a direct sum† of the *irreducible representations*, $\mathbf{\Gamma}^{(\gamma)}$. The coefficient $n^{(\gamma)}$ indicates the number of times that the γth irreducible representation is included in $\mathbf{\Gamma}$.

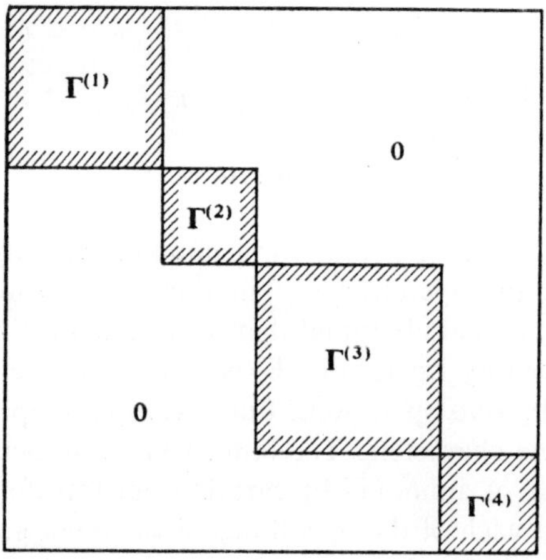

FIG. 9. Block-diagonal form of a transformation matrix: The irreducible representations.

Before going on to consider the application of group–theoretical methods to the molecular vibrational problem, it is necessary to discuss several general properties of irreducible representations. First, suppose that a group is of order g, and that these g operations have been collected into k different classes of mutually conjugate operations, following the principles outlined earlier. It can be shown that the group $\mathfrak{G}$ possesses precisely k nonequivalent irreducible representations, $\mathbf{\Gamma}^{(1)}, \mathbf{\Gamma}^{(2)}, \ldots, \mathbf{\Gamma}^{(k)}$, whose dimensions $d_1, d_2, \ldots, d_k$, satisfy the relation

$$d_1^{\,2} + d_2^{\,2} + \ldots + d_k^{\,2} = g. \tag{9}$$

† The direct sum, $\mathbf{C}$, of two square matrices $\mathbf{A}$ and $\mathbf{B}$ is obtained by writing one after the other along the diagonal of $\mathbf{C}$; thus,

$$\mathbf{C} = \left(\begin{array}{c:c} \mathbf{A} & \mathbf{O} \\ \hdashline \mathbf{O} & \mathbf{B} \end{array}\right).$$

This statement is often taken as a basic theorem of representation theory. It is found that for any point group there is only one set of k integers, the sum of whose squares is equal to g. Hence, from Eqn (9), the number of irreducible representations, as well as their dimensions, can be determined for any group. It was stated earlier that all operations belonging to the same class have the same characters. The classes are denoted $\mathscr{K}_1, \mathscr{K}_2, \ldots, \mathscr{K}_k$ and contain $g_1, g_2, \ldots, g_k$ operations, respectively. Finally, the symbol $\chi_j^{(\gamma)}$ is used to represent the character of the γth representation of an operation of class $\mathscr{K}_j$.

The above definitions allow the property of orthogonality of the characters to be stated in the form

$$\frac{1}{g} \sum_{j=1}^{k} g_j \chi_j^{(\gamma')*} \chi_j^{(\gamma)} = \delta_{\gamma',\gamma}, \tag{10}$$

where $\chi_j^{(\gamma')*}$ is the complex conjugate of $\chi_j^{(\gamma')}$ and $\delta_{\gamma',\gamma}$ is the Kronecker delta. The orthogonality of the characters will be used repeatedly in the group-theoretical applications which follow†. Because of the property expressed by Eqn (10), each irreducible representation can be specified by its set of characters. These quantities are conveniently listed in a *character table*.

VIII. Character Tables

The tables of the characters have the general form shown in Table 3. Each column represents a class of symmetry operations, while the rows designate the different irreducible representations. The entries in the table are simply the characters (traces) of the corresponding submatrices. Two specific properties of character tables will now be considered.

As shown earlier, the operation E is always in a class by itself, and it is by convention identified with $\mathscr{K}_1$, the arbitrarly-chosen first class of operations. In a given representation, the operation E corresponds to a unit matrix whose rank is equal to the dimension of the representation. Hence, the resulting character, the sum of the diagonal elements, is also equal to the dimension of the representation. The dimension of each representation can thus be easily determined by inspection of the corresponding entry in the first column of the character table.

A further property of character tables arises from the fact that every point group contains an irreducible representation which is invariant under all group operations. This irreducible representation consists of a one-by-one unit matrice for every class of operation, the characters thus being always

† Equations (9) and (10) can both be derived from a general orthogonality theorem. See, for example, reference (**5**), Sec. 3–2.

equal to unity. As this irreducible representation is by convention taken to be $\Gamma^{(1)}$, the first row of any character table consists solely of ones. The significance of character tables will become more apparent by consideration of an example.

TABLE 3

$\mathfrak{G}$	$\mathscr{K}_1 = E$	$\mathscr{K}_2$	$\mathscr{K}_3$	...	$\mathscr{K}_k$
$\Gamma^{(1)}$	$\chi_1^{(1)}$	$\chi_2^{(1)}$	$\chi_3^{(1)}$	...	$\chi_k^{(1)}$
$\Gamma^{(2)}$	$\chi_1^{(2)}$				
$\Gamma^{(3)}$	$\chi_1^{(3)}$				
$\vdots$	$\vdots$				
$\Gamma^{(k)}$	$\chi_1^{(k)}$	...	...	...	$\chi_k^{(k)}$

It was shown earlier that the $g = 6$ symmetry operations of the point group $\mathscr{C}_{3v}$ can be divided into three classes: E, $2C_3$, and $3\sigma_v$. Thus, $k = 3$ and Eqn (9) becomes

$$d_1^2 + d_2^2 + d_3^2 = 6. \tag{11}$$

Only the set of integers 1, 1, 2, satisfies this relation, the order being arbitrary. Therefore, in the point group $\mathscr{C}_{3v}$ there are two different irreducible representations of order one and one of order two, and the characters appearing in the first column of the character table (Table 4) are accounted for. Furthermore, all of the characters in the first row are unity, identifying $\Gamma^{(1)}$ as the totally symmetric irreducible representation.

TABLE 4

$\mathscr{C}_{3v}$	E	$2C_3$	$3\sigma_v$
$\Gamma^{(1)}$	1	1	1
$\Gamma^{(2)}$	1	1	-1
$\Gamma^{(3)}$	2	-1	0

The construction of the remainder of Table 4 is accomplished by application of the orthogonality property, Eqn (10). These results can be easily verified. For example, from the second line, the characters of $\mathbf{\Gamma}^{(2)}$, one finds

$$\tfrac{1}{6}[1(1)^2 + 2(1)^2 + 3(-1)^2] = 1, \tag{12}$$

and from the first and second lines

$$\tfrac{1}{6}[1(1 \times 1) + 2(1 \times 1) + 3(-1 \times 1)] = 0, \tag{13}$$

in agreement with Eqn (10). In this way the tables of characters for the various point groups can be easily developed [See (**2**), Chapter VI]. Character tables for most of the common point groups are given in Appendix C.

In molecular spectroscopy the term *symmetry species* is usually substituted for the rather unwieldy expression "irreducible representation". Therefore, in the following discussions, the $\mathbf{\Gamma}^{(\gamma)}$'s will often be referred to as symmetry species, while the spectroscopic terms *degeneracy* will sometimes be used to describe their dimensions. In addition, a more-or-less standard notation has been adopted for the various symmetry species (see Appendix C).

IX. "Magic Formula"

It has been argued above that the number of irreducible representations in any point group is equal to the number of classes, k. Thus, Eqn (8) becomes

$$\mathbf{\Gamma} = \sum_{\gamma=1}^{k} n^{(\gamma)} \mathbf{\Gamma}^{(\gamma)}. \tag{14}$$

Consider now, the matrix $\mathbf{\Gamma}$ which represents the operation R. The trace of this matrix, χ_R, is the character of R in $\mathbf{\Gamma}$, a quantity which is independent of the choice of basis coordinates. Since χ_R is merely the sum of the diagonal elements of $\mathbf{\Gamma}$, it is also equal to the sum of the traces of the individual submatrices $\mathbf{\Gamma}^{(\gamma)}$, each multiplied by $n^{(\gamma)}$, the number of times that each $\mathbf{\Gamma}^{(\gamma)}$ appears in $\mathbf{\Gamma}$. It follows that

$$\chi_R = \sum_{\gamma=1}^{k} n^{(\gamma)} \chi_R{}^{(\gamma)}, \tag{15}$$

an expression which holds for each operation R of $\mathfrak{G}$. However, as all operations of a given class have the same characters, only k different equations of the type of Eqn (15) can be formed, one for each class. Hence,

$$\chi_j = \sum_{\gamma=1}^{k} n^{(\gamma)} \chi_j{}^{(\gamma)}, \qquad j = 1, 2, \ldots, k. \tag{16}$$

where $\chi_j^{(\gamma)}$ is the character in $\Gamma^{(\gamma)}$ of an operation of class $\mathscr{K}_j$, and the $\chi_j^{(\gamma)}$'s are the characters of the irreducible representations $\Gamma^{(\gamma)}$ of the jth class.

The set of simultaneous equations (16) can be solved using the orthogonality properties of the characters [Eqn (10)]. Multiplication of both sides of Eqn (16) by $\chi_j^{(\gamma')*}$ weighted by the factor g_j, and summation over the class index j, yield the expression

$$\sum_{j=1}^{k} g_j \chi_j^{(\gamma')*} \chi_j = \sum_{j=1}^{k} g_j \chi_j^{(\gamma')*} \left[\sum_{\gamma=1}^{k} n^{(\gamma)} \chi_j^{(\gamma)} \right]$$

$$= \sum_{\gamma=1}^{k} n^{(\gamma)} \left[\sum_{j=1}^{k} g_j \chi_j^{(\gamma')*} \chi_j^{(\gamma)} \right]. \tag{17}$$

However, from Eqn (10) the quantity in brackets is equal to $g\delta_{\gamma'\gamma}$ and Eqn (17) becomes

$$\sum_{j=1}^{k} g_j \chi^{(\gamma')*} \chi_j = \sum_{\gamma=1}^{k} n^{(\gamma)} g\delta_{\gamma'\gamma} = gn^{(\gamma)} \tag{18}$$

or

$$n^{(\gamma)} = \frac{1}{g} \sum_{j=1}^{k} g_j \chi_j^{(\gamma)*} \chi_j. \tag{19}$$

This result allows $n^{(\gamma)}$, the number of times that the γth irreducible representation appears in a reducible representation whose characters are the χ_j's, to be calculated using the elements $\chi_j^{(\gamma)}$ of the character table of the point group. Equation (19) is of such widespread applicability that it is referred to by many students of group theory as the "magic formula".

X. Classification of Molecular Motions

One obvious application of Eqn (19) is the determination of the symmetry species (irreducible representations) corresponding to the normal modes, and, hence, the vibrational fundamentals, of a polyatomic molecule. As an example, consider again the ammonia molecule, NH_3, whose symmetry is completely described by the operations of point group $\mathscr{C}_{3v}$.

In order to apply Eqn (19) to this problem, it is necessary to evaluate χ_j for each class of operations. As this quantity is independent of the choice of basis coordinates, it can be easily evaluated by generalizing Eqn (5). This transformation corresponds to the effect of any proper operation on a given atom which is not shifted by the symmetry operation. For improper operations, $z \to -z$, and the general transformation is given by

$$\mathbf{R}_j = \begin{pmatrix} \cos\phi_j & -\sin\phi_j & 0 \\ \sin\phi_j & \cos\phi_j & 0 \\ 0 & 0 & \pm 1 \end{pmatrix} \tag{20}$$

for each unshifted atom.

Hence,

$$\chi_j = m_j(2\cos\phi_j \pm 1). \tag{21}$$

where m_j is the number of atoms which are not shifted by an operation of class j, and the sign is positive or negative as the operation is, respectively, proper or improper. As $3N$ basis coordinates have been used in this case, the six external modes of translation and rotation have been automatically included in the problem. The application of the above principles to the case of ammonia results in the characters, $\chi_j(\text{tot})$, given in Table 5. Then, using the magic formula, Eqn (19), the values of $n_{\text{tot}}^{(\gamma)}$ given in the first column to the right of the character table of Table 5 are easily calculated. Here, the irreducible representations are designated by the standard symbols defined in Appendix C.

TABLE 5

$\mathscr{C}_{3v}$	E	$2C_3$	$3\sigma_v$	$n_{\text{tot}}^{(\gamma)}$	$n_{\text{trans}}^{(\gamma)}$	$n_{\text{rot}}^{(\gamma)}$	$n_{\text{vib}}^{(\gamma)}$
A_1	1	1	1	3	1	0	2
A_2	1	1	−1	1	0	1	0
E	2	−1	0	4	1	1	2
m_j	4	1	2				
χ_j (tot)	12	0	2				
χ_j (trans)	3	0	1				
χ_j (rot)	3	0	−1				

From the above results, given as $n_{\text{tot}}^{(\gamma)}$ of Table 5, must be subtracted the contributions of the translational and rotational degrees of freedom. The displacement of the center of gravity, g, of the molecule (See Fig. 1 in Chapter 1), can be specified by a vector **T**, which transforms as Eqn (20). Thus, the characters for translation are given by

$$\chi_j(\text{trans}) = 2\cos\phi_j \pm 1. \tag{22}$$

To find the corresponding results for the three rotational degrees of freedom, consider the angular momentum vector

$$\mathfrak{M} = \mathbf{T} \times \mathbf{p} = \begin{vmatrix} \Delta x & \Delta y & \Delta z \\ p_x & p_y & p_z \\ \mathbf{i} & \mathbf{j} & \mathbf{k} \end{vmatrix}. \tag{23}$$

After a symmetry operation R_j, this vector becomes

$$\mathfrak{M}' = \mathbf{T}' \times \mathbf{p}', \tag{24}$$

with $\mathbf{T}' = \mathbf{R}_j\mathbf{T}$ and $\mathbf{p}' = \mathbf{R}_j\mathbf{p}$. The effect of the symmetry operation R_j on the angular momentum vector can be expressed as

$$\mathfrak{M}' = \mathbf{P}_j\mathfrak{M}, \tag{25}$$

where $\mathbf{P}_j$ can be found by expansion of Eqns (23) and (24). The result

$$\mathbf{P}_j = \begin{pmatrix} \pm\cos\phi_j & \mp\sin\phi_j & 0 \\ \pm\sin\phi_j & \pm\cos\phi_j & 0 \\ 0 & 0 & 1 \end{pmatrix},$$

leads to the characters

$$\chi_j(\text{rot}) = \text{Trace}\,\mathbf{P}_j = \pm 2\cos\phi_j + 1 \tag{26}$$

for the three rotational degrees of freedom. For proper operations, this result is identical to that found for translation [Eqn (22)], while for improper operations it takes the opposite sign.†

Returning now to the ammonia example, the characters corresponding to the external degrees of freedom can be determined from the expressions found above. These results, which are given in Table 5, have been used with Eqn (19) to calculate $n_{\text{trans}}^{(\gamma)}$ and $n_{\text{rot}}^{(\gamma)}$. For each symmetry species the number of fundamental vibrations is then found by difference, as given in the last column.

† The method outlined in this paragraph is applicable only to nonlinear molecules. In the case of linear molecules, the angular momentum has only two components. [See Chapter 4 and reference (6)].

That is, the structure of the reduced representation of the vibrational degrees of freedom of ammonia is given by†

$$\mathbf{\Gamma} = 2A_1 + 2E. \tag{27}$$

The method outlined above for determining the distribution of the $3N - 6$ vibrational modes among the symmetry species is based on the use of Cartesian coordinates. As these $3N$ coordinates are always linearly independent and form a complete set, this method is completely general. However, it has the disadvantage that it always includes the six external degrees of freedom which must then be subtracted from the complete problem. An alternative method depends on the use of a set of internal coordinates as the basis.

XI. Internal Coordinates and Redundancy

As an illustration, define a set of internal coordinates for ammonia consisting of the three changes in bond lengths, Δr_1, Δr_2, Δr_3, and the interbond angles, $\Delta\alpha_{12}$, $\Delta\alpha_{23}$, and $\Delta\alpha_{31}$, as shown in Fig. 10. These six coordinates,

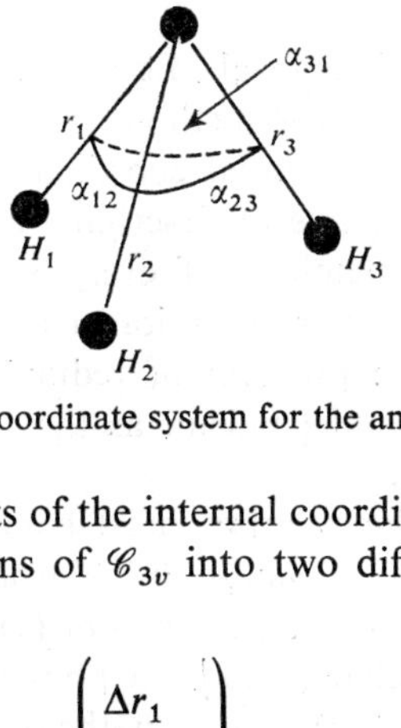

FIG. 10. Internal coordinate system for the ammonia molecule.

which form the components of the internal coordinate matrix, **S**, are divided by the symmetry operations of $\mathscr{C}_{3v}$ into two different symmetrically equivalent sets; that is,

$$\mathbf{S} = \begin{pmatrix} \Delta r_1 \\ \Delta r_2 \\ \Delta r_3 \\ \hdashline \Delta\alpha_{12} \\ \Delta\alpha_{23} \\ \Delta\alpha_{31} \end{pmatrix} = \begin{pmatrix} \mathbf{S}_r \\ \hdashline \mathbf{S}_\alpha \end{pmatrix}. \tag{28}$$

† Following the conventional notation (Appendix C), the symbols for the irreducible erpresentations of Eqn (27) are not shown in bold-face type, even though they represent, in general, matrix quantities. The plus sign in this case indicates that the direct sum is to be taken.

The characters, χ_j, for each symmetrical set can be found by observing the effect of each of the group operations on the coordinates of the set. Those coordinates which are shifted by a symmetry operation do not contribute to the character, while each of those which remains fixed contributes the value +1. Thus, for the coordinates $\mathbf{S}_r$ of ammonia $\chi_E = 3$, $\chi_{2C_3} = 0$, and $\chi_{3\sigma_v} = 1$. Applying Eqn (19) to this result, and using the character table (Table 6), the structure of the representation of $\mathbf{S}_r$ is found to be

$$\mathbf{\Gamma}(r) = A_1 + E. \tag{29}$$

Consideration of the angular coordinates, $\mathbf{S}_\alpha$, leads to the identical result,

$$\mathbf{\Gamma}(\alpha) = A_1 + E. \tag{30}$$

Hence, the structure of the reduced representation of the six vibrational degrees of freedom of ammonia becomes

$$\mathbf{\Gamma}(\text{vib}) = \mathbf{\Gamma}(r) + \mathbf{\Gamma}(\alpha) = 2A_1 + 2E, \tag{31}$$

in agreement with the previous result, Eqn (27).

In the application of the above method, care must be taken in the choice of internal coordinates. All essential coordinates must be included, in the sense that they must be capable of describing all possible atomic displacements. Furthermore, all members of every symmetrically equivalent set must be represented. The latter condition may require more than $3N - 6$ coordinates, leading to the problem of redundancy. In this case the coordinates are not linearly independent, as they are interrelated by certain redundancy conditions (7).

The problem of redundancy arises most commonly when four or more bonds are attached to a single atom. In the chloroform molecule, for example, a logical set of internal coordinates consists of (1) the variation in the C—H bond length, (2) the three changes in C—Cl bond lengths. (3) changes in the three Cl—C—Cl angles, and (4) changes in the three angles H—C—Cl. Thus,

$$\mathbf{S}^\dagger = (\Delta R\ \Delta r_1\ \Delta r_2\ \Delta r_3\ \Delta\alpha_{12}\ \Delta\alpha_{23}\ \Delta\alpha_{31}\ \Delta\beta_1\ \Delta\beta_2\ \Delta\beta_3), \tag{32}$$

where the coordinates are defined in Fig. 11. Ten internal coordinates have in this case been chosen to describe the $3 \times 5 - 6 = 9$ vibrational degrees of freedom.

Using the method described earlier, based on Cartesian coordinates, the $3N = 15$ total degrees of freedom of this molecule are found to correspond to the structure

$$\Gamma(\text{tot}) = 4A_1 + A_2 + 5E, \tag{33}$$

and, hence,

$$\Gamma(\text{vib}) = 3A_1 + 3E. \tag{34}$$

However, the distribution in the various symmetry species of the internal coordinates defined by Eqn (32) is found as given in Table 7. The sum, based on these internal coordinates, is given by

$$\Gamma(\text{int}) = 4A_1 + 3E. \tag{35}$$

Comparison of this result with Eqn (34) shows that one fictitious degree of freedom of species A_1 has been included in the internal coordinate system **S**.

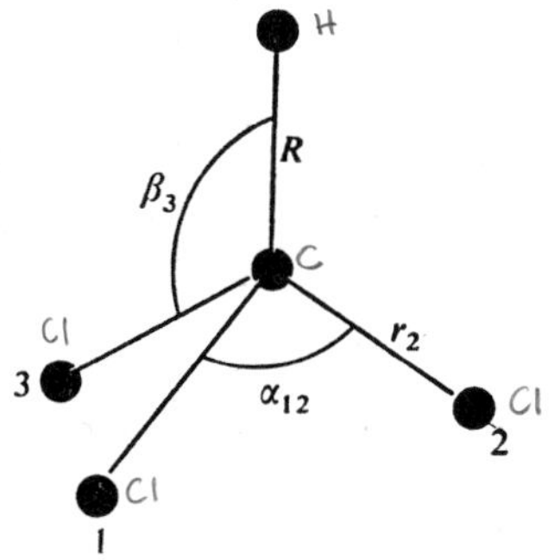

FIG. 11. Internal coordinate system for the chloroform molecule.

This redundancy results from the lack of independence of the six interbond angles, the specific redundancy condition corresponding to the relation

$$\Delta\alpha_{12} + \Delta\alpha_{23} + \Delta\alpha_{31} + \Delta\beta_1 + \Delta\beta_2 + \Delta\beta_3 = 0. \tag{36}$$

TABLE 6

$\mathscr{C}_{3v}$	E	$2C_3$	$3\sigma_v$	$n_r^{(\gamma)}$	$n_\alpha^{(\gamma)}$	$n_{\text{vib}}^{(\gamma)}$
A_1	1	1	1	1	1	2
A_2	1	1	−1	0	0	0
E	2	−1	0	1	1	2
$\chi_j(\mathbf{S}_r)$	3	0	1			
$\chi_j(\mathbf{S}_\alpha)$	3	0	1			

In setting up the vibrational problem in internal coordinates it is generally simpler to use symmetrically complete sets. As will be seen later, the advantages provided by the resulting symmetry properties of the coordinates more than outweigh the inconvenience caused by the addition of the few extra redundant coordinates.

TABLE 7

$\mathscr{C}_{3v}$	E	$2C_3$	$3\sigma_v$	$n_{tot}^{(\gamma)}$	$n_{vib}^{(\gamma)}$	$n_{int}^{(\gamma)}$	$n_{red}^{(\gamma)}$
A_1	1	1	1	4	3	4	1
A_2	1	1	−1	1	0	0	0
E	2	−1	0	5	3	3	0
χ_j (tot)	15	0	3	cart.			
χ_j (R)	1	1	1	A_1			
χ_j (r)	3	0	1	A_1+E			
χ_j (α)	3	0	1	A_1+E			
χ_j (β)	3	0	1	A_1+E			

XII. SYMMETRY COORDINATES

The general problem of the vibrations of polyatomic molecules was treated in Chapter 1 using the **F**- and **G**-matrix method. It was shown that when the problem is formulated in terms of internal coordinates, the vibrational frequencies can be found by solution of the secular equation, Eqn (36) in Chapter 1. In the first part of the present chapter, the question of molecular symmetry has been discussed. It is now appropriate to consider the role of molecular symmetry in the formulation of the vibration problem.

As a simple example, consider again the water molecule, having $3 \times 3 - 6 = 3$ vibrational degrees of freedom and belonging to point group $\mathscr{C}_{2v}$. The internal coordinates chosen in Chapter 1 are changes in the two O—H bond lengths and the variation in the bond angle. Thus, $\mathbf{S}^\dagger = (\Delta r_1\ \Delta r_2\ \Delta\alpha)$ and the symmetry properties of these coordinates are found by consideration of the character table for $\mathscr{C}_{2v}$, as shown in Table 8.

The coordinate $\Delta\alpha$, which alone forms a complete symmetrically equivalent set, corresponds to symmetry species A_1. That is, it is invariant under all four operations of the point group. On the other hand, the coordinates Δr_1 and Δr_2 together form a second complete set, and, although neither of these two coordinates has symmetry properties which are consistent with the point group, there exist certain linear combinations of them which do possess such properties. The results given in Table 8 indicate that one such combination must be totally symmetric (species A_1), while the other has the symmetry of species B_2. It is easily verified that the linear combinations $\Delta r_1 + \Delta r_2$ and $\Delta r_1 - \Delta r_2$ have the desired properties. The first is clearly unchanged by the four symmetry operations of $\mathscr{C}_{2v}$, while the second is modified by neither the operations E nor $\sigma_v(yz)$, but is multiplied by -1 by the operations C_2 or $\sigma_v(zx)$, in agreement with the characters given in row B_2 of the character table.

TABLE 8

$\mathscr{C}_{2v}$	E	C_2	$\sigma_v(zx)$	$\sigma_v(yz)$	$n_r^{(\gamma)}$	$n_\alpha^{(\gamma)}$
A_1	1	1	1	1	1	1
A_2	1	1	-1	-1	0	0
B_1	1	-1	1	-1	0	0
B_2	1	-1	-1	1	1	0
$\chi(r)$	2	0	0	2		
$\chi(\alpha)$	1	1	1	1		

Coordinates such as these described above, which have the symmetry properties of the point group, are called *symmetry coordinates.* As they transform in the same manner as the irreducible representations when used as basis coordinates, they factor the secular determinant into block–diagonal form, by analogy with Fig. 9. Thus, while normal coordinates must be found in order to diagonalize completely the secular determinant, the factorization resulting from the use of symmetry coordinates often provides considerable simplification of the vibrational problem. Furthermore, symmetry coordinates can be chosen *a priori* by analysis of the molecular symmetry alone.

In the example considered above, $\Delta r_1 - \Delta r_2$ is the only symmetry coordinate of species B_2. Thus, it results in a factor of degree one in the completely reduced secular determinant, and, therefore, must also be a normal coordinate. On the other hand, the two normal coordinates of species A_1 are certain linear combinations of the symmetry coordinates $\Delta\alpha$ and $\Delta r_1 + \Delta r_2$ which can only be found by solution of the secular equations.

It is usually found to be convenient to normalize the symmetry coordinates. Hence, for water, the three symmetry coordinates are given by

$$\text{Species } A_1: \begin{cases} \mathrm{S}_1 = \Delta\alpha \\ \mathrm{S}_2 = \dfrac{1}{\sqrt{2}}(\Delta r_1 + \Delta r_2), \end{cases} \qquad (37)$$

$$\text{Species } B_2: \mathrm{S}_3 = \frac{1}{\sqrt{2}}(\Delta r_1 - \Delta r_2),$$

where the normalizing constant $\mathfrak{N}_p$ is determined by the condition

$$\mathfrak{N}_p{}^2 \sum_t S_t{}^2 = 1 \qquad (38)$$

for each symmetry coordinate S_p. In general form, the relations of Eqn (37) are expressed by the matrix transformation

$$\mathbf{S} = \mathbf{U}\mathrm{S}, \qquad (39)$$

which for the above example becomes

$$\begin{pmatrix} \mathrm{S}_1 \\ \mathrm{S}_2 \\ \mathrm{S}_3 \end{pmatrix} = \begin{pmatrix} 0 & 0 & 1 \\ 1/\sqrt{2} & 1/\sqrt{2} & 0 \\ 1/\sqrt{2} & -1/\sqrt{2} & 0 \end{pmatrix} \begin{pmatrix} \Delta r_1 \\ \Delta r_2 \\ \Delta\alpha \end{pmatrix}. \qquad (40)$$

It should be noted that when the symmetry coordinates have been normalized, $\mathbf{U}$ is orthogonal; that is, $\mathbf{U}^\dagger = \mathbf{U}^{-1}$.

A general rule for the construction of symmetry coordinates can be written in the form [(1), p. 119]

$$\mathrm{S}_p{}^{(\gamma)} = \mathfrak{N}_p \sum_R \chi_R{}^{(\gamma)}(RS_1), \qquad (41)$$

where (RS_1) represents the coordinate which results when the operation R acts on a chosen internal coordinate, S_1. The summation in Eqn (41) is taken over those operations which are necessary to generate all internal coordinates in the equivalent set.

In order to show how use is made of symmetry coordinates as the bases of the vibrational problem, reconsider the kinetic and potential energies as given earlier,

$$2T = \dot{\mathbf{S}}^{\dagger}\,\mathbf{G}^{-1}\,\dot{\mathbf{S}} = \dot{\mathbf{Q}}^{\dagger}\,\mathbf{E}\,\dot{\mathbf{Q}} \tag{42}$$

and

$$2V = \mathbf{S}^{\dagger}\,\mathbf{F}\,\mathbf{S} = \mathbf{Q}^{\dagger}\,\mathbf{\Lambda}\,\mathbf{Q}. \tag{43}$$

From Eqn (39),

$$\mathbf{S} = \mathbf{U}^{-1}\,\boldsymbol{\mathsf{S}} = \mathbf{U}^{\dagger}\,\boldsymbol{\mathsf{S}}, \tag{44}$$

hence,

$$2T = \dot{\boldsymbol{\mathsf{S}}}^{\dagger}\,\mathbf{U}\mathbf{G}^{-1}\,\mathbf{U}^{\dagger}\,\dot{\boldsymbol{\mathsf{S}}} \tag{45}$$

and

$$2V = \boldsymbol{\mathsf{S}}^{\dagger}\,\mathbf{U}\,\mathbf{F}\,\mathbf{U}^{\dagger}\,\boldsymbol{\mathsf{S}}, \tag{46}$$

which suggests the definition of kinetic and potential energy matrices based on symmetry coordinates, namely,

$$\boldsymbol{\mathsf{G}}^{-1} = \mathbf{U}\,\mathbf{G}^{-1}\,\mathbf{U}^{\dagger}, \tag{47}$$

or

$$\boldsymbol{\mathsf{G}} = \mathbf{U}\,\mathbf{G}\,\mathbf{U}^{\dagger}, \tag{48}$$

and

$$\boldsymbol{\mathsf{F}} = \mathbf{U}\,\mathbf{F}\,\mathbf{U}^{\dagger}. \tag{49}$$

Thus, the kinetic and potential energies become

$$2T = \dot{\boldsymbol{\mathsf{S}}}^{\dagger}\,\boldsymbol{\mathsf{G}}^{-1}\,\dot{\boldsymbol{\mathsf{S}}} = \dot{\mathbf{Q}}^{\dagger}\,\mathbf{E}\,\dot{\mathbf{Q}} \tag{50}$$

and

$$2V = \boldsymbol{\mathsf{S}}^{\dagger}\boldsymbol{\mathsf{F}}\,\boldsymbol{\mathsf{S}} = \mathbf{Q}^{\dagger}\,\mathbf{\Lambda}\,\mathbf{Q}, \tag{51}$$

respectively.

Equations (50) and (51) lead to the secular determinant as before,

$$|\boldsymbol{\mathsf{G}}\,\boldsymbol{\mathsf{F}} - \mathbf{E}\lambda_{k}| = 0. \tag{52}$$

The eigenvectors, $\boldsymbol{\mathsf{L}}$, can be found, since from Eqns (48) and (49),

$$\boldsymbol{\mathsf{GF}} = \mathbf{U}\mathbf{G}\mathbf{U}^{\dagger}\,\mathbf{U}\mathbf{F}\mathbf{U}^{\dagger} = \mathbf{U}\mathbf{G}\mathbf{F}\mathbf{U}^{-1}; \tag{53}$$

hence,

$$\mathbf{U}^{-1}\,\mathbf{GFU} = \mathbf{GF} \tag{54}$$

and

$$\mathbf{L}^{-1}\,\mathbf{U}^{-1}\,\mathbf{GFUL} = \mathbf{L}^{-1}\,\mathbf{GFL} = \mathbf{\Lambda}. \tag{55}$$

$$\therefore \mathbf{L} = \mathbf{UL}. \tag{56}$$

Returning now to the example of the water molecule, and using **F** given by Eqn (53), Chap. 1, and **U** from Eqn (40), the transformation $\mathbf{UFU}^\dagger$ gives

$$\mathbf{F} = \left(\begin{array}{cc|c} f_r + f_{rr} & \sqrt{2}f_{r\alpha} & 0 \\ \sqrt{2}f_{r\alpha} & f_\alpha & 0 \\ \hline 0 & 0 & f_r - f_{rr} \end{array}\right). \tag{57}$$

Similarly, taking **G** from Eqn (52), Chap. 1, the transformation $\mathbf{UGU}^\dagger$ becomes

$$\mathbf{G} = \left(\begin{array}{cc|c} \dfrac{2\mu_H + 2\mu_O(1 - \cos\alpha)}{r^2} & \dfrac{2\mu_O \sin\alpha}{r} & 0 \\ \dfrac{2\mu_O \sin\alpha}{r} & \mu_H + \mu_O(1 + \cos\alpha) & 0 \\ \hline 0 & 0 & \mu_H + \mu_O(1 - \cos\alpha) \end{array}\right). \tag{58}$$

As both **F** and **G** are factored by the use of symmetry coordinates, the secular determinant is similarly factored. The problem of calculating the vibrational frequencies is thus divided into two parts: solution of a linear equation for the single frequency of species B_2 and of a quadratic equation for the pair of frequencies of species A_1.

The symmetry coordinates introduced in this section are examples of internal symmetry coordinates. In some applications, particularly to crystals, it is found to be convenient to define external symmetry coordinates which are suitable linear combinations of Cartesian coordinates. The method of constructing such external symmetry coordinates is completely analogous to that of Eqn (41).

XIII. Vibrational Selection Rules

In order to determine the selection rules governing vibrational transitions in a polyatomic molecule, it is necessary to consider the symmetry properties of the vibrational wave functions. For example, in the harmonic approximation the wave function corresponding to the kth normal mode is essentially the Hermite polynomial $H_{v_k}(Q_k)$†. Since $H_0(Q_k) = 1$, the ground state has all of the symmetry of the undistorted molecule, that is, the symmetry described by the molecular point group. For the first excited state of vibration,

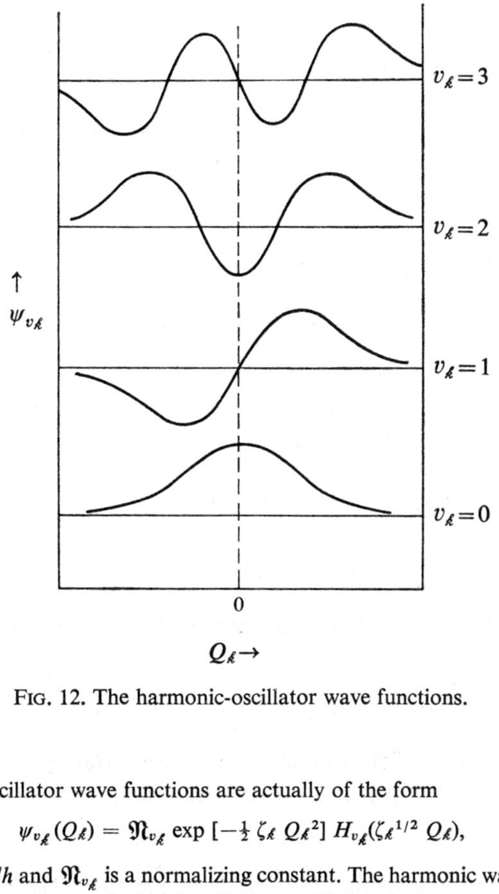

FIG. 12. The harmonic-oscillator wave functions.

† The harmonic-oscillator wave functions are actually of the form

$$\psi_{v_k}(Q_k) = \mathfrak{N}_{v_k} \exp\left[-\tfrac{1}{2}\zeta_k Q_k^2\right] H_{v_k}(\zeta_k^{1/2} Q_k),$$

where $\zeta_k = 4\pi^2 \nu_k/h$ and $\mathfrak{N}_{v_k}$ is a normalizing constant. The harmonic wave function for a molecule with 3N–6 degrees of vibration freedom will then be the product of 3N–6 such functions and the exponential factor will be of the form

$$\exp\left[-\tfrac{1}{2}\sum_{k=1}^{3N-6} \zeta_k Q_k^2\right].$$

As this factor is invariant under all operations of the molecular symmetry group, the symmetry of ψ_{v_k} is determined by the corresponding Hermite polynomial, $H_{v_k}(\zeta_k^{1/2} Q_k)$.

$v_k = 1$ for one value of k. And, since $H_1(Q_k) = 2Q_k$,† this state has just the symmetry of the normal coordinate Q_k, as shown in Fig. 12.

The principle illustrated above can be generalized by considering the effect of any symmetry operation R on a given vibrational wave function. If n represents the ground vibrational state, the result of a symmetry operation R can be expressed as

$$\psi_n \xrightarrow{R} \psi_n \tag{59}$$

for all operations R of the point group, because, as indicated above, the ground state always belongs to the totally symmetric representation.

For any other vibrational state m, symmetry coordinates can be used as the basis set. That is, the wave function can always be written as some linear combination of $\psi^{(\gamma)}$'s, each of which belongs to an irreducible representation γ of the point group. If the γth symmetry species is non-degenerate, the effect of an operation R of class j can be represented by

$$\psi^{(\gamma)} \xrightarrow{R} \chi_j^{(\gamma)} \psi^{(\gamma)}. \tag{60}$$

This statement is the origin of the coefficients ± 1 used earlier in the construction of symmetry coordinates for the water molecule. In the event γ is a degenerate species, further considerations enter.

The dipole moment components can also be expressed on the basis of symmetry coordinates. Thus, for example,

$$\mathfrak{m}_x^{(\gamma')} \xrightarrow{R} \chi_j^{(\gamma')} \mathfrak{m}^{(\gamma')}. \tag{61}$$

and the effect of R on the dipole moment matrix elements can be written

$$\int \psi_n^* \mathfrak{m}_x^{(\gamma')} \psi_m^{(\gamma)} \, d\tau \xrightarrow{R} \chi_j^{(\gamma')} \chi_j^{(\gamma)} \int \psi_n^* \mathfrak{m}_x^{(\gamma')} \psi_m^{(\gamma)} \, d\tau = \int \psi_n^* \mathfrak{m}_x^{(\gamma')} \psi_m^{(\gamma)} \, d\tau, \tag{62}$$

as the matrix elements of the dipole moment are certainly invariant under the group operations. Multiplying by g_j and summing over all of the classes j, Eqn (62) becomes

$$\sum_j g_j \chi_j^{(\gamma')} \chi_j^{(\gamma)} \int \psi_n^* \mathfrak{m}_x^{(\gamma')} \psi_m^{(\gamma)} d\tau = g \int \psi_n^* \mathfrak{m}_x^{(\gamma')} \psi_m^{(\gamma)} \, d\tau. \tag{63}$$

From the orthogonality of the characters [Eqn (10)], the summation over j vanishes unless $\gamma' = \gamma$, in which case the dipole moment component being considered and the wave function of the upper state, m, belong to the same

† See reference (1), p. 289, for a tabulation of the Hermite polynomials.

species. This result, then, constitutes the selection rule for the transition $n \rightarrow m$ [See Eqn (68) of Chapter 1].

A more general statement of the selection rule developed above is that the nonvanishing of the integral of Eqn (63) requires that the direct product of the irreducible representations corresponding to the various factors in the integrand must contain the totally symmetric representation.† Thus, the reduction of the direct product,

$$\mathbf{\Gamma}^{(\gamma'')} \times \mathbf{\Gamma}^{(\gamma')} \times \mathbf{\Gamma}^{(\gamma)} = \sum_{l} n^{(l)} \mathbf{\Gamma}^{(l)} \tag{64}$$

can be used to determine the selection rule for any transition $n \rightarrow m$, where the state n, belonging to species γ'', is now not necessarily the ground state. The corresponding matrix elements will not necessarily be zero, then, if $n^{(1)} \neq 0$.

As an illustration of this general selection rule, consider the staggered configuration of ethane shown in Fig. 3. This structure is of point group $\mathscr{D}_{3d}$, thus the vibrational analysis can be carried out using Table 9 in the manner described earlier. The resulting structure of the representation for the 18 vibrational degrees of freedom is given in the last column. As the symmetry properties of the various normal modes are now specified, there remains the problem of determining the symmetry species of the various components of the molecular dipole moment. This problem has, in fact, already been solved, as it is evident that any vector, such as $\mathfrak{m}$, with components $\mathfrak{m}_x$, $\mathfrak{m}_y$, and $\mathfrak{m}_z$ in the molecule-fixed coordinate system, transforms exactly as the translation vector, $\mathbf{T}$. The equation derived earlier for χ_j(trans), Eqn (22), is hence directly applicable to the determination of the characters for the dipole moment components. The symmetry species of a given component of $\mathfrak{m}$ is then the same as that for the corresponding component of $\mathbf{T}$. From Table 9 it is seen that in $\mathscr{D}_{3d}$, T_z falls in species A_{2u}, while T_x and T_y are members of the doubly degenerate species E_u.

As the symmetry species of the components of the dipole moment are now known, the selection rule governing any optical transition can be determined by application of Eqn (64). For the vibrational fundamentals, infrared activity can be determined by inspection. If any component of the dipole moment appears in the same species as a given normal coordinate, Q_k, the fundamental transition corresponding to that coordinate can be active in the infrared spectrum. For combinations, the direct product of the symmetry

† If $\mathbf{A}$ is a square matrix of rank s and $\mathbf{B}$ is a square matrix of rank t, their direct product $\mathbf{A} \times \mathbf{B}$ is a square matrix of rank st whose elements consist of the products of all possible pairs of elements, one member taken each from $\mathbf{A}$ and $\mathbf{B}$.

species of the initial and final states must fall in the same species as a component of the dipole moment for infrared absorption to be allowed.

A completely analogous argument holds for the case of the Raman effect. However, in this case the dipole moment is induced in the system by the electric field of the exciting radiation; thus,

$$\mathfrak{m}_{\text{ind}} = \boldsymbol{\alpha}\mathscr{E}, \tag{65}$$

where $\mathscr{E}$ is the electric vector of the radiation and $\boldsymbol{\alpha}$ is the polarizability tensor,

$$\boldsymbol{\alpha} = \begin{pmatrix} \alpha_{xx} & \alpha_{xy} & \alpha_{xz} \\ \alpha_{yx} & \alpha_{yy} & \alpha_{yz} \\ \alpha_{zx} & \alpha_{zy} & \alpha_{zz} \end{pmatrix}, \tag{66}$$

which can usually be assumed to be symmetric [See, however, (**8**)]. The selection rules for Raman spectra are determined from the symmetry properties of the components of $\boldsymbol{\alpha}$ and the vibrational wave functions.

The effect of a symmetry operation on the induced moment and the electric field can be expressed as †

$$\mathbf{R}_j\mathfrak{m} = \mathfrak{m}' \tag{67}$$

and

$$\mathbf{R}_j\mathscr{E} = \mathscr{E}', \tag{68}$$

respectively, where the orthogonal matrix $\mathbf{R}_j$ is given by Eqn (20). Then, since

$$\mathfrak{m}' = \boldsymbol{\alpha}'\mathscr{E}', \tag{69}$$

$$\mathbf{R}_j\mathfrak{m} = \mathfrak{m}' = \boldsymbol{\alpha}'\mathbf{R}_j\mathscr{E} = \boldsymbol{\alpha}'\mathbf{R}_j\boldsymbol{\alpha}^{-1}\mathfrak{m}. \tag{70}$$

$$\therefore \mathbf{R}_j = \boldsymbol{\alpha}'\mathbf{R}_j\boldsymbol{\alpha}^{-1} \tag{71}$$

and

$$\boldsymbol{\alpha}' = \mathbf{R}_j\boldsymbol{\alpha}\mathbf{R}_j^{-1} = \mathbf{R}_j\boldsymbol{\alpha}\mathbf{R}_j^{\dagger}. \tag{72}$$

The last transformation, which can be evaluated using Eqns (20) and (66), leads to the characters

$$\chi_j(\boldsymbol{\alpha}) = 2\cos\phi_j[\pm 1 \pm 2\cos\phi_j]. \tag{73}$$

The resulting symmetry properties of the components of $\boldsymbol{\alpha}$, or certain of their linear combinations, are included in the character tables of Appendix C (See also Appendix G).

† The subscript ind of Eqn (65) has been dropped.

TABLE 9

$\mathscr{D}_{3d}$	E	$2C_3$	$3C_2$	i	$2S_6$	$3\sigma_d$		$n_{tot}^{(\gamma)}$	$n_{vib}^{(\gamma)}$
A_{1g}	1	1	1	1	1	1	$\alpha_{xx} + \alpha_{yy}, \alpha_{zz}$	3	3
A_{2g}	1	1	−1	1	1	−1	R_z	1	0
E_g	2	−1	0	2	−1	0	$(R_x, R_y)(\alpha_{xx} - \alpha_{yy}, \alpha_{xy})(\alpha_{yz}, \alpha_{zx})$	4	3
A_{1u}	1	1	1	−1	−1	−1		1	1
A_{2u}	1	1	−1	−1	−1	1	T_z	3	2
E_u	2	−1	0	−2	1	0	(T_x, T_y)	4	3
χ_j (tot)	24	0	0	0	0	4		24	18

As pointed out earlier in this section, selection rules for vibrational fundamentals can be determined by inspection of the character table for the molecule symmetry group. Coincidence of a component of the dipole moment or the polarizability, and hence their derivatives, with a given vibrational mode in the same symmetry species is a necessary condition for infrared or Raman activity, respectively. The selection rules for the overtones of nondegenerate vibrations can be derived by arguments similar to those presented for the fundamentals. It is found that all levels involving even values of the quantum number v_k are totally symmetric, while those associated with odd values of v_k have the same symmetry as the corresponding normal mode. For excited levels of degenerate vibrations, the problem is somewhat more difficult. It can be shown, for example, that the characters corresponding to an excited vibration of a doubly degenerate mode with quantum number v_k are given by [(**1**), Section 7–3)]

$$\chi_{v_k}(R) = \tfrac{1}{2}[\chi(R)\chi_{v_k-1}(R) + \chi(R^{v_k})]. \tag{74}$$

For triply degenerate frequencies, the analogous expression is

$$\chi_{v_k}(R) = \tfrac{1}{3}(2\chi(R)\chi_{v_k-1}(R) + \tfrac{1}{2}\{\chi(R^2) - [\chi(R)]^2\}\chi_{v_k-2}(R) + \chi(R^{v_k})). \tag{75}$$

In using the above equations, the conventions $\chi_1(R) = \chi(R)$, $\chi_0(R) = 1$, and $\chi_{-n}(R) = 0$ are adopted.

To illustrate the application of the above rules, consider the methane molecule, of symmetry $\mathcal{T}_d$, whose character table is given in Table 10. The structure of the representation of the fundamentals is given by

$$\Gamma_{(\mathrm{vib})} = A_1 + E + 2F_2. \tag{76}$$

From Table 10 it is apparent that all fundamentals are Raman-active, while only the two of species F_2 are active in the infrared absorption spectrum.

The selection rules for combinations are determined from the characters which are the corresponding products of the characters of the species involved. For a binary combination of a frequency of species E with another of species F_2, element-by-element multiplication of the appropriate rows yields the values of $\chi_j(E \times F_2)$ shown in Table 10. Application of the magic formula then gives the result

$$E \times F_2 = F_1 + F_2. \tag{77}$$

Thus, since Eqn (77) contains species F_2, the combination $E \times F_2$ is active in the infrared spectrum.

TABLE 10

$\mathscr{T}_d$	E	$8C_3$	$3C_2$	$6S_4$	$6\sigma_d$		$n^{(\gamma)}_{\text{tot}}$	$n^{(\gamma)}_{\text{vib}}$	$n^{(\gamma)}_{E\times F_2}$	$n^{(\gamma)}_{F_2\times F_2}$	$n^{(\gamma)}_{F_2^2}$
A_1	1	1	1	1	1	$\alpha_{xx}+\alpha_{yy}+\alpha_{zz}$	1	1	0	1	1
A_2	1	1	1	−1	−1		0	0	0	0	0
E	2	−1	2	0	0	$(2\alpha_{zz}-\alpha_{xx}-\alpha_{yy}, \alpha_{xx}-\alpha_{yy})$	1	1	0	1	1
F_1	3	0	−1	1	−1	**R**	1	0	1	1	0
F_2	3	0	−1	−1	1	**T** $(\alpha_{xy}, \alpha_{yz}, \alpha_{zx})$	3	2	1	1	1
χ_j (tot)	15	0	−1	−1	3						
$\chi_j(E\times F_2)$	6	0	−2	0	0						
$\chi_j(F_2\times F_2)$	9	0	1	1	1						
$\chi_j(F_2{}^2)$	3	0	3	1	1						

For combination of the two different vibrations of species F_2, the characters are simply the squares of the corresponding characters of F_2, which are represented by $\chi_j(F_2 \times F_2)$ in Table 10. Thus,

$$F_2 \times F_2 = A_1 + E + F_1 + F_2. \tag{78}$$

The overtones of nondegenerate fundamentals can be treated in a similar manner, or using the general principle stated above. In the present example, all overtones of the A_1 species vibration are forbidden in the infrared, although they are active in the Raman spectrum.

The selection rules for the overtones of either vibration of species F_2 require application of Eqn (75) for their determination. The results shown in Table 10 correspond to the reduction

$$(F_2)^2 = A_1 + E + F_2. \tag{79}$$

It should be noted that this result is quite different from the combination represented by Eqn (78); that is, $(F_2)^2 \neq F_2 \times F_2$.

XIV. Subgroups and Correlation Diagrams

It was pointed out earlier in this chapter that if the symmetry of a molecule is characterized by a point group $\mathfrak{G}$, certain of the symmetry operations of $\mathfrak{G}$ form a subgroup $\mathfrak{H}$. If each class of operations of $\mathfrak{H}$ contains all of the operations of the corresponding class of $\mathfrak{G}$, the group $\mathfrak{H}$ is called an *invariant subgroup* (or, sometimes, a *self-conjugate subgroup*) of $\mathfrak{G}$. Any irreducible representation of a group $\mathfrak{G}$ is a representation, but not necessarily an irreducible representation, of the subgroup $\mathfrak{H}$.

Consider as an example, the methane molecule, CH_4. Its molecular symmetry group is $\mathscr{T}_d$. If one of the hydrogen atoms is replaced by a deuterium atom, forming CH_3D, the symmetry is lowered to that of $\mathscr{C}_{3v}$. Substitution of a second atom of deuterium yields the compound CH_2D_2, whose symmetry is of point group $\mathscr{C}_{2v}$. In the analyses of the normal vibrations of the isotopic species in this series it is of interest to determine the reduction of the irreducible representations of $\mathscr{T}_d$ in the groups $\mathscr{C}_{3v}$ and $\mathscr{C}_{2v}$. In general, this type of "mapping" of the irreducible representations of a group $\mathfrak{G}$ on one of its subgroups $\mathfrak{H}$ is found by considering the characters associated with those symmetry operations of the group $\mathfrak{G}$ which are also symmetry operations of $\mathfrak{H}$.

Returning now to the methane example, the operations of $\mathscr{C}_{3v}$ are E, $2C_3$, and $3\sigma_v$. Only two of the eight C_3 operations of $\mathscr{T}_d$ are contained in $\mathscr{C}_{3v}$; thus, $\mathscr{C}_{3v}$ is not an invariant subgroup of $\mathscr{T}_d$. Similarly, only three of the reflections, σ_d, are symmetry operations of $\mathscr{C}_{3v}$. They become the operations of the class $3\sigma_v$ of the subgroup $\mathscr{C}_{3v}$. Those characters of $\mathscr{T}_d$ which are needed to find the mapping of $\mathscr{T}_d$ on $\mathscr{C}_{3v}$, which are given in Table 11, are copied from the character table of $\mathscr{T}_d$. If the entries in each line are compared successively with the elements of the character table for $\mathscr{C}_{3v}$, it is easily found that A_1 of $\mathscr{T}_d$ (Table 11) is identical to the line labelled A_1 in $\mathscr{C}_{3v}$. Similarly, species A_2 and E in Table 11 correspond, respectively, to A_2 and E of $\mathscr{C}_{3v}$. Thus, these representations of $\mathscr{T}_d$ are also irreducible representations of $\mathscr{C}_{3v}$.

TABLE 11

		E	$2C_3$	$3\sigma_v$	Species of $\mathscr{C}_{3v}$	Species of $\mathscr{C}_s$
$\mathscr{C}_{3v}$	A_1	1	1	1		A'
	A_2	1	1	-1		A''
	E	2	-1	0		$A' + A''$
$\mathscr{T}_d$	A_1	1	1	1	A_1	
	A_2	1	1	-1	A_2	
	E	2	-1	0	E	
	F_1	3	0	-1	$A_2 + E$	
	F_2	3	0	1	$A_1 + E$	

Consider now the characters 3, 0, and -1 of F_1 shown in Table 11. These values do not correspond to a species of $\mathscr{C}_{3v}$. Thus, although these values are characters of a representation in $\mathscr{C}_{3v}$, the representation is not an irreducible one. It can, however, be reduced in the usual way using the "magic formula". It is easily found that this representation has the structure

$$\Gamma_{F_1 \text{ in } \mathscr{T}_d} = A_2 + E \tag{80}$$

In other words $F_1 \Rightarrow A_2 + E$ forms part of the mapping of $\mathcal{T}_d$ on $\mathcal{C}_{3v}$. This result can be easily determined by inspection if it is noted that the sums taken column by column of the characters of A_2 and E in $\mathcal{C}_{3v}$ are equal to the entries following F_1 in Table 11. The complete mapping of a group onto one of its subgroups is often given in the form of a correlation diagram. For the present example the correlation between $\mathcal{T}_d$ and $\mathcal{C}_{3v}$ is shown on the left-hand side of Table 12.

TABLE 12

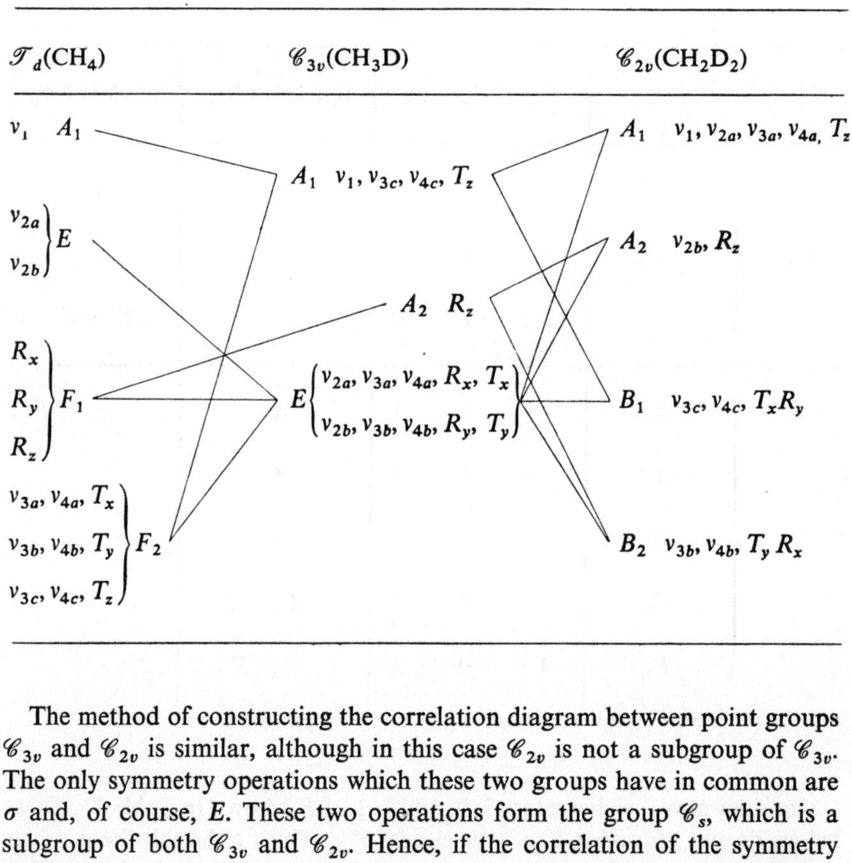

The method of constructing the correlation diagram between point groups $\mathcal{C}_{3v}$ and $\mathcal{C}_{2v}$ is similar, although in this case $\mathcal{C}_{2v}$ is not a subgroup of $\mathcal{C}_{3v}$. The only symmetry operations which these two groups have in common are σ and, of course, E. These two operations form the group $\mathcal{C}_s$, which is a subgroup of both $\mathcal{C}_{3v}$ and $\mathcal{C}_{2v}$. Hence, if the correlation of the symmetry species of $\mathcal{C}_{3v}$ and $\mathcal{C}_{2v}$ is needed, it can be found by mapping each of these groups onto its common subgroup, $\mathcal{C}_s$. The result for $\mathcal{C}_{3v}$ is summarized in Table 11.

Since the group $\mathcal{C}_s$ is associated with no chemical species in the series, it is of no direct significance and can be suppressed. The direct correlation is then made between the groups $\mathscr{C}_{3v}$ and $\mathscr{C}_{2v}$, as shown on the right-hand side of Table 12. However, it must be remembered that the construction of a correlation diagram of this type requires the passage between the two groups via their common subgroup.

In Chapters 1 and 2 the well-known theory of the vibrational spectra of polyatomic molecules has been summarized. The dynamical methods introduced in the treatment of the molecular vibration problem will now be extended to include the vibrations of crystal lattices. Later, in Chapter 4, crystal symmetry and the theory of space groups will be applied to the dynamics of crystal vibrations.

References

1. Wilson, E. B. Jr., Decius, J. C. and Cross, P. C. "Molecular Vibrations", McGraw-Hill, New York (1955).
2. Barchewitz, M.-P. "Spectroscopie infrarouge I. Vibrations moléculaires", Gauthier-Villars, Paris (1961).
3. Cotton, F. A. "Chemical Applications of Group Theory", Interscience, New York (1963).
4. Schonland, D. S. "Molecular Symmetry", Van Nostrand, London (1965).
5. Tinkham, M. "Group Theory and Quantum Mechanics", McGraw-Hill, New York (1964).
6. Mitra, S. S. *Zeit. Krist.* **116,** 149 (1961).
7. Decius, J. C. *J. Chem. Phys.* **17,** 1315 (1949).
8. Loudon, R. *Advan. Phys.* **13,** 423 (1964).

Chapter 3

Lattice Dynamics

Virtually every discussion of lattice dynamics begins with an analysis of the motion of a one-dimensional array of identical particles. This problem is one of the oldest in the history of mathematical physics, having been first treated by Newton in 1686. The propagation of an elastic wave along the chain of point masses was used by Newton as a model for the propagation of a sound wave in air. However, the sound velocity which he computed did not agree with the experimental value due to his use of the isothermal, rather than the adiabatic, bulk modulus of air. This error was corrected by Laplace in 1822, who showed that the use of the adiabatic constant in Newton's formula resulted in excellent agreement with experiment.

The question of the vibrations of one-dimensional lattices received considerable attention from theoreticians in the eighteenth century. Such great names as Taylor, Euler, the Bernoullis, Fourier, and Lagrange figure among the contributors to this problem. Their work provided the basis for numerous modern applications, including eigenvalue problems, Fourier expansion, and wave propagation, as well as the lattice theory of crystals.

The treatment of the dynamics of crystal lattices (**1–3**) is usually considered to originate with Born's one-dimensional model of sodium chloride (1912). The analysis of this model, which consisted of a linear chain of two kinds of particles arranged alternately, depended heavily on earlier work, particularly that of Lord Kelvin. It was applied by Born, and others, to the problem of the specific heats of crystals. This model, sometimes referred to as the diatomic chain, will be treated later in this chapter.

I. Infinite Monatomic Chain

Consider an infinite linear array of identical atoms of mass M separated at equilibrium by distance d, as shown in Fig. 1. The quantity d is effectively the length of the unit cell in the one-dimensional crystal. Individual particles are identified by the index n, so that $n = 0$ defines the origin of the one-dimensional coordinate system X. If the longitudinal displacement from equilibrium of the nth particle is denoted by ρ_n, the coordinate of that particle becomes

$$X_n = nd + \rho_n. \tag{1}$$

The distance between any two particles n and $n+m$, whose equilibrium separation is equal to md, is given by

$$r_{n,n+m} = X_{n+m} - X_n = md + \rho_{n+m} - \rho_n. \tag{2}$$

As the potential energy of interaction between the two particles is a function of $|r_{n+m}|$, the total potential energy of the lattice becomes

$$V = \sum_n \sum_{m>0} V(X_{n+m} - X_n). \tag{3}$$

The index m is restricted to positive integers so that each pair interaction is counted but once. The argument of V in Eqn (3) is then always positive.

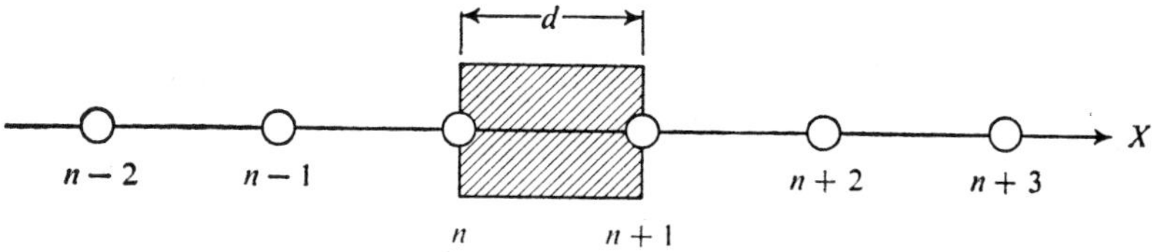

FIG. 1. Linear array of identical atoms.

If it is assumed that the displacements, ρ_n, are small, the potential energy can be expanded in a Taylor's series. Thus, each term in Eqn (3) is of the form

$$V(X_{n+m} - X_n) = V(md) + V'(md)(\rho_{n+m} - \rho_n) + \tfrac{1}{2}V''(md)(\rho_{n+m} - \rho_n)^2 + \ldots, \tag{4}$$

where a prime represents differentiation with respect to $r_{n,n+m}$ and the various coefficients have been evaluated at equilibrium ($r_{n,n+m} = md$). The total potential energy is then found by substituting Eqn (4) into Eqn (3). The Taylor-series expansion of the potential energy carried out here is completely analogous to that of the molecular potential energy in Chapter 1, Eqn (23). The force acting on a given atom l is now given by

$$F_l \equiv -\frac{\partial V}{\partial \rho_l} = -\frac{\partial}{\partial \rho_l}\sum_n \sum_{m>0} [V'(md)(\rho_{n+m} - \rho_n) + \tfrac{1}{2}V''(md)(\rho_{n+m} - \rho_n)^2], \tag{5}$$

where cubic and higher terms in ρ_{n+m} have been neglected in the harmonic approximation. When carrying out the differentiation indicated in Eqn (5) it is seen that all terms vanish except those in which $n = l$ and $n + m = l$. Thus,

$$F_l = -\frac{\partial}{\partial \rho_l} \sum_{m>0} [V'(md)(\rho_{l+m} - \rho_l) + \tfrac{1}{2} V''(md)(\rho_{l+m} - \rho_l)^2$$

$$+ V'(md)(\rho_l - \rho_{l-m}) + \tfrac{1}{2} V''(md)(\rho_l - \rho_{l-m})^2] \tag{6}$$

$$= - \sum_{m>0} [-V'(md) - V''(md)(\rho_{l+m} - \rho_l)$$

$$+ V'(md) + V''(md)(\rho_l - \rho_{l-m})] \tag{7}$$

and

$$F_l = \sum_{m>0} V''(md)(\rho_{l+m} - 2\rho_l + \rho_{l-m}). \tag{8}$$

In Eqn (7) the first term inside the brackets, $-V'(md)$, is the force of atom $l + m$ on atom l. This term is in this case exactly compensated by the third term inside the brackets, $+V'(md)$, which is the force of atom $l - m$ on atom l. This cancellation is the direct result of the translational symmetry of the infinite lattice. As will be seen later, in a finite chain atoms are distinguishable by their positions in the chain. Thus, forces such as those appearing in Eqn (7) will not cancel and rather complicated equilibrium conditions arise near the ends of the chain.

Using Newton's second law, the equations of motion for an infinite monatomic chain become

$$F_l = M\ddot{\rho}_l = \sum_{m>0} f_m(\rho_{l+m} - 2\rho_l + \rho_{l-m}), \tag{9}$$

where the quantity $V''(md)$ of Eqn (7) has now been replaced by f_m, as it represents the force constant for the interaction of two atoms separated by m unit-cell dimensions. If a wave solution of the form

$$\rho_l = A\,\mathrm{e}^{-2\pi i(\nu t - kld)} \tag{10}$$

is assumed, Eqn (8) becomes, after division by ρ_l,

$$-4\pi^2 \nu^2 M = \sum_{m>0} f_m[\mathrm{e}^{2\pi ikmd} - 2 + \mathrm{e}^{-2\pi ikmd}] \tag{11}$$

$$= -2 \sum_{m>0} f_m(1 - \cos 2\pi kmd), \tag{12}$$

and the positive root is given by

$$\nu = \frac{1}{\pi}\left[\sum_{m>0} \frac{f_m}{2M}(1 - \cos 2\pi kmd)\right]^{1/2}. \tag{13}$$

Hence, ν is a periodic function of the wave number k, the period being equal to $1/d$. If only forces between nearest neighbors are included, $m = 1$, and Eqn (13) becomes

$$\nu = \frac{1}{\pi}\left[\frac{f_1}{2M}(1 - \cos 2\pi kd)\right]^{1/2} \tag{14}$$

$$= \frac{1}{\pi}\sqrt{\frac{f_1}{M}}\,|\sin \pi kd|. \tag{15}$$

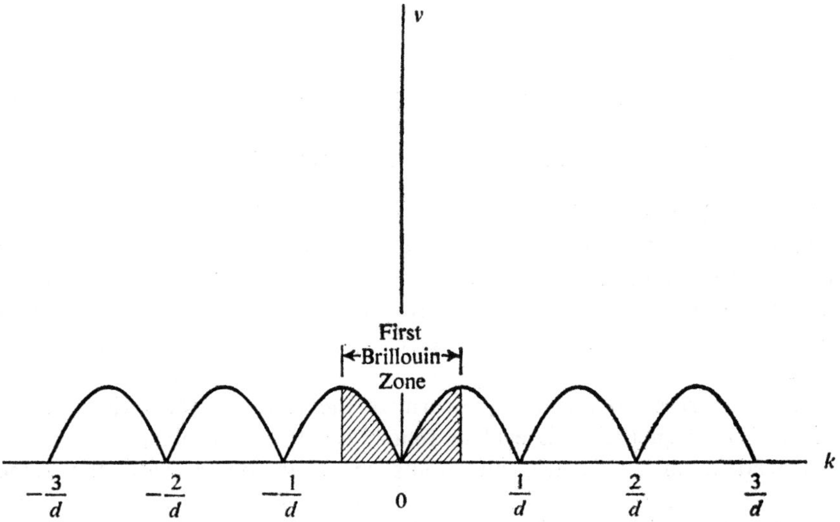

FIG. 2. Plot of Eqn (15) for the monatomic chain.

Equation (15) is plotted in Fig. 2, where it is seen that ν does indeed have a period equal to $1/d$. The series of maxima occurs at a frequency

$$\nu_c = \frac{1}{\pi}\sqrt{\frac{f_1}{M}}, \tag{16}$$

which is sometimes referred to as the cut-off frequency because of the analogy between the mechanical system considered here and an electrical low-pass filter.

In general, it is not necessary to include in Eqn (15) the entire range of values of k. For a given value of ν there is a solution of the form

$$\rho_l = A\,\mathrm{e}^{-2\pi i(\nu t - k'ld)}, \tag{17}$$

where $|k'| \leqslant 1/2d$. However, the solutions corresponding to

$$k_n = k' + n/d, \tag{18}$$

are equally valid. If $k = k_n$,

$$\rho_l = A\,\mathrm{e}^{-2\pi i(\nu t - k'ld + nl)} = A\,\mathrm{e}^{-2\pi i(\nu t - k'ld)}, \tag{19}$$

and the solutions for various values of n are entirely equivalent. Hence, there remain but two values of k that need be considered, namely, $k = \pm k'$, corresponding to the two particular solutions of the second-order differential equations for the displacements [Eqn (9)]. The general solution for $\nu \leqslant \nu_c$ becomes

$$\rho_l = A_+\,\mathrm{e}^{-2\pi i(\nu t - kld)} + A_-\,\mathrm{e}^{-2\pi i(\nu t + kld)} \tag{20}$$

$$= (A_+\,\mathrm{e}^{2\pi ikld} + A_-\,\mathrm{e}^{-2\pi ikld})\,\mathrm{e}^{-2\pi i\nu t}, \tag{21}$$

where $-1/2d \leqslant k \leqslant 1/2d$.† The two terms on the right-hand side of Eqn (20) represent waves travelling in opposite directions along the chain. The range of values specified above is referred to as the first *Brillouin zone* of the linear lattice. Often, only the positive half of the first Brillouin zone is represented, as in Fig. 3. This plot of frequency ν versus the wave number, k, is known as the *dispersion relation* for the linear monatomic lattice.

It is important to visualize the form of the vibration for various values of k. From Eqn (21) it is evident that if $k = 0$, the displacements ρ_l are independent of k. Hence, all atoms move in phase with the same

† For frequencies higher than the cut-off frequency, ν_c, Eqn (15) becomes

$$\nu = \frac{1}{\pi}\sqrt{\frac{f_1}{M}}\,|\sinh \pi\kappa d\,|,$$

where $\kappa = ik$ is real, and the solutions are of the form

$$\rho = A\,\mathrm{e}^{\pm 2\pi(i\nu t - \kappa ld)}.$$

These solutions, which represent exponentially attenuated waves, would have to be included in a completely general solution of the problem.

amplitude and direction, as shown in Fig. 3. At the zone edge, where $k = 1/2d$, Eqn (21) becomes

$$\rho_l = A_+ \, e^{\pi i l} \, . \, e^{2\pi i \nu t} = A_+ \, (-1)^l \, e^{-2\pi i \nu t}, \tag{22}$$

which represents a standing wave in which all particles have the same amplitude of oscillation, but alternate ones are out of phase. This mode of motion is sometimes referred to as the π mode because neighboring atoms differ in phase by π radians. For intermediate values of k the system exhibits travelling-wave motion in which the wavelength $\lambda = 1/k$ of the so-called *descriptive wave* lies somewhere between infinity (for $k = 0$) and its lower limit ($2d$) at the cut-off frequency† (See Fig. 3).

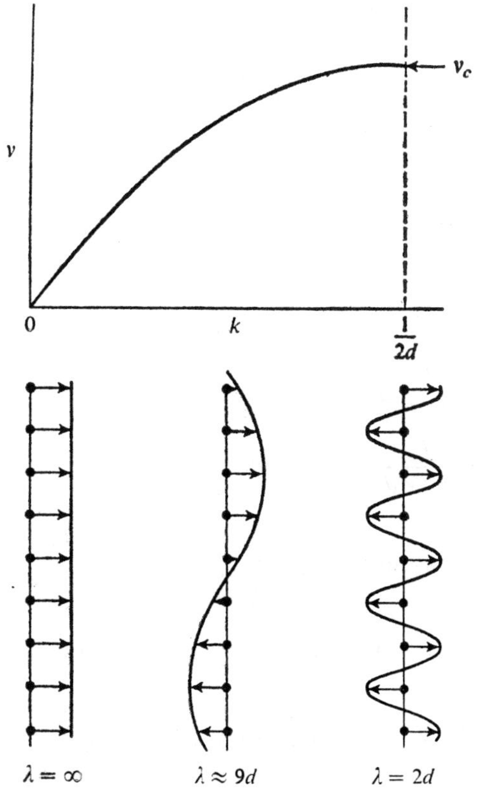

FIG. 3. Dispersion relation for the monatomic chain showing forms of the vibrational modes for certain values of k.

† It should be noted that the wave number is defined here as $k = 1/\lambda$. However, in the literature of lattice dynamics the definition $k = 2\pi/\lambda$ is often used. The latter definition would require k to be replaced by $k/2\pi$ throughout this text.

The results developed above were based on the simplified force field in which only interactions between nearest neighbors were considered. Indeed, for repulsive forces, which are normally of very short range, it is usually not necessary to include interactions beyond those between nearest neighbors. However, very long-range forces resulting, for example, from Coulombic interactions in ionic crystals, may extend over distances involving many unit cells. Hence, in the general relation of Eqn (12),

$$\nu^2(k) = \frac{1}{2\pi^2 M} \sum_{m>0} f_m(1 - \cos 2\pi kmd), \tag{23}$$

the limit on the summation may vary considerably depending on the nature of the forces involved.

A general method (4) of determining the range of forces in crystals can be developed from Eqn (23). Multiplying both sides by $\cos 2\pi knd$, where n is an integer, and integrating over the wave numbers, k, one finds,

$$\int_{-1/2d}^{1/2d} \nu^2(k) \cos(2\pi kn)\, dk$$

$$= \frac{1}{2\pi^2 M} \sum_{m>0} f_m \int_{-1/2d}^{1/2d} (1 - \cos 2\pi kmd) \cos(2\pi knd)\, dk \tag{24}$$

$$= \begin{cases} 0 & \text{if} \quad n \neq m \\ -\dfrac{f_m}{4\pi^2 Md} & \text{if} \quad n = m. \end{cases} \tag{25}$$

Hence, the force constants are given by

$$f_m = -4\pi^2 Md \int_{-1/2d}^{1/2d} \nu^2(k) \cos(2\pi kmd)\, dk, \tag{26}$$

which is essentially the cosine transform of $\nu^2(k)$. Equation (26) provides a general solution to the problem of determining the force constants in a one-dimensional crystal from a knowledge of the dispersion relations.† This method has been applied to several simple three-dimensional crystals.

† The development of neutron scattering techniques provides an experimental means of determining dispersion curves for real crystals. Thus, the method of Eqn (26) is likely to become very useful in the treatment of the force constant problem.

II. Boundary Conditions

When the chain of Fig. 1 is of finite length, suitable *boundary conditions* must be imposed on the general solution of Eqn (20). If the chain consists of $\mathcal{N}$ atoms that are free to oscillate, the ends of the chain can be fixed, for example by imposing the conditions†

$$\rho_0 = \rho_{\mathcal{N}+1} = 0. \tag{27}$$

Thus, for $l = 0$, Eqn (21) becomes $A_+ + A_- = 0$, and the general solution can be written

$$\rho_l = B \sin(2\pi k l d)\, e^{2\pi i \nu t}, \tag{28}$$

where B is a constant. The second condition of Eqn (27) is satisfied by

$$\rho_{\mathcal{N}+1} = B \sin[2\pi k(\mathcal{N}+1)d] = 0, \tag{29}$$

with

$$2\pi k(\mathcal{N}+1)d = 2\pi k L = s\pi, \tag{30}$$

where $L = (\mathcal{N}+1)d$ is the length of the chain and s is an integer. Equation (29) is simply a statement of the fact that there must be an integral number of half wavelengths in length L of the chain. The only values of s which yield independent solutions are $s = 1, 2, \ldots, \mathcal{N}$, corresponding to the $\mathcal{N}$ normal modes of vibration of the system.

An expression for the normal frequencies is obtained by substituting Eqn (30) into Eqn (14). Thus,

$$\nu_s = \frac{1}{\pi}\sqrt{\frac{f_1}{M}} \sin\frac{\pi s}{2(\mathcal{N}+1)} = \nu_c \sin\frac{\pi s}{2(\mathcal{N}+1)}, \qquad s = 1, 2, \ldots, \mathcal{N} \tag{31}$$

and the corresponding displacements are given by

$$\rho_{l,s} = B_s \sin\left(\frac{\pi l s}{\mathcal{N}+1}\right) e^{2\pi i \nu t}, \qquad s = 1, 2, \ldots, \mathcal{N}. \tag{32}$$

The general solution to the problem of the vibration of the finite chain can be written in the form

$$\rho_l = \sum_s B_s \sin\left(\frac{\pi l s}{\mathcal{N}+1}\right) e^{2\pi i \nu t}, \qquad s = 1, 2, \ldots, \mathcal{N}. \tag{33}$$

† As pointed out earlier in this chapter, the cancellation of certain forces depends on the translational symmetry of the chain [See Eqn (7)]. Hence, in the finite chain the equilibrium interatomic distance, d, is perturbed near the ends of the chain. This effect is neglected in the following treatment [See reference (**1**)].

These results will be reconsidered in Chapter 7 in connection with the vibrations of polymers of finite length.

Another boundary condition which is often applied to the linear chain problem is the *cyclic* or periodic condition proposed by Born and von Kármán (5). In this case the chain is considered to be essentially unbounded, but a requirement of long-range periodicity is introduced. That is,

$$\rho_l = \rho_{l+\mathcal{N}}, \tag{34}$$

where $\mathcal{N} \gg 1$. There is thus a running wave solution, as for the infinite chain, but the values of k are restricted to

$$k = \frac{s}{\mathcal{N}\,d} = \frac{s}{L}, \tag{35}$$

where s is zero or a positive or negative integer. This result is found by imposing the condition of Eqn (34) on the solutions given in Eqn (19). The cyclic boundary condition can be visualized by arranging the particles

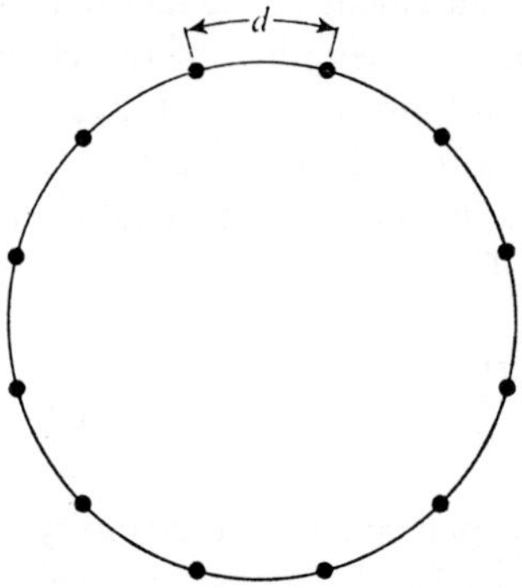

FIG. 4. Schematic representation of the cyclic boundary condition for $\mathcal{N} = 12$.

around a ring of circumference L, as shown in Fig. 4. The solutions must then be periodic in X with period L. The range of the integer s is specified by

$$s = 0, \pm 1, \pm 2, \ldots, \eta, \tag{36}$$

where $\eta = +\mathcal{N}/2$ if $\mathcal{N}$ is even, the single mode corresponding to $s = \mathcal{N}/2$ being the π-mode, and $\eta = \pm(\mathcal{N}-1)/2$ if $\mathcal{N}$ is odd. Thus, there are $\mathcal{N}$ independent solutions, as there must be for the one-dimensional motion of $\mathcal{N}$ particles.

III, Diatomic Chain

Now consider a chain consisting of two types of atoms arranged alternatively, as shown in Fig. 5. This one-dimensional model might be used to represent a simple ionic crystal, for example. The distance, d, is then the dimension of the unit cell which contains two particles. Proceeding as in the case of the monatomic chain, the forces on the atoms of type 1 and 2 are given respectively, by

$$F_{2n} = -f'(\rho_{2n} - \rho_{2n-1}) + f(\rho_{2n+1} - \rho_{2n}) = M_1 \ddot{\rho}_{2n} \tag{37}$$

and

$$F_{2n+1} = -f(\rho_{2n+1} - \rho_{2n}) + f'(\rho_{2n+2} - \rho_{2n+1}) = M_2 \ddot{\rho}_{2n+1}, \tag{38}$$

where, for simplicity, only nearest-neighbor forces have been included. The force constants f and f' have been used to represent bonds of types

- - - - -o——O- - - - - and ——O- - - - -o——, respectively.

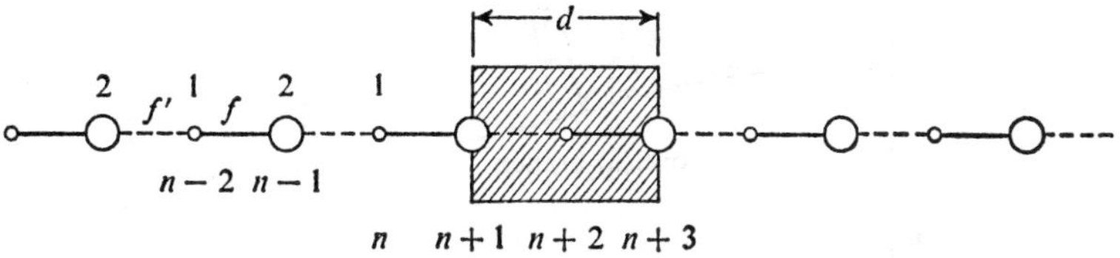

Fig. 5. Linear diatomic chain.

As in the case of the monatomic chain, periodic solutions in time and space are proposed of the form

$$\rho_{2n} = A_1 \, e^{-2\pi i(\nu t - nkd)} \tag{39}$$

and

$$\rho_{2n+1} = A_2 \, e^{-2\pi i[\nu t - (n+\frac{1}{2})kd]}. \tag{40}$$

Substitution of these expressions in Eqns (37) and (38) leads to the pair of linear, homogeneous equations for the amplitudes A_1 and A_2,

$$\begin{aligned} &[4\pi^2 \nu^2 M_1 - (f + f')] A_1 + [f e^{\pi ikd} + f' e^{-\pi ikd}] A_2 = 0 \\ \text{and} \\ &[f' e^{\pi ikd} + f e^{-\pi ikd}] A_1 + [4\pi^2 \nu^2 M_2 - (f' + f)] A_2 = 0. \end{aligned} \tag{41}$$

A nontrivial solution of these equations exists only if the determinant of the coefficients vanishes. Thus, the secular determinant becomes

$$\begin{vmatrix} f + f' - 4\pi^2 \nu^2 M_1 & -(f e^{\pi ikd} + f' e^{-\pi ikd}) \\ -(f e^{-\pi ikd} + f' e^{\pi ikd}) & f' + f - 4\pi^2 \nu^2 M_2 \end{vmatrix} = 0. \tag{42}$$

The roots of this determinant are given by

$$\nu^2 = \frac{f+f'}{8\pi^2}\left(\frac{1}{M_1}+\frac{1}{M_2}\right)$$
$$\pm\frac{1}{4\pi^2}\left[\left(\frac{f+f'}{2}\right)^2\left(\frac{1}{M_1}+\frac{1}{M_2}\right)^2-\frac{4ff'}{M_1M_2}\sin^2\pi kd\right]^{1/2}. \quad (43)$$

For the special case in which all bonds are identical, $f = f'$, which implies equal spacing of the atoms, and Eqn (43) reduces to

$$\nu^2 = \frac{f}{4\pi^2}\left\{\left(\frac{1}{M_1}+\frac{1}{M_2}\right)\pm\left[\left(\frac{1}{M_1}+\frac{1}{M_2}\right)^2-\frac{4}{M_1M_2}\sin^2\pi kd\right]^{1/2}\right\}. \quad (44)$$

This model is the one used originally by Born and von Kármán to represent a simple ionic crystal (**2**, **5**).

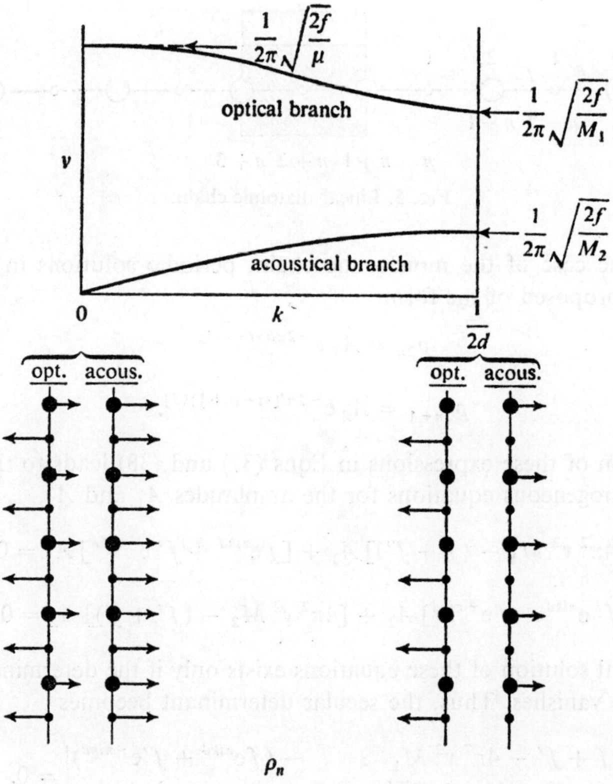

FIG. 6. Dispersion curves for the longitudinal vibrations of the diatomic chain.

For $k = 0$, the roots of Eqn (44) which are of physical interest are 0 and $\frac{1}{2\pi}\sqrt{\frac{2f}{\mu}}$, where μ is the reduced mass defined by $\frac{1}{\mu} = \frac{1}{M_1} + \frac{1}{M_2}$. The positive roots of Eqn (44) have been plotted in the positive half of the first Brillouin zone in Fig. 6 assuming that $M_1 < M_2$. The curve which passes through the origin is called the *acoustical branch*, because the frequencies fall in the region of sonic or ultrasonic waves. The upper curve, referred to as the *optical branch*, represents frequencies in the optical (infrared) spectral region.

The forms of the normal vibrations for each branch are given schematically in Fig. 6 for the limiting values $k = 0$ and $k = 1/2d$. It should be noted that at $k = 0$, the optical branch represents a simple stretching of the bond between two given atoms in which their center of mass remains fixed. As the two atoms are assumed to be different, the oscillating dipole moment which results can interact with incident radiation, producing absorption in the infrared region. The acoustic vibration, on the other hand, does not result in a change in dipole moment at $k = 0$.

From Fig. 6 it is seen that wave solutions of the type of Eqns (39) and (40) do not exist for frequencies between $\frac{1}{2\pi}\sqrt{\frac{2f}{M_1}}$ and $\frac{1}{2\pi}\sqrt{\frac{2f}{M_2}}$. In this *frequency gap*, k is imaginary, hence waves having frequencies in this region are damped.

The general diatomic chain of Fig. 5 can also be used as a model of a simple molecular crystal, if, for example, f is associated with a covlaent bond and f' is used to represent the much weaker intermolecular forces. If $f \gg f'$, the roots of Eqn (43) are given approximately by

$$\nu_{\text{int}} \approx \frac{1}{2\pi}\sqrt{\frac{f}{\mu}} \tag{45}$$

and

$$\nu_{\text{ext}} \approx \frac{1}{\pi}\sqrt{\frac{f'}{M_1 + M_2}}\,|\sin \pi k d|. \tag{46}$$

The first solution, ν_{int}, which is independent of k, is the vibrational frequency of an isolated diatomic molecule, as in the ideal-gas phase. And, for all values of k the center of gravity of each molecule remains fixed in its unit cell. The second solution arises from translational motion of the rigid diatomic molecules. Note that the latter result agrees with Eqn (15) for a monatomic chain, where the effective mass is the total mass of the molecule.

The approximate solutions represented by Eqns (45) and (46) are associated with the limiting case, $f \gg f'$. In general, ν_{int} exhibits some dependence on k. Nevertheless, the above approximation is the basis of the separation which is often made of internal and external motion in crystals containing molecules or polyatomic ions (See Chapter 6).

In the above discussion of the diatomic chain, only longitudinal vibrations have been treated. If the potential energy of displacement of the atoms perpendicular to the chain is determined by a bending force constant, f_α, the force on atom $2n$ is given by

$$F_{2n} = f_\alpha(\Delta\alpha_{2n-1} + \Delta\alpha_{2n+1} - 2\,\Delta\alpha_{2n}), \tag{47}$$

where $\Delta\alpha_{2n}$ is the change in the angle between the two bonds joining atom $2n$ to its nearest neighbors. Replacing each $\Delta\alpha_{2n}$, etc., by

$$\Delta\alpha_{2n} = \frac{2}{d}(\sigma_{2n+1} + \sigma_{2n-1} - 2\,\sigma_{2n}), \tag{48}$$

where σ_{2n} is the perpendicular displacement of the $2n$th atom, Eqn (47) yields the equations of motion

$$F_{2n} = \frac{2f_\alpha}{d}[\sigma_{2n-2} - 4\,\sigma_{2n-1} + 6\,\sigma_{2n} - 4\,\sigma_{2n+1} + \sigma_{2n+2}] = M_1\,\ddot{\sigma}_{2n} \tag{49}$$

and

$$F_{2n+1} = \frac{2f_\alpha}{d}[\sigma_{2n-1} - 4\,\sigma_{2n} + 6\,\sigma_{2n+1} - 4\,\sigma_{2n+2} + \sigma_{2n+3}] = M_2\,\ddot{\sigma}_{2n+1}, \tag{50}$$

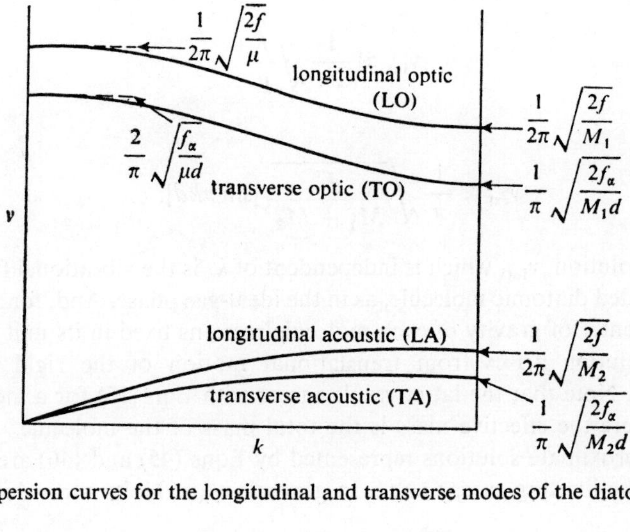

FIG. 7. Dispersion curves for the longitudinal and transverse modes of the diatomic chain.

which lead, as before, to the secular determinant. The solutions can be written in the form

$$4\pi^2 \nu^2 = \frac{2f_\alpha}{M_1 M_2 d}\{(M_1 + M_2)(3 + \cos 2\pi kd)$$

$$\pm [(M_1 + M_2)^2(3+\cos 2\pi kd)^2 - 4M_1 M_2(\cos^2 2\pi kd - 2\cos 2\pi kd + 1)]^{1/2}\}. \tag{51}$$

Thus, for the three-dimensional motion of a diatomic chain there is one pair of dispersion curves (one acoustical and one optical branch) for each direction in space. In the present example, as the two transverse directions are equivalent, each of these branches corresponds to doubly generate frequencies. These results are presented in Fig. 7.

IV. Distribution of Lattice Frequencies

Let $\omega(k)$ be the number of lattice frequencies in a unit range of the wave number, k. This quantity is sometimes called the density of states in k space. For the one-dimensional, finite, monatomic chain there is one vibrational mode in each interval $\Delta k = 1/2L$. Hence, from Eqn (30)

$$\omega(k) = \begin{cases} 2L & \text{for} \quad k \leqslant 1/2d \\ 0 & \text{for} \quad k > 1/2d. \end{cases} \tag{52}$$

In the case of the cyclic lattice of Fig. 4 the analogous result, which is obtained from Eqn (35), is

$$\omega(k) = \begin{cases} L & \text{for} \quad -1/2d \leqslant k \leqslant 1/2d \\ 0 & \text{otherwise.} \end{cases} \tag{53}$$

On a scale of absolute values of k Eqn (53) becomes identical to Eqn (52).

Now, define the number of vibrational states per unit frequency range by $g(\nu)$. Then, the number of frequencies in the interval $\nu + d\nu$ is given by

$$g(\nu)\, d\nu = \omega(k)\frac{dk}{d\nu}\, d\nu = \omega(k)\left(\frac{d\nu}{dk}\right)^{-1} d\nu. \tag{54}$$

The quantity $\dfrac{d\nu}{dk}$ can be obtained directly from the dispersion relation. For the monatomic chain the dispersion relation for positive k is given by Eqn (15),

$$\nu = \frac{1}{\pi}\sqrt{\frac{f_1}{M}}\sin \pi kd = \nu_c \sin \pi kd. \tag{55}$$

Hence,

$$k = \frac{1}{\pi d} \sin^{-1}\left(\frac{\nu}{\nu_c}\right) \tag{56}$$

and

$$\frac{dk}{d\nu} = \frac{1}{\pi d} \frac{1}{(\nu_c^2 - \nu^2)^{1/2}}. \tag{57}$$

This result is plotted as the solid curve in Fig. 8.

It is interesting to compare the above result with that obtained from the continuum model of Debye (**2, 6**), which is often used in calculating specific heats of crystals. Debye assumed the relation

$$\nu\lambda = \nu/k = v, \tag{58}$$

which is appropriate for a sound wave propagating in a continuous medium with velocity v. In this case, using Eqns (52) and (54), the frequency distribution becomes

$$g(\nu) = 2\,L/v, \tag{59}$$

which is independent of ν. However, this distribution must cut off at a frequency $\nu_D = v/2d = \pi\nu_c/2$, as shown by the dashed line in Fig. 8. According to the Debye model, all frequencies up to the maximum value, ν_D, are equally probable. For the lattice model, on the other hand, the frequencies are crowded near the cut-off frequency, ν_c. This singularity arises from the vanishing of the group velocity, $d\nu/dk$, of the descriptive wave at the boundary of the first zone, $k = 1/2d$.

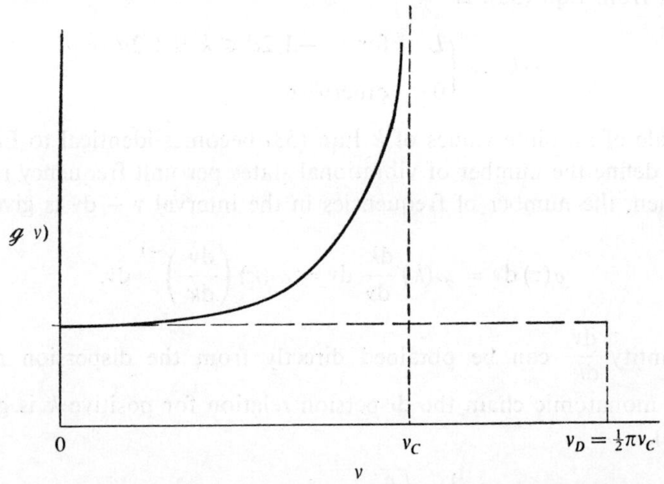

FIG. 8. Frequency distribution for monatomic chain. Solid curve represents the results for the Born model [Eqn (57)]. Dashed curve is the Debye distribution [Eqn (59)].

For the longitudinal vibrations of the diatomic lattice it was shown that two frequency branches are obtained. By applying the method illustrated here for calculating the frequency distribution it is found that the optical branch has singularities at each limiting frequency determined by the zone boundaries. The frequency distribution for the acoustical branch is very similar to that found above for the monatomic lattice. The complete distribution is shown in Fig. 9.

It was shown earlier that if intramolecular forces are much stronger than the forces between molecules, the optical frequency becomes nearly independent of k. In this case the two singularities associated with the optical branch become pushed together and can be approximated by a delta function (**7, 8**). The delta-function distribution is that which is used in the Einstein model of specific heats.

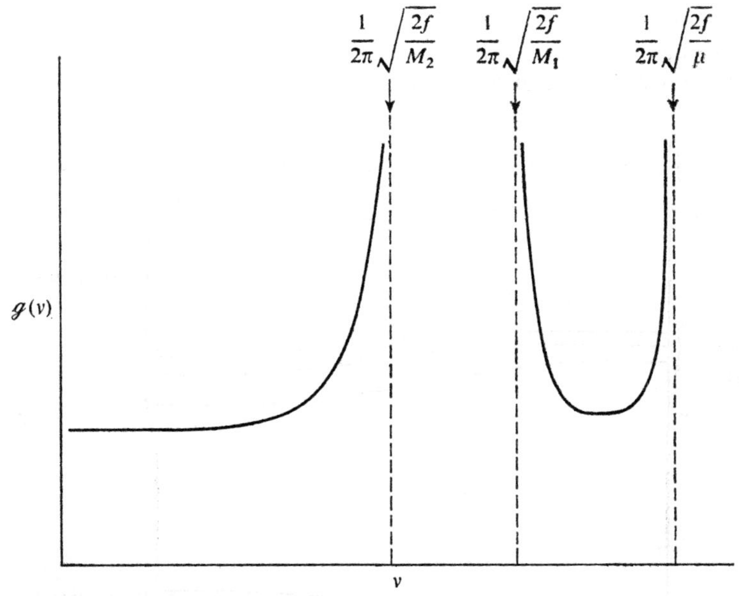

FIG. 9. Frequency distribution for the diatomic chain (Born model).

V. VIBRATIONS OF THREE-DIMENSIONAL LATTICES

The generalization to three dimensions of equations of motion such as Eqns (37) and (38) is straightforward, but complicated. This problem, which was first formulated by Born and von Kármán, is discussed in detail by Born and Kun Huang (**2**). There, it is evident that the mathematical complexity of the problem increases very rapidly with the number of atoms in the unit cell. Hence, serious calculations have for the most part been limited to very simple systems which are of little practical interest, particularly to molecular spectroscopists. However, such calculations are important because the qualitative behavior of more complex systems can often be developed from them by extrapolation.

Proceeding as in the one-dimensional cases, the solutions in the harmonic approximation are of the form

$$\boldsymbol{\rho}_n = \mathbf{A}_n \exp\left[-2\pi i(\nu t - \mathbf{k}\cdot\mathbf{x}_l)\right], \tag{60}$$

in direct analogy with Eqn (10), for example. Now, $\mathbf{k}$ is a vector quantity known as the *wave vector*. The equilibrium position of each atom is located with respect to a suitable space-fixed origin by the vector $\mathbf{x}_l$. As before, solution of the secular determinant yields the frequencies as functions of the wave vector $\mathbf{k}$, which has the magnitude of the reciprocal wavelength and the direction of propagation of the descriptive wave. As in the one-dimensional case, the determination of the dispersion relations allows the distribution of lattice frequencies to be calculated.

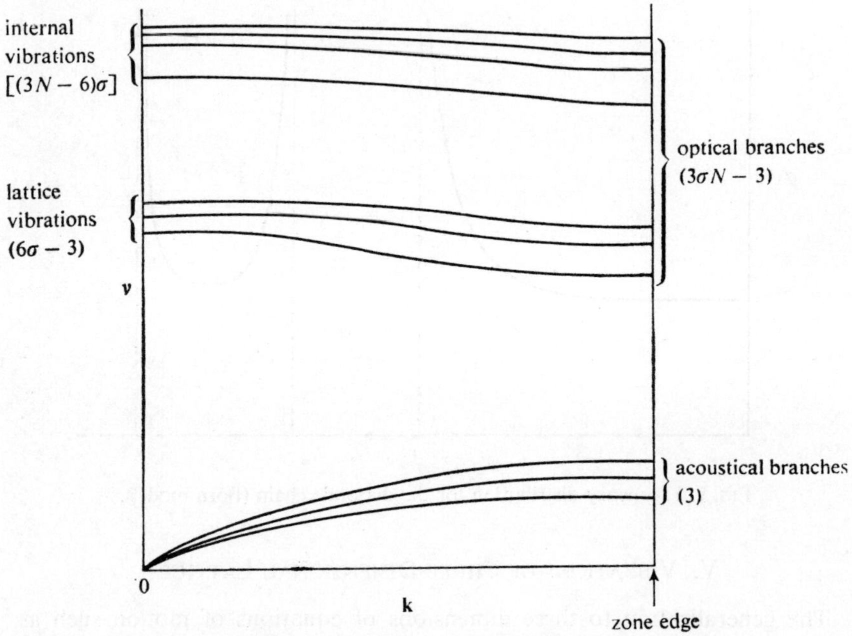

FIG. 10. General form of dispersion curves for a molecular crystal.

As might be expected, there are in general three branches for each particle in the unit cell. Thus, for a simple example of a molecular crystal whose unit cell contains σ molecules, each of which is composed of N atoms, as many as $3N\sigma$ separate branches can exist. Of course the actual number of branches may be reduced due to degeneracies in the system. Three of the branches correspond to acoustical modes, hence, there are up to $3N\sigma - 3$ optical branches. The general form of the dispersion relations for such a system is sketched in Fig. 10 for a given direction of the wave vector. The various branches have, in general, different shapes depending on the direction of **k**.

As pointed out earlier in connection with the vibration of the diatomic chain, intramolecular forces are relatively strong in a molecular crystal. Hence, the $3N - 6$ vibrational fundamentals of a molecule in the gas phase are not severely modified when the molecule is in a crystalline environment. If there are σ molecules per unit cell, there will be $\sigma(3N - 6)$ possible "internal" frequencies, although degeneracies are usually present to reduce this number. As in the corresponding limiting case of the diatomic chain, these "internal" frequencies are usually only slightly dependent on **k**.

The remaining degrees of freedom are derived from the 3σ translational and 3σ rotational motions of the molecules. However, three of these degrees of freedom are responsible for the acoustic modes of the crystal, leaving $6\sigma - 3$ as the maximum number of optical external modes. These latter, modes which are referred to by infrared and Raman spectroscopists as the "lattice vibrations", usually fall in the far-infrared region of the absorption spectrum of a crystal and close to the exciting line in its Raman spectrum.

VI. Phonons

In the discussion of the dynamics of crystal lattices which has been presented thus far in this chapter, emphasis has been placed on the vibrational frequencies and their distributions. The approach has been entirely classical. However, as in the molecular case, which was treated in Chapter 1, the laws of quantum mechanics must be called upon to provide a correct description of the dynamical system. In crystals, as in isolated molecules, there is a principle of correspondence. In fact, in the harmonic approximation the vibrational energy levels again take the form

$$E_{v\ell} = h\nu_{\ell}(v_{\ell} + \tfrac{1}{2}), \qquad v_{\ell} = 0, 1, 2, \ldots \tag{61}$$

as in Eqn (55) Chap. 1, and the quantity ν_{ℓ} can be identified with the classical vibrational frequency of the ℓth normal mode. However, in crystals the frequency ν_{ℓ} is a function of the wave vector, as determined by the dispersion relations.

The quantum of lattice-vibrational (sound) energy, $h\nu_{\ell}$, is often referred to as a *phonon*, by analogy with the term photon for a quantum of electromagnetic (light) energy. Phonons have directional properties, as they depend directly on the wave vector, **k**. The effective momentum of a phonon can be represented by $h\mathbf{k}$. Hence, its interaction with a photon of electromagnetic radiation, for example, is governed by the law of conservation of momentum, as well as the usual principle of energy conservation.

Consider the case of inelastic scattering of a photon by a crystal lattice. If the incident photon with wave vector **K** interacts with a lattice to form a phonon plus a scattered photon, momentum conservation requires that

$$\mathbf{K} = \mathbf{K}' + \mathbf{k} + \mathbf{k}_h, \qquad (62)$$

where $\mathbf{K}'$ is the wave vector for the scattered photon and $\mathbf{k}_h$ is a reciprocal lattice vector [See Eqn (69)]. In the event the incident photon is completely absorbed by the lattice, $\mathbf{K}' = 0$, and Eqn (62) becomes

$$\mathbf{K} = \mathbf{k} + \mathbf{k}_h. \qquad (63)$$

The principle illustrated by Eqn (62) is also directly applicable to the inelastic scattering of thermal neutrons by phonons. In this case **K** and $\mathbf{K}'$ are, respectively, the wave vectors of the incident and scattered neutron. Energy conservation is assured by the requirement

$$\frac{h^2\mathbf{K}^2}{2M_n} = \frac{h^2\mathbf{K}'^2}{2M_n} + h\nu_k(\mathbf{k}), \qquad (64)$$

where M_n is the neutron mass. Hence, an experimental determination of the energy of the scattered neutrons as a function of the direction $\mathbf{K} - \mathbf{K}'$ provides a direct method of studying phonon dispersion in crystals.

Examples of phonon spectra of cubic crystals, as determined by inelastic neutron scattering (**9**), are shown in Fig. 11. In cubic crystals the results are comparatively easy to interpret, because in the [100], [110], or [111] directions, for example, the descriptive waves are polarized in either the

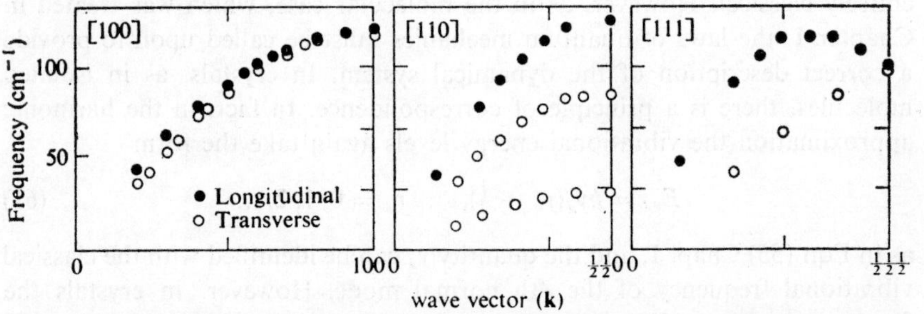

FIG. 11. Phonon dispersion curves for sodium in the [001], [110], and [111] directions at 90°K. Neutron-diffraction data from reference (**9**). Note that crystalline sodium has a body-centred cubic structure. As its primitive unit cell contains but one atom (see Fig. 7, Chap. 4), the dispersion curves consist only of acoustic branches. This figure is reproduced from reference (**10**).

longitudinal or transverse sense. The separate dispersion curves for the longitudinal and transverse directions are clearly evident in Fig. 11. Furthermore, for waves propagating in the [110] direction, two transverse branches are found, indicating the inequivalence of the two directions perpendicular to the [110] direction (the C_2 axis of the cube).

In general, studies of thermal-neutron scattering by phonons have proven the correctness of Born's theory of lattice vibrations. However, to provide quantitative agreement with the shapes of the experimental dispersion curves, it has been found necessary to introduce additional parameters in the theoretical model. One approach is the so-called shell model (**11**), which attempts to take into account the effect of atomic polarization by introducing additional force constants between the outer shells of valence electrons and the inner atomic cores.

VII. Reciprocal Lattices and Brillouin Zones

As in the case of linear chains, not all values of **k** correspond to distinct solutions. This result can be better appreciated after some simple properties of three-dimensional crystal lattices have been considered.

The equilibrium configuration of an infinite three-dimensional lattice is described by a primitive unit cell which generates the lattice by simple translations. This transitional symmetry will be considered in detail in the following chapter. The primitive cell can be represented by three *basic lattice vectors*, $\mathbf{t}_1$, $\mathbf{t}_2$, and $\mathbf{t}_3$, as shown in Fig. 12. The position of each atom m in a given cell is specified with respect to the origin o_n of the cell by a vector $\mathbf{r}_m$. The position of each origin o_n is determined by a vector

$$\boldsymbol{\tau}_n = n_1 \mathbf{t}_1 + n_2 \mathbf{t}_2 + n_3 \mathbf{t}_3, \tag{65}$$

where n_1, n_2, and n_3 are integers. The equilibrium position of each atom in the lattice can now be specified by the vector

$$\mathbf{x}_l = \boldsymbol{\tau}_n + \mathbf{r}_m. \tag{66}$$

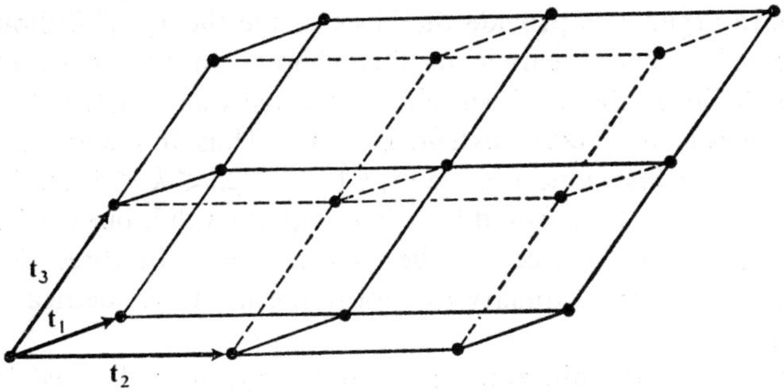

FIG. 12. Crystal lattice constructed from three primitive vectors.

Hence, introducing Eqn (66) into the solutions such as Eqn (60), it is seen that the factor $\exp(2\pi i\, \mathbf{k}\cdot\boldsymbol{\tau}_n)$ represents the phase difference between motions in different primitive cells.

It is convenient to define by the relations

$$\mathbf{b}_1 \equiv \frac{\mathbf{t}_2 \times \mathbf{t}_3}{|\mathbf{t}_1\cdot(\mathbf{t}_2 \times \mathbf{t}_3)|},\quad \mathbf{b}_2 \equiv \frac{\mathbf{t}_3 \times \mathbf{t}_1}{|\mathbf{t}_2\cdot(\mathbf{t}_3 \times \mathbf{t}_1)|},\quad \text{and}\quad \mathbf{b}_3 \equiv \frac{\mathbf{t}_1 \times \mathbf{t}_2}{|\mathbf{t}_3\cdot(\mathbf{t}_1 \times \mathbf{t}_2)|} \tag{67}$$

reciprocal lattice vectors, which have the property

$$\mathbf{b}_i\cdot\mathbf{t}_j = \delta_{ij},\qquad i, j = 1, 2, 3. \tag{68}$$

The vectors defined in Eqn (67) form the basis of a lattice in **k** space, the *reciprocal lattice*. Any lattice vector in reciprocal space can be defined by

$$\mathbf{k}_h = h_1\,\mathbf{b}_1 + h_2\,\mathbf{b}_2 + h_3\,\mathbf{b}_3, \tag{69}$$

where h_1, h_2, and h_3 are integers. Thus, the scalar product

$$\boldsymbol{\tau}_n\cdot\mathbf{k}_h = n_1 h_1 + n_2 h_2 + n_3 h_3 \tag{70}$$

is also an integer. Clearly, the addition of any reciprocal lattice vector $\mathbf{k}_h$ to the wave vector **k** does not change the phase factor, as

$$\exp[2\pi i(\mathbf{k}+\mathbf{k}_h)\cdot\boldsymbol{\tau}_n] = \exp[2\pi i\,\mathbf{k}\cdot\boldsymbol{\tau}_n]\exp[2\pi i\,\mathbf{k}_h\cdot\boldsymbol{\tau}_n] = \exp[2\pi i\,\mathbf{k}\cdot\boldsymbol{\tau}_h]. \tag{71}$$

Therefore, all independent solutions of the form of Eqn (60) are taken into account if the values of **k** are limited to one primitive cell in **k** space. This conclusion follows from the fact that for any point outside of a given cell, there is a point within the cell which is connected to it by a lattice vector $\mathbf{k}_h$.

The above arguments provide the basis of the theory of Brillouin zones, which were introduced earlier in this chapter in the treatment of the vibrations of linear chains. It should be pointed out that the choice of the Brillouin zones is, to a certain extent, arbitrary. Thus, in the one-dimensional case (Sec. I) the first zone was defined by $-1/2d \leqslant k \leqslant 1/2d$. The choice of the region $0 \leqslant k \leqslant 1/d$ would have been equally valid, but would not have been symmetric with respect to the origin, $k = 0$. In three dimensional reciprocal space, it is customary to choose the most symmetrical system of Brillouin zones.

The choice of Brillouin zones in a given crystal is dictated by several criteria. First, the volume of the first zone must be equal to the volume of the primitive cell in reciprocal space, which is given by

$$V_c' = |\mathbf{b}_1 \cdot (\mathbf{b}_2 \times \mathbf{b}_3)| = \frac{(\mathbf{t}_2 \times \mathbf{t}_3) \cdot [(\mathbf{t}_3 \times \mathbf{t}_1) \times (\mathbf{t}_1 \times \mathbf{t}_2)]}{|\mathbf{t}_1 \cdot (\mathbf{t}_2 \times \mathbf{t}_3)|^3}$$

$$= \frac{1}{|\mathbf{t}_1 \cdot (\mathbf{t}_2 \times \mathbf{t}_3)|} = \frac{1}{V_c}, \tag{72}$$

where $V_c = |\mathbf{t}_1 \cdot (\mathbf{t}_2 \times \mathbf{t}_3)|$ is the volume of the primitive cell in the space of the direct lattice; that is, the volume of the first Brillouin zone is equal to the reciprocal of the volume of the primitive cell. Secondly, no two points

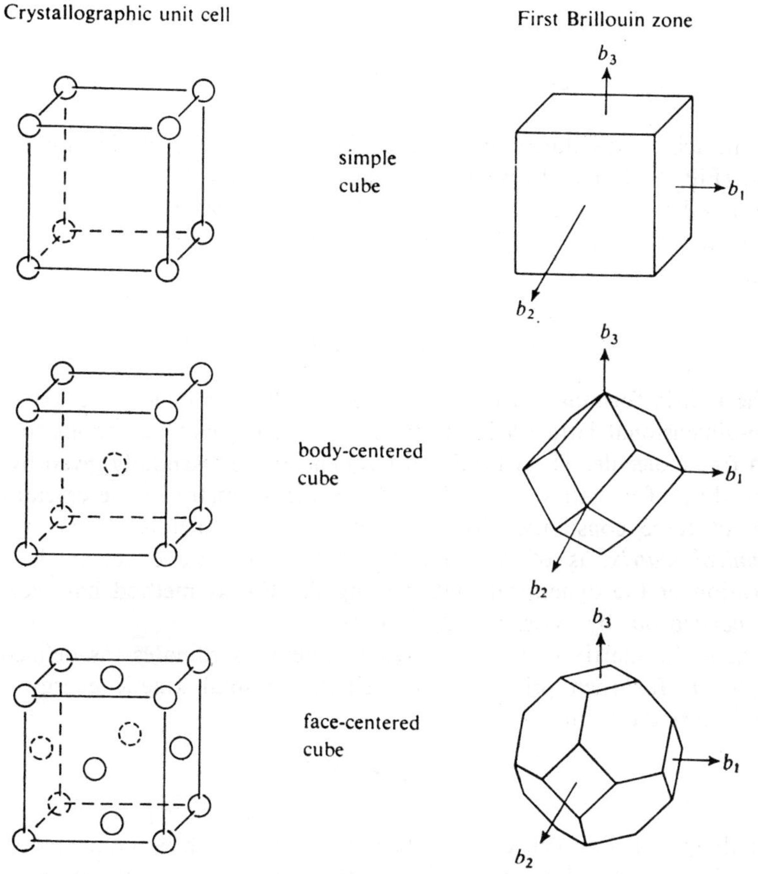

FIG. 13. The cubic lattices and their first Brillouin zones.

in the zone are related by a reciprocal lattice vector, $\mathbf{k}_h$. Finally, as pointed out above, it is customary to choose the zone in such a manner that it has all of the point symmetry of the original lattice. Some examples of Brillouin zones in three dimensions are shown in Fig. 13, all of which have the symmetry of the original cubic lattices.

It is often convenient to introduce cyclic boundary conditions, as was done for the linear chain. Then, by analogy with Eqn (53), the number of lattice frequencies per unit volume in reciprocal space becomes

$$\omega(k) = L_1 L_2 L_3 = \mathcal{N}_1 \mathcal{N}_2 \mathcal{N}_3 V_c \tag{73}$$

where L_1, L_2, and L_3 are the dimensions of the three-dimensional cyclic unit consisting of $\mathcal{N}_1\mathcal{N}_2\mathcal{N}_3$ primitive cells. The total number of values of $\mathbf{k}$ in the first Brillouin zone is then

$$\mathcal{N}_1 \mathcal{N}_2 \mathcal{N}_3 V_c V_c' = \mathcal{N}_1 \mathcal{N}_2 \mathcal{N}_3. \tag{74}$$

This principle was illustrated in one dimension by the cyclic monatomic chain (Fig. 4). For real crystals the quantity $\mathcal{N}_1 \mathcal{N}_2 \mathcal{N}_3$ represents a very large number of unit cells. Hence, the density of $\mathbf{k}$ values becomes so great that sums in $\mathbf{k}$ space can be replaced by integrals over the first Brillouin zone.

VIII. Dynamical Matrix

The matrix formulation of the problem of the harmonic vibrations of a three-dimensional lattice leads to a secular determinant analogous to that for a free molecule. The vibrational frequencies are essentially given by the eigenvalues of a matrix which depends on the geometry of the crystal and a set of force constants. However, this matrix, which is known as the *dynamical matrix*, is of course a function of the wave vector, $\mathbf{k}$. The derivation of the dynamical matrix using the $\mathbf{F}-\mathbf{G}$ method has recently been carried out by Piseri and Zerbi (**12**).

If $\boldsymbol{\xi}_n$ is the matrix of Cartesian displacement coordinates (as defined in Chapter 1) for each primitive unit cell n, the total kinetic energy of a crystal takes the form

$$2T = \sum_n \dot{\boldsymbol{\xi}}_n^{\dagger} \mathbf{M} \dot{\boldsymbol{\xi}}_n, \tag{75}$$

by analogy with Eqn (9), Chap 1. The matrix $\mathbf{M}$, which is the reciprocal $\mathbf{G}$ matrix in the Cartesian system, is the diagonal matrix of atomic masses, ordered as explained on page 4. If a complete set of internal coordinates, S_n, for the nth unit cell can be chosen, the potential energy of the crystal becomes, in the harmonic approximation,

$$2V = \sum_n \sum_{n'} \mathbf{S}_n^{\dagger} \mathbf{F}_{nn'} \mathbf{S}_{n'}. \tag{76}$$

Each matrix $\mathbf{F}_{nn'}$ is a $3N\sigma \times 3N\sigma$ force-constant matrix. When $n = n'$, it is simply the force constant matrix of the nth unit cell, which contains $N\sigma$ atoms. When $n \neq n'$, $\mathbf{F}_{nn'}$ is a matrix of interaction force constants which characterize forces between atoms of different unit cells. Since the translational symmetry of the crystal requires that $\mathbf{F}'_{nn'}$ depends only on the difference $|n - n'| = s$, one can write $\mathbf{F}_{nn'} = \mathbf{F}_s$ and $\mathbf{F}_{n'n} = \mathbf{F}_s^{\dagger}$.

It is convenient to introduce a set of Cartesian "phonon coordinate vectors" by the expression†

$$\Xi(\mathbf{k}) = \frac{1}{\sqrt{\mathcal{N}}} \sum_n \xi_n \exp\left[-2\pi i\, \mathbf{k} \cdot \boldsymbol{\tau}_n\right], \tag{77}$$

where $\mathcal{N}$ is the number of unit cells in the crystal. Analogously, internal "phonon coordinate vectors" can be defined by

$$\Sigma(\mathbf{k}) = \frac{1}{\sqrt{\mathcal{N}}} \sum_n \mathbf{S}_n \exp\left[-2\pi i\, \mathbf{k} \cdot \boldsymbol{\tau}_n\right]. \tag{78}$$

When the summations in Eqns (75) and (76) are replaced by integrals over the first Brillouin zone, the kinetic and potential energies become

$$2T = \int \dot{\Xi}^{\dagger}(\mathbf{k})\, \mathbf{M}\, \dot{\Xi}(\mathbf{k})\, d\mathbf{k} \tag{79}$$

and

$$2V = \int \Sigma^{\dagger}(\mathbf{k})\, \mathbf{F}_{\Sigma}(\mathbf{k})\, \Sigma(\mathbf{k})\, d\mathbf{k}, \tag{80}$$

respectively, where $\mathbf{F}_{\Sigma}(\mathbf{k})$ is a matrix defined by

$$\mathbf{F}_{\Sigma}(\mathbf{k}) \equiv \mathbf{F}_0 + \sum_s \left(\mathbf{F}_s^{\dagger} \exp\left[-2\pi i\, \mathbf{k} \cdot \boldsymbol{\tau}_s\right] + \mathbf{F}_s \exp\left[2\pi i\, \mathbf{k} \cdot \boldsymbol{\tau}_s\right]\right). \tag{81}$$

The Cartesian and internal phonon coordinate systems are related by a linear transformation. The transformation matrix can be found by considering

† The "phonon coordinate vectors" are essentially symmetry coordinates based on the translational symmetry of the unit cell [see Chapter 4].

first the transformation

$$\mathbf{S}_n = \mathbf{B}_{-l}\,\xi_{n-l} + \dots + \mathbf{B}_{-1}\,\xi_{n-1} + \mathbf{B}_0\,\xi_n + \mathbf{B}_1\,\xi_{n+1} + \dots + \mathbf{B}_l\,\xi_{n+l} \tag{82}$$

$$= \mathbf{B}_0\,\xi_n + \sum_{l=1}^{m} (\mathbf{B}_{-l}\,\xi_{n-l} + \mathbf{B}_l\,\xi_{n+l}). \tag{83}$$

Each term in Eqn (83) is analogous to Eqn (12), Chap. 1. Substituting Eqn (83) into Eqn (78), and using the definition of $\mathbf{\Xi}(\mathbf{k})$ [Eqn (77)], one finds

$$\mathbf{\Sigma}(\mathbf{k}) = \sum_{m=-l}^{l} \mathbf{B}_l \exp\left[-2\pi i\,\mathbf{k}\cdot\boldsymbol{\tau}_m\right]\mathbf{\Xi}(\mathbf{k}). \tag{84}$$

Thus, the desired transformation matrix takes the form

$$\mathbf{B}(\mathbf{k}) = \sum_{m=-l}^{l} \mathbf{B}_m \exp\left[-2\pi i\,\mathbf{k}\cdot\boldsymbol{\tau}_m\right]. \tag{85}$$

The potential-energy matrix $\mathbf{F}_{\Sigma}(\mathbf{k})$ can now be transformed to Cartesian coordinates, viz.

$$\mathbf{F}_{\Xi}(\mathbf{k}) = \mathbf{B}^{\dagger}(\mathbf{k})\,\mathbf{F}_{\Sigma}(\mathbf{k})\,\mathbf{B}(\mathbf{k}), \tag{86}$$

and the vibrational frequencies found from the eigenvalues of the product

$$\mathbf{G}_{\Xi}\,\mathbf{F}_{\Xi}.$$

The secular equations take the form

$$\mathbf{G}_{\Xi}\,\mathbf{F}_{\Xi}(\mathbf{k})\,\mathbf{L}(\mathbf{k}) = \mathbf{\Lambda}(\mathbf{k})\,\mathbf{L}(\mathbf{k}), \tag{87}$$

where $\mathbf{G}_{\Xi} = \mathbf{M}^{-1}$ and the eigenvalues are given by the roots of

$$|\mathbf{G}_{\Xi}\,\mathbf{F}_{\Xi}(\mathbf{k}) - \mathbf{E}\lambda_{\ell}(\mathbf{k})| = 0 \tag{88}$$

for each value of $\mathbf{k}$. However, since $\mathbf{G}_{\Xi}$ is diagonal, the secular determinant can be rewritten in the form

$$|\mathbf{G}_{\Xi}^{\,1/2}\,\mathbf{F}_{\Xi}(\mathbf{k})\,\mathbf{G}_{\Xi}^{\,1/2} - \mathbf{E}\lambda_{\ell}(\mathbf{k})| = 0, \tag{89}$$

as indicated in Chapter 1 [See Eqn (46)]. The Hermitian matrix

$$\mathbf{D}(\mathbf{k}) \equiv \mathbf{G}_{\Xi}^{\,1/2}\,\mathbf{F}_{\Xi}(\mathbf{k})\,\mathbf{G}_{\Xi}^{\,1/2} \tag{90}$$

is the *dynamical matrix*.

The secular determinant in the form of Eqn (89) represents a general formulation of the problem of the harmonic vibrations of a crystal. In principle, a force field $\mathbf{F}_s$ can be determined for a particular value of $\mathbf{k}$, say, $\mathbf{k} = 0$, from a set of observed vibrational frequencies. Then, the form of the family of dispersion curves can be traced out by successive solution of Eqn (89) for various values of $\mathbf{k}$. The application of this method to the vibrations of polymer chains will be discussed in Chapter 7.

As an example of the method outlined above, reconsider the diatomic chain of Fig. 5. For simplicity assume nearest-neighbor interactions and $f = f'$. In this case Eqn (81) becomes

$$\mathbf{F}_\Sigma(\mathbf{k}) = \mathbf{F}_0 = \begin{pmatrix} f & 0 \\ 0 & f \end{pmatrix}. \tag{91}$$

The transformation from Cartesian displacement coordinates to internal coordinates has the form

	Δx_1^{j-1}	Δx_2^{j-1}	Δx_1^{j}	Δx_2^{j}	Δx_1^{j+1}	Δx_2^{j+1}
Δr_1^{j-1}	-1	1	0	0	0	0
Δr_2^{j-1}	0	-1	1	0	0	0
Δr_1^{j}	0	0	-1	1	0	0
Δr_2^{j}	0	0	0	-1	1	0
Δr_1^{j+1}	0	0	0	0	-1	1
Δr_2^{j+1}	0	0	0	0	0	-1

The matrices appearing in Eqn (83) are then:

$$\mathbf{B}_{-1} = \begin{pmatrix} 0 & 0 \\ 0 & 0 \end{pmatrix}, \quad \mathbf{B}_0 = \begin{pmatrix} -1 & 1 \\ 0 & -1 \end{pmatrix}, \quad \text{and} \quad \mathbf{B}_1 = \begin{pmatrix} 0 & 0 \\ 1 & 0 \end{pmatrix}, \tag{92}$$

and the transformation matrix of Eqn (85) is

$$\mathbf{B}(k) = \begin{pmatrix} -1 & 1 \\ e^{2\pi i k d} & -1 \end{pmatrix}. \tag{93}$$

Substitution of Eqns (91) and (93) into Eqn (86) yields

$$\mathbf{F}(k) = \begin{pmatrix} 2f & -f(1+e^{-2\pi ikd}) \\ -f(1+e^{2\pi ikd}) & 2f \end{pmatrix}, \tag{94}$$

and with

$$\mathbf{M} = \mathbf{G}^{-1} = \begin{pmatrix} M_1 & 0 \\ 0 & M_2 \end{pmatrix} \tag{95}$$

the dynamical matrix becomes

$$\mathbf{D}(k) = \begin{pmatrix} \dfrac{2f}{M_1} & -\dfrac{f}{\sqrt{M_1 M_2}}(1+e^{-2\pi ikd}) \\ -\dfrac{f}{\sqrt{M_1 M_2}}(1+e^{2\pi ikd}) & \dfrac{2f}{M_2} \end{pmatrix}. \tag{96}$$

Solution of the resulting secular determinant gives the dispersion relation found earlier for this system [Eqn (44)].

References

1. Brillouin, L. "Wave Propagation in Periodic Structures", Second edition, Dover, New York (1953).
2. Born, M. and Kun Huang, "Dynamical Theory of Crystal Lattices", Oxford (1954).
3. Maradudin, A. A., Montroll, E. W. and Weiss, G. H. "Theory of Lattice Dynamics in the Harmonic Approximation", Academic Press, New York (1963).
4. Forman, A. J. E. and Lomer, W. M. *Proc. Roy. Soc. (London)* **70B,** 1143 (1957).
5. Born, M. and von Kármán, Th. *Physik Zeit.* **13,** 297 (1912).
6. Debye, P. *Ann. Physik* **39,** 789 (1912).
7. Smith, R. A. "Wave Mechanics of Crystalline Solids", Chapman and Hall, London (1961).
8. Einstein, A. *Ann. Physik* **22,** 180 (1907); **34,** 170 (1911).
9. Woods, A. D. B., Brockhouse, B. N,. March, R. H. and Bowers, R. *Proc. Phys. Soc. (London)* **79B,** 440 (1962).
10. Kittel, C. Chap. 1 in "Phonons in Perfect Lattices and in Lattices with Point Imperfections" (R. W. H. Stevenson, ed.), Oliver and Boyd, Edinburgh (1966).
11. Dick, B. J. and Overhauser, A. W. *Phys. Rev.* **112,** 90 (1958).
12. Piseri, L. and Zerbi, G. *J. Mol. Spectry.* **26,** 254 (1968).

Chapter 4

Crystal Symmetry

The subject of molecular symmetry was treated in Chapter 2. There it was shown that the various symmetry operations of identity, rotation, reflection, inversion, and rotation-reflection form a point group which characterizes the symmetry of the equilibrium configuration of any isolated molecule. Furthermore, it was found to be possible to classify the normal modes of vibration of the molecule on the basis of the irreducible representations of its point group. For a molecule or complex ion in a crystal, however, certain additional operations must be defined which serve to characterize its translational symmetry in the lattice. The addition of these translational operations results in the formation of the space groups, which are the bases of classification of all crystals.

I. Crystal Classes

The point-symmetry operations, E, C_n^k, i, S_n^k, and σ were described in Chapter 2. These same operations can be used to specify the symmetry of a given unit cell in a crystal. However, as the unit cells of a crystal must completely fill space, certain restrictions are placed on the operations C_n^k and S_n^k. It can be easily shown (**1, 2**) that only proper and improper rotations by 60° and 90° are possible, corresponding to $n = 1, 2, 3, 4$, and 6. With this limitation it is found that only 32 groups can be formed from the operations that describe the symmetry of a unit cell. These 32 point groups constitute the 32 *crystal classes.*

Crystallographers prefer to replace the rotation-reflection operation, S_n^k, by rotation-inversion. However, the two types of operations are found to be equivalent. In discussions of molecular symmetry the Schönflies notation is almost invariably used. In crystallography, on the other hand, the International, or Hermann–Mauguin system is usually preferred. In the latter system a proper rotation about an n-fold axis is represented by n, and the corresponding improper rotation is written $\bar{n}$. In addition, reflection with respect to a plane of symmetry is represented by the letter m. Thus, $m \equiv \bar{2} = \sigma$. The two systems of notation are compared in Table 1, along

TABLE 1

Type	Operation		Element	Symbol		
				Schönflies	International (Hermann–Mauguin)	Graphic
proper (first kind)	rotation by $\phi =$	$\frac{2\pi}{1}$	1-fold (monad)	$C_1 \equiv E$	1	None
		$\frac{2\pi}{2}$	2-fold (diad)	C_2	2	⬮
		$\frac{2\pi}{3}$	3-fold (triad)	C_3	3	▲
		$\frac{2\pi}{4}$	4-fold (tetrad)	C_4	4	◆
		$\frac{2\pi}{6}$	6-fold (hexad)	C_6	6	⬢
improper (second kind)	rotation by $\phi =$	$\frac{2\pi}{1}$	1-fold inversion (center)	i	$\overline{1}$	○
		$\frac{2\pi}{2}$	2-fold inversion (plane)	σ	$m \equiv \overline{2}$	— [a]
		$\frac{2\pi}{3}$	3-fold inversion	S_6	$\overline{3}$	△
		$\frac{2\pi}{4}$	4-fold inversion	S_4	$\overline{4}$	◈
		$\frac{2\pi}{6}$	6-fold inversion	S_3	$\overline{6}$	⬣

with the corresponding graphic symbols which will be used in later examples.

The 32 crystal classes are identified in Table 2, where they are grouped within the seven crystal systems. The International point-group notation has been introduced in addition to that of Schönflies. It is evident that the International notation for each group is derived directly from the symbols for the characteristic operations of that group. For a detailed description of the Hermann–Mauguin system in its various forms, the reader is referred to the "International Tables for X-Ray Crystallography"(**3**).

II. Properties of Space Groups

As pointed out in Chapter 2, each point-symmetry operation can be represented by a linear orthogonal transformation, $\mathbf{R}$. If the vector $\mathbf{x}$ locates any point in the system, the effect of the symmetry operation R can be written in the form

$$\mathbf{x}' = \mathbf{R}\,\mathbf{x}, \tag{1}$$

where $\mathbf{x}'$ is the point into which $\mathbf{x}$ is transformed by $\mathbf{R}$. It is always possible to choose a coordinate system such that $\mathbf{R}$ is given by

$$\mathbf{R} = \begin{pmatrix} \cos\phi & -\sin\phi & 0 \\ \sin\phi & \cos\phi & 0 \\ 0 & 0 & \pm 1 \end{pmatrix}, \tag{2}$$

where ϕ is the angle of rotation and $+1$ or -1 is chosen for proper or improper operations, respectively. Equation (2) was used to deduce the expression for the character per unshifted atom [Eqn (21), Chap. 2].

Equation (1) can be generalized to include the possibility of translation of a point in space by writing

$$\mathbf{x}' = \mathbf{R}\,\mathbf{x} + \boldsymbol{\tau}, \tag{3}$$

where the vector $\boldsymbol{\tau}$ represents a displacement or *translation* of the point $\mathbf{x}$. In the notation of Seitz (**4**), Eqn (3) takes the form

$$\mathbf{x}' = \{\mathbf{R}|\boldsymbol{\tau}\}\mathbf{x}. \tag{4}$$

Some general properties of operators of the type $\{\mathbf{R}|\boldsymbol{\tau}\}$ can be easily developed.

Consider the transformation of Eqn (3) followed by

$$\mathbf{x}'' = \mathbf{R}'\,\mathbf{x}' + \boldsymbol{\tau}' = \{\mathbf{R}'|\boldsymbol{\tau}'\}\mathbf{x}'. \tag{5}$$

TABLE 2

System	Axial and angular relationships	Axial ratios and angles to be specified for each substance	Essential symmetry	Point groups or crystal classes
Triclinic	$a \neq b \neq c$ $\alpha \neq \beta \neq \gamma \neq 90°$	$a : b : c$ α, β, γ	No planes or axes	$\mathscr{C}_1$ $\mathscr{C}_i(\mathscr{S}_2)$ 1 $\bar{1}$
Monoclinic	1st setting[a] $a \neq b \neq c$ $\alpha = \beta = 90° \neq \gamma$ —— 2nd setting[a] $a \neq b \neq c$ $\alpha = \gamma = 90° \neq \beta$	$a : b : c$ γ —— $a : b : c$ β	One twofold axis or one plane	$\mathscr{C}_2$ $\mathscr{C}_s$ $\mathscr{C}_{2h}$ 2 m $2/m$
Orthorhombic	$a \neq b \neq c$ $\alpha = \beta = \gamma = 90°$	$a : b : c$	Three mutually perpendicular twofold axes or two perpendicular planes	$\mathscr{D}_2(\mathscr{V})$ $\mathscr{D}_{2h}(\mathscr{V}_h)$ $\mathscr{C}_{2v}$ 222 mmm $2mm$

System	Axial and angular relationships	Axial ratios and angles to be specified for each substance	Essential symmetry	Point groups or crystal classes
Tetragonal	$a = b \neq c$ $\alpha = \beta = \gamma = 90°$	$c : a$	One fourfold axis	$\mathscr{S}_4$ $\mathscr{C}_4$ $\mathscr{C}_{4v}$ $\mathscr{C}_{4h}$ $\mathscr{D}_4$ $\mathscr{D}_{4h}$ $\mathscr{D}_{2d}$ $\bar{4}$ 4 4*mm* 4/*m* 422 4*mmm* $\bar{4}2m$
Trigonal (may be taken as subdivision of hexagonal)	(Rhombohedral axes) $a = b = c$ $\alpha = \beta = \gamma < 120° \neq 90°$ —— (Hexagonal axes) $a = b \neq c$ $\alpha = \beta = 90°, \gamma = 120°$	α —— $c : a$	One threefold axis	$\mathscr{C}_3$ $\mathscr{C}_{3v}$ $\mathscr{C}_{3h}$ $\mathscr{D}_3$ $\mathscr{D}_{3h}$ $\mathscr{D}_{3d}$ 3 3*m* $\bar{6}$ 32 $\bar{6}m2$ $\bar{3}m$
Hexagonal	$a = b \neq c$ $\alpha = \beta = 90°$ $\gamma = 120°$	$c : a$	One sixfold axis	$\mathscr{C}_6$ $\mathscr{C}_{6v}$ $\mathscr{C}_{6h}$ $\mathscr{D}_6$ $\mathscr{D}_{6h}$ $\mathscr{S}_6$ 6 6*mm* 6/*m* 622 6/*mmm* $\bar{3}$
Cubic	$a = b = c$ $\alpha = \beta = \gamma = 90°$	None	Four threefold axes	$\mathscr{I}$ $\mathscr{I}_d$ $\mathscr{I}_h$ $\mathscr{O}$ $\mathscr{O}_h$ 23 $\bar{4}3m$ *m*3 432 *m*3*m*

[a] See reference (3)

The result of the successive operations is then given by

$$\mathbf{x}'' = \mathbf{R}'(\mathbf{R}\mathbf{x} + \boldsymbol{\tau}) + \boldsymbol{\tau}' \tag{6}$$

$$= \mathbf{R}'\mathbf{R}\,\mathbf{x} + (\mathbf{R}'\boldsymbol{\tau} + \boldsymbol{\tau}'), \tag{7}$$

where the operation $(\mathbf{R}'\boldsymbol{\tau} + \boldsymbol{\tau}')$ is also a translation. Thus, in operator notation the product of two operations is defined by

$$\{\mathbf{R}'|\boldsymbol{\tau}'\}\{\mathbf{R}|\boldsymbol{\tau}\} = \{\mathbf{R}'\mathbf{R}|\mathbf{R}'\boldsymbol{\tau} + \boldsymbol{\tau}'\}, \tag{8}$$

and the identity operation is represented by $\{\mathbf{E}|\mathbf{0}\}$.

The inverse of the operator $\{\mathbf{R}|\boldsymbol{\tau}\}$ can be found from Eqn (8) by imposing the conditions $\mathbf{R}'\mathbf{R} = \mathbf{E}$ and $\mathbf{R}'\boldsymbol{\tau} + \boldsymbol{\tau}' = \mathbf{0}$. Then,

$$\{\mathbf{R}|\boldsymbol{\tau}\}^{-1} = \{\mathbf{R}^{-1}|-\mathbf{R}^{-1}\boldsymbol{\tau}\}. \tag{9}$$

From the above properties it is evident that the set of operations $\{\mathbf{R}|\boldsymbol{\tau}\}$ forms a group. Furthermore, the set of all pure translations $\{\mathbf{E}|\boldsymbol{\tau}\}$ forms an invariant subgroup, since if $\{\mathbf{E}|\boldsymbol{\tau}\}$ is a pure translation in the group, the operation

$$\{\mathbf{R}|\boldsymbol{\tau}\}^{-1}\{\mathbf{E}|\boldsymbol{\tau}'\}\{\mathbf{R}|\boldsymbol{\tau}\} = \{\mathbf{R}^{-1}|-\mathbf{R}^{-1}\boldsymbol{\tau}\}\{\mathbf{E}|\boldsymbol{\tau}'\}\{\mathbf{R}|\boldsymbol{\tau}\} = \{\mathbf{E}|\mathbf{R}^{-1}\boldsymbol{\tau}'\}, \tag{10}$$

where $\{\mathbf{E}|\mathbf{R}^{-1}\boldsymbol{\tau}'\}$ is also a pure translation.

If the pure translation operations considered above are restricted to the primitive translations, $\{\mathbf{E}|\boldsymbol{\tau}_n\}$, where

$$\boldsymbol{\tau}_n = n_1\mathbf{t}_1 + n_2\mathbf{t}_2 + n_3\mathbf{t}_3 \tag{11}$$

and n_1, n_2, and n_3 are integers, the group of operations $\{\mathbf{R}|\boldsymbol{\tau}\}$ forms $\mathfrak{S}$, the *space group* of the crystal. The vectors $\mathbf{t}_1$, and $\mathbf{t}_2$, and $\mathbf{t}_3$ are the three linearly independent primitive lattice vectors. The points generated by the vectors $\boldsymbol{\tau}_n$ constitute the lattice. The set of translations $\{\mathbf{E}|\boldsymbol{\tau}_n\}$ forms a group $\mathfrak{T}$, known as the *translation group*, an invariant, Abelian subgroup of $\mathfrak{S}$.

In an infinite crystal the group $\mathfrak{S}$ is an infinite group. However, if the cyclic boundary conditions are assumed, each lattice vector satisfies the relation

$$(n_i + \mathcal{N}_i)\,\mathbf{t}_i = n_i\,\mathbf{t}_i \qquad i = 1, 2, 3, \tag{12}$$

where $\mathcal{N}_1$, $\mathcal{N}_2$, and $\mathcal{N}_3$ are the numbers of primitive cells in each crystallographic direction. Thus, $\mathcal{N}_1\mathcal{N}_2\mathcal{N}_3$ is the total number of primitive cells in the cyclic unit and is equal to the order of the translation group, $\mathfrak{T}$.

With $\mathfrak{T}$ a subgroup of $\mathfrak{S}$ it is possible to form the factor group, represented symbolically by $\mathfrak{S}/\mathfrak{T}$, which is isomorphic with the point group of operations $\{\mathbf{R}|\mathbf{0}\}$, which defines the crystal class. Formally, the space group is written

$$\mathfrak{S} = \mathfrak{U} \otimes \mathfrak{T}, \tag{13}$$

where the symbol $\otimes$ stands for the semi-direct product and $\mathfrak{U}$ is the *unit cell group*, or *factor group*, of the space group (**5**). The irreducible representations of $\mathfrak{U}$ include all irreducible representations of $\mathfrak{S}$ which are invariant under primitive translations. The importance of the factor group will become apparent in the application of group-theoretical methods to the problem of crystal vibrations.

As the lattice generated by the primitive translations of a space group is invariant under the operations of its factor group, the operations $\{\mathbf{E}|\boldsymbol{\tau}_n\}$ of the translation group are also restricted. Only 14 different lattices, the *Bravais lattices*, can be generated by the various translation operations. Hence the possible lattices associated with a space group depend only on the crystal class to which the space group belongs. A complete list of the 230 space groups is given in Appendix E, along with the corresponding Bravais lattices.

It has been seen that space groups are partially characterized by the invariant subgroups of primitive translations, $\{\mathbf{E}|\boldsymbol{\tau}_n\}$. Furthermore, the operations $\{\mathbf{R}|\mathbf{0}\}$, which are rotations in the general sense, form one of the 32 point groups. However, the space groups corresponding to a given point group and Bravais lattice are further distinguished by the form of the translational parts of the operations $\{\mathbf{R}|\boldsymbol{\tau}\}$ of the space group. When $\mathbf{R} = \mathbf{E}$, the operations $\{\mathbf{E}|\boldsymbol{\tau}\}$ are necessarily primitive translations, $\{\mathbf{E}|\boldsymbol{\tau}_n\}$, the operations of the translation group. However, when $\mathbf{R} \neq \mathbf{E}$ the translational parts of $\{\mathbf{R}|\boldsymbol{\tau}\}$ are not necessarily primitive. All operators can, however, be written in the form

$$\{\mathbf{R}|\mathbf{v}(\mathbf{R}) + \boldsymbol{\tau}_n\} = \{\mathbf{E}|\boldsymbol{\tau}_n\}\{\mathbf{R}|\mathbf{v}(\mathbf{R})\}, \tag{14}$$

where $\boldsymbol{\tau}_n$ represents all primitive translations and $\mathbf{v}(\mathbf{R})$ is either zero or a nonprimitive translation. For 73 of the 230 space groups $\mathbf{v}(\mathbf{R}) = \mathbf{0}$, and for each point group operation, $\mathbf{R}$, there is a space group operation $\{\mathbf{R}|\mathbf{0}\}$. Thus, in these simple, or *symmorphic*, space groups the entire point group is a subgroup of $\mathfrak{S}$.

When $\mathbf{v}(\mathbf{R}) \neq \mathbf{0}$ in Eqn (14), the operations $\{\mathbf{R}|\mathbf{v}(\mathbf{R})\}$ are nonprimitive translations followed by proper or improper rotations. The operations are the glide reflections and screw rotations of crystallography, which will be illustrated in the following section.

The irreducible representations of the space group can be found by considering the symmetry properties of the wave vector, **k**. For each particular **k** there will be some factor-group operations which leave it invariant. These operations form a group, $\mathfrak{W}$, called the *group of the wave vector* [a little group of the first kind (**5**) whose irreducible representations are sometimes referred to as the *small representations*]. If **k** is allowed to take on all possible values, corresponding to all points within and on the surface of the first Brillouin zone, the irreducible representations of $\mathfrak{S}$ can be found from the irreducible representations of the $\mathfrak{W}$'s.

If a given wave vector, **k**, is subjected in turn to each of the factor-group operations which is not contained in its $\mathfrak{W}$, it will generate a set of vectors which, with **k** itself, form the *star* of **k**. The irreducible representations of the symmorphic space groups can be found from a knowledge of the groups of the wave vectors and their stars. This method will be illustrated in Chapter 8 in the development of vibrational selection rules for impure crystals.

III. Screw Axes and Glide Planes

In order to account for the translational symmetry of crystals it is necessary to define two additional operations: Rotation-translation and reflection-translation. The corresponding symmetry elements are, respectively, the *screw axis* and the *glide plane*.

Rotation-translation, or screw rotation, involves a clockwise rotation about an axis by an angle of $\phi = 2\pi/n$, where $n = 1, 2, 3, 4,$ or 6, followed by a nonprimitive translation in the direction of the rotation axis. The international symbol for a screw rotation is n_p, where $p = 1, 2, 3, \ldots (n-1)$, and p/n is the fraction of a primitive translation involved in the operation. A simple example to illustrate this type of operation is shown in Fig. 1. The complete notation for the rotation operations and the corresponding symmetry elements is given in Table 3.

The reflection-translation, or glide reflection, involves reflection in a plane, followed by a nonprimitive translation along a direction parallel to the plane. A glide reflection is represented by a lower-case letter *a*, *b*, *c*, *n*, or *d*,

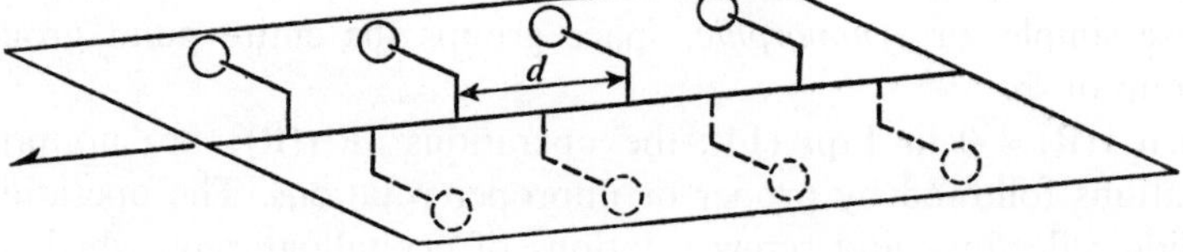

FIG. 1. Rotation–translation (screw) operation $2_1 \equiv C_2^s$.

TABLE 3

Symbols: Schönflies	Symbols: International	Symmetry axis	Graphical symbol	Nature of right-handed screw translation along the axis
E	1	Rotation monad	None	None
i	$\bar{1}$	Inversion monad	○	None
C_2	2	Rotation diad	⬮ (normal to paper)	None
			⟶ (parallel to paper)	
C_2^s	2_1	Screw diad	(normal to paper)	$c/2$
			(parallel to paper)	Either $a/2$ or $b/2$
			Normal to paper	
C_3	3	Rotation triad	▲	None
C_3^s	3_1	Screw triads	▲	$c/3$
$(C_3^s)^2$	3_2		▲	$2c/3$
S_6	$\bar{3}$	Inversion triad	△	None

Symbols: Schönflies	Symbols: International	Symmetry axis	Graphical symbol (normal to plane of paper)	Nature of right-handed screw translation along the axis
C_4	4	Rotation tetrad	◆	None
C_4^s	4_1	Screw tetrads		$c/4$
$(C_4^s)^2$	4_2			$2c/4$
$(C_4^s)^3$	4_3			$3c/4$
S_4	$\bar{4}$	Inversion tetrad		None
C_6	6	Rotation hexad	⬢	None
C_6^s	6_1	Screw hexads		$c/6$
$(C_6^s)^2$	6_2			$2c/6$
$(C_6^s)^3$	6_3			$3c/6$
$(C_6^s)^4$	6_4			$4c/6$
$(C_6^s)^5$	6_5			$5c/6$
S_3	$\bar{6}$	Inversion hexad		None

TABLE 4

Symbols		Symmetry plane	Graphical symbol		Nature of glide translation
Schönflies	International		Normal to plane of projection	Parallel to plane of projection	
σ	m	Reflection plane (mirror)		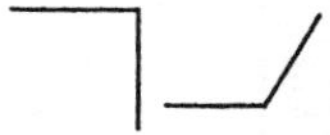	None (Note. If the plane is at $z = \frac{1}{4}$ this is shown by printing $\frac{1}{4}$ beside the symbol.)
$\sigma^{g(a \text{ or } b)}$	a, b	Axial glide plane		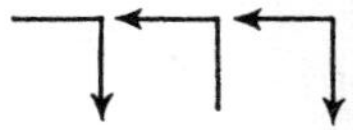	$a/2$ along [100] or $b/2$ along [010]; or along ⟨100⟩.
$\sigma^{g(c)}$	c			None	$c/2$ along z-axis; or $(a + b + c)/2$ along [111] on rhombohedral axes.
$\sigma^{g(n)}$	n	Diagonal glide plane (net)			$(a + b)/2$ or $(b + c)/2$ or $(c + a)/2$; or $(a + b + c)/2$ (tetragonal and cubic).
$\sigma^{g(d)}$	d	"Diamond" glide plane			$(a \pm b)/4$ or $(b \pm c)/4$ or $(c \pm a)/4$; or $(a \pm b \pm c)/4$ (tetragonal and cubic).

depending on the crystal symmetry. The various types of glide planes are represented graphically by different broken lines, as shown in Table 4. The glide-reflection operation is illustrated in Fig. 2.

With the addition of screw rotations and glide reflections to the five point-group operations, it is now possible to specify the operations of the factor group. As an example, consider a crystal belonging to the monoclinic system. According to Table 2 the crystal classes in this system are $\mathscr{C}_2 \equiv 2$, $\mathscr{C}_s \equiv m$, and $\mathscr{C}_{2h} \equiv 2/m$, whose essential symmetry elements are, respectively, a two-fold axis, a plane, or both. In the last case the plane of symmetry is perpendicular to the axis and a center of inversion occurs where these elements intersect. The rotation axis may be associated with either a simple 180° rotation ($C_2 \equiv 2$) or a 180° screw rotation (2_1). Similarly, the plane or reflection may be either a simple plane or a glide plane. It can be shown that there exist only six different combinations of these elements. Hence, the class $\mathscr{C}_{2h} \equiv 2/m$ contains six different space groups (See Appendix E).

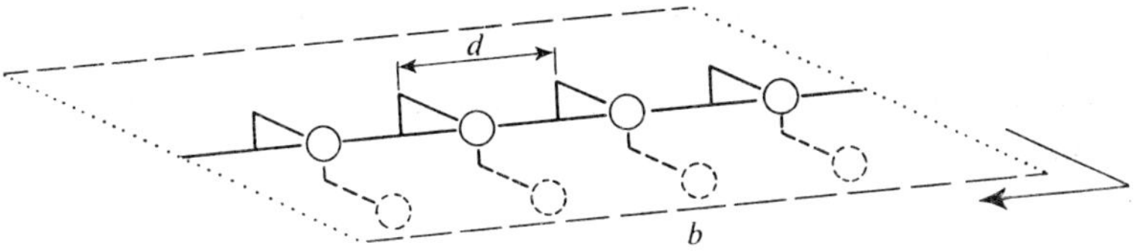

FIG. 2. Reflection–translation (glide) operation $b \equiv \sigma^{g(b)}$.

One combination of the above symmetry elements characterizes the space group $P\,2_1/b \equiv \mathscr{C}_{2h}{}^5$. The symmetry operations are: $E \equiv 1$, $i \equiv \bar{1}$, $C_2{}^s \equiv 2_1$ and $\sigma^{g(b)} \equiv b$, where the symbol b has been used to specify the direction of the glide operation. The unit cell of a crystal of space group $P\,2_1/b$ is shown in Fig. 3. Figure 3a, which is taken directly from the International Tables (p. 98), is a projection of the cell on the a, b plane. The perspective drawing of Fig. 3b shows the positions and directions of the glide planes. This figure will be referred to later in this chapter in connection with the vibrational analysis of naphthalene. The space group $P\,2_1/b \equiv \mathscr{C}_{2h}{}^5$ is important, as a number of organic crystals, particularly those of fused polynuclear aromatic compounds, exhibit this symmetry.

The Hermann–Mauguin, or international, space-group notation used in the above paragraph is derived in the following manner. First, a capital letter indicates the nature of the Bravais lattice. A simple, or primitive, lattice is denoted by *P*, body centered by *I* (for inner), side-centered by *A*, *B*, or *C*, and face-centered (all faces) by *F*. The trigonal or rhombohedral lattice is denoted by *R*. Following the capital letter for the lattice type comes a symbol representing the generating axis, and, if there is a plane of symmetry or a glide plane perpendicular to it, the two symbols are combined, as in the symbol $P\,2_1/b$ used above. Additional symbols are added to represent further elements of symmetry, but only to the extent necessary to characterize the space group. These symbols are sometimes referred to as the *abbreviated symbols.*

The Schönflies notation for the space groups follows directly the point-group notation used in Chapter 2. An arbitrary superscript is added to distinguish the various space groups of a given class. Thus, a convenience of the Schönflies symbolism is that the crystal class, which is the point-group that is isomorphic with the factor group of the space group, can be found by simply dropping the superscript on the space-group symbol. On the other hand, the Schönflies symbol provides no information regarding the nature of the lattice. As will be illustrated later, it is extremely important to know whether or not a given crystallographic unit cell is a primitive cell. The symbol P in the international notation immediately provides this information. Furthermore, identification of a nonprimitive Bravais lattice by a suitable capital letter other than P or R allows the proper division of the unit cell to be made in order to find the primitive cell. The 14 Bravais lattices and their primitive cells are illustrated in Appendix F.

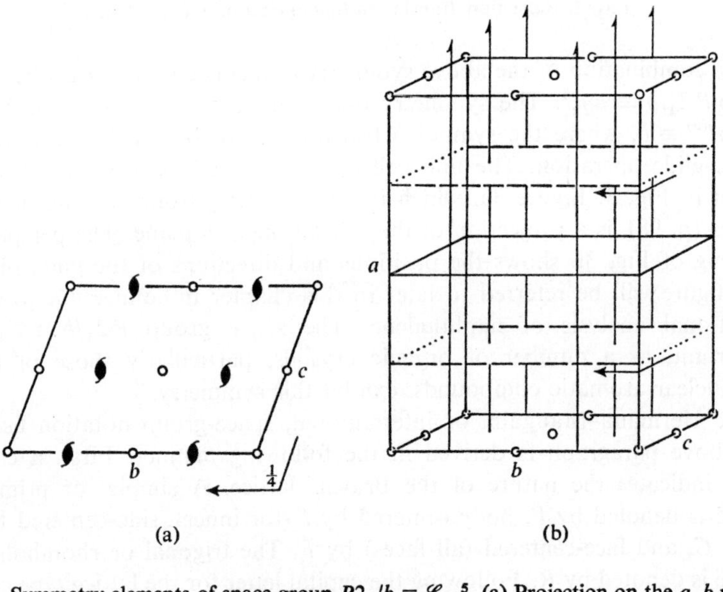

FIG. 3. Symmetry elements of space group $P2_1/b \equiv \mathscr{C}_{2h}{}^5$. (a) Projection on the a, b plane. (b) Crystallographic unit cell.

IV. Site Symmetry

Consider an arbitrary point in a given unit cell. If it is subjected to one of the operations of the factor group, it will generate another, or the same, point. For example, if a six-fold symmetry axis is present, successive rotations about it generate six symmetrically equivalent points, unless the initial point happens to lie on the axis. If, in addition to the six-fold axis, there is a plane of symmetry perpendicular to it, a point contained in neither symmetry element will generate twelve equivalent points under the corresponding operations. In crystallography these points are known as *equivalent positions* in the unit cell, while spectroscopists refer to them as *equivalent sites*. The number of equivalent sites which can be generated by a given site is the *multiplicity* of the site.

A given site in a crystal can be further characterized by the group of point symmetry operations which leave it invariant. This group, which is known as the *site group*, is necessarily a subgroup of the factor group. In general a unit cell contains sites of several different symmetries and sometimes several distinct sets of sites having the same symmetry.

The site group characterizes the symmetry of the crystalline field which surrounds the site in question. Thus, the geometrical structure of a molecule or complex ion which occupies a given site is determined by the symmetry of that site. A molecule may exhibit lower symmetry in the crystal than in the gas phase, because in the crystalline solid it must conform to the symmetry of its site. In this case the site group is necessarily a subgroup of the molecular point group which describes the symmetry of the free molecule.

The 230 space groups are listed in Appendix E using both the international and the Schönflies notations. For each space group is given the number of distinct sets of sites of a given symmetry, the site symmetry, and the multiplicity of each set.

V. Primitive Cells

Consider the unit cell of a simple cubic lattice (Fig. 4). In this case the primitive vectors $\mathbf{t}_1$, $\mathbf{t}_2$, and $\mathbf{t}_3$ are given by

$$\mathbf{t}_1 = \mathbf{i}\,a, \quad \mathbf{t}_2 = \mathbf{j}\,a, \quad \text{and} \quad \mathbf{t}_3 = \mathbf{k}\,a, \tag{15}$$

where $\mathbf{i}$, $\mathbf{j}$, and $\mathbf{k}$ are unit vectors along the X, Y, and Z axes. respectively, and a is the dimension of the cube. Lattice points, which are generated by simple translations of the cube are found at $X = n_1 a$, $Y = n_2 a$, and $Z = n_3 a$, where n_1, n_2, and n_3 are integers. The unit cell can of course be chosen with the origin at a corner of the cube. However, it is often more conveneint to place the origin at the center of the cube so that a single cell is invariant under all operations of the point group which describe the cubic symmetry.

In the case of a body-centered cube of dimension a with the origin at a corner, lattice points are produced at $X = (n_1 + \frac{1}{2})\,a$, $Y = (n_2 + \frac{1}{2})\,a$, and $Z = (n_3 + \frac{1}{2})\,a$. Thus, two lattice points exist for each set of integers n_1, n_2, and n_3. Therefore, the body-centered cubic cell chosen here cannot be a primitive cell, as it contains two equivalent lattice points. The primitive cell for this lattice must have half of the volume of the body-centered cube.

For the body-centered cubic lattice, the primitive cell can be found by constructing the basis vectors, $\mathbf{t}_1$, $\mathbf{t}_2$, and $\mathbf{t}_3$ from the central lattice point to three of its nearest neighbors, as shown in Fig. 5. In this case these vectors are given by

$$\mathbf{t}_1 = \frac{a}{2}(-\mathbf{i} + \mathbf{j} + \mathbf{k}), \quad \mathbf{t}_2 = \frac{a}{2}(\mathbf{i} - \mathbf{j} + \mathbf{k}), \quad \text{and} \quad \mathbf{t}_3 = \frac{a}{2}(\mathbf{i} + \mathbf{j} - \mathbf{k}). \tag{16}$$

A parallelepiped constructed with sides parallel to $\mathbf{t}_1$, $\mathbf{t}_2$, and $\mathbf{t}_3$ has a volume

$$|\mathbf{t}_1 \cdot (\mathbf{t}_2 \times \mathbf{t}_3)| = \frac{a^3}{8}(2 + 2) = \frac{a^3}{2}, \tag{17}$$

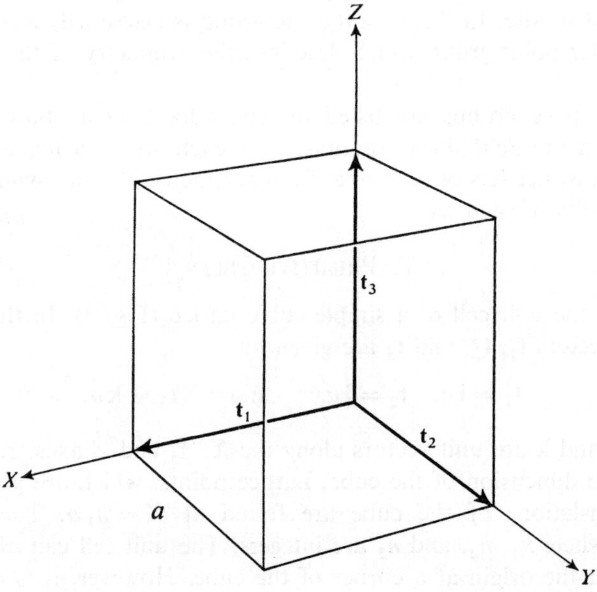

FIG. 4. Unit cell of simple cubic lattice showing primitive vectors.

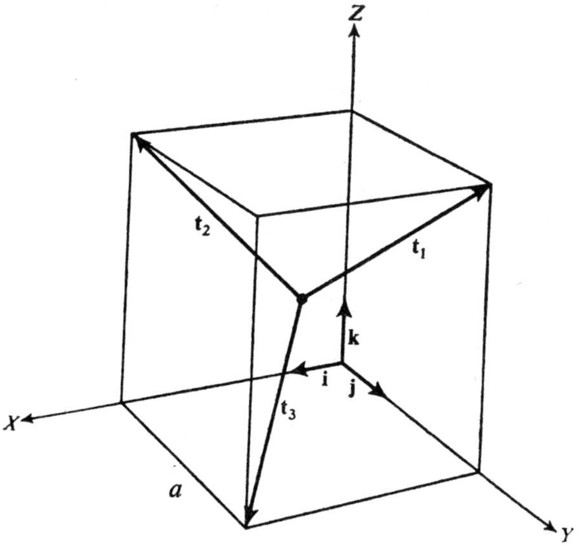

FIG. 5. Unit cell of body-centered cubic lattice showing primitive vectors.

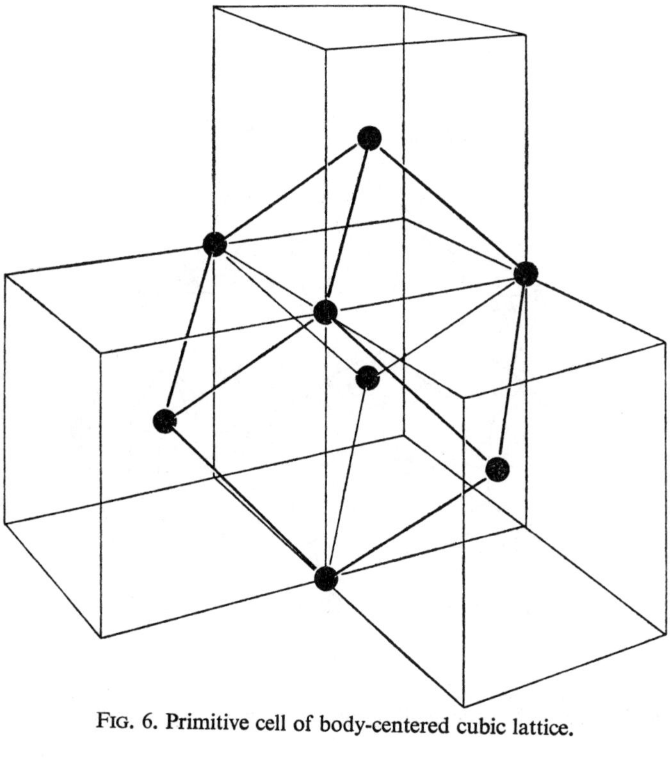

FIG. 6. Primitive cell of body-centered cubic lattice.

which is indeed half of the volume of the cube. The cell defined by Eqn (16), and shown in Fig. 6, contains but a single lattice point and is, therefore, a primitive cell.

The primitive cell found above for the body-centered cubic lattice does not have cubic symmetry. A little study of Fig. 6 will show that it has the symmetry of point group $\mathscr{D}_{3d}$. In order to find a primitive cell having cubic symmetry, consider the original body-centered cube. Now, draw lines from the central lattice point to each of its eight nearest neighbors (the corners of

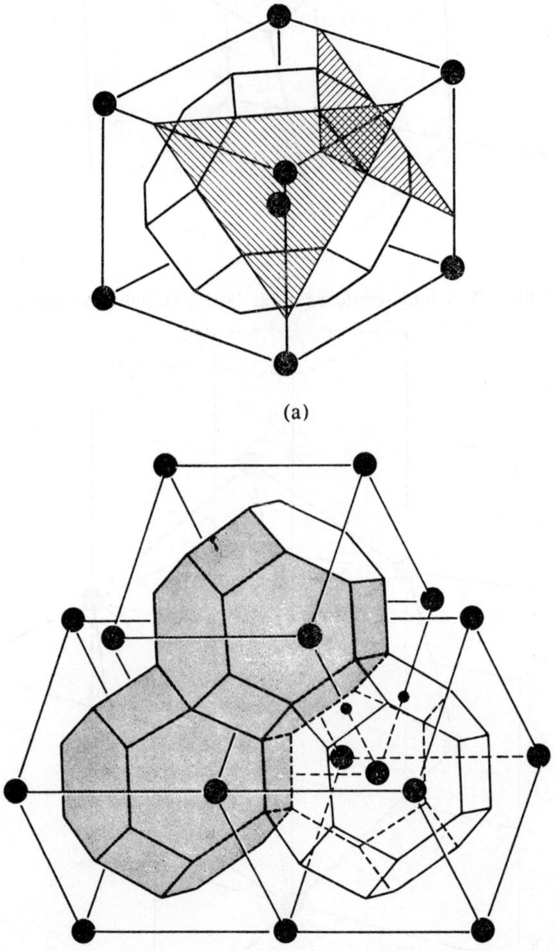

(a)

(b)

FIG. 7. Wigner–Seitz cell for theb ody-centered cubic lattice. (a) Construction of Wigner–Seitz cell within cube. (b) Method of filling space using Wigner–Seitz cells. [See reference (7)].

the cube). Construct the planes which are the perpendicular bisectors of each of these eight lines. These planes, plus portions of the faces of the original cube, define the structure shown in Fig. 7. This primitive cell is known as the *Wigner–Seitz cell* for the body-centered cubic lattice (**6**). It exhibits the complete symmetry of a cube in the sense that it is transformed into itself by all of the operations of the cubic point groups, $\mathcal{T}$, $\mathcal{T}_h$, $\mathcal{T}_d$, $\mathcal{O}$, and $\mathcal{O}_h$. This method has been used to find the Wigner–Seitz cells for the 14 Bravais lattices, as given in Appendix F.

VI. Irreducible Representations of the Translation Group

For simplicity consider a symmorphic space group, whose translational subgroup, $\mathfrak{T}$, is the group of primitive translations $\{\mathbf{E}|\boldsymbol{\tau}_n\}$. The cyclic boundary conditions can be introduced in the form

$$\{\mathbf{E}|\mathbf{t}_1\}^{\mathcal{N}_1} = \{\mathbf{E}|\mathbf{t}_2\}^{\mathcal{N}_2} = \{\mathbf{E}|\mathbf{t}_3\}^{\mathcal{N}_3} = \{\mathbf{E}|\mathbf{0}\}. \tag{18}$$

In this case the full translational group, $\mathfrak{T}$, is simply the direct product

$$\mathfrak{T} = \mathfrak{T}_1 \times \mathfrak{T}_2 \times \mathfrak{T}_3 \tag{19}$$

of subgroups $\mathfrak{T}_1$, $\mathfrak{T}_2$, and $\mathfrak{T}_3$ formed, respectively, by the operations $\{\mathbf{E}|\mathbf{t}_1\}$ and its powers, $\{\mathbf{E}|\mathbf{t}_2\}$ and its powers, and $\{\mathbf{E}|\mathbf{t}_3\}$ and its powers. This factorization of $\mathfrak{T}$ is possible, because, as can be seen from Eqn (8), the product of two operations corresponds to addition of their primitive vectors. Hence, all of the primitive translations commute, and the three groups commute with each other.

Since all of the operations of $\mathfrak{T}$ are commutative, it is an Abelian group, each operation forming a class by itself. Thus the number of classes, and hence the number of irreducible representations, is equal to the order of the group, $\mathcal{N}_1\mathcal{N}_2\mathcal{N}_3$. From Eqn (9), Chap. 2, it is evident that all of the irreducible representation matrices $\Gamma^{(\gamma)}$ are of rank one, that is, they are simply numbers. These numbers are just the characters, or,

$$\Gamma^{(\gamma)} = \chi^{(\gamma)}_{100}, \chi^{(\gamma)}_{200}, \dots \chi^{(\gamma)}_{\mathcal{N}_100}; \dots ; \chi^{(\gamma)}_{\mathcal{N}_1\mathcal{N}_2\mathcal{N}_3}. \tag{20}$$

These quantities must obey the same rules of multiplication as the group operations, namely

$$\chi^{(\gamma)}_{n_1n_2n_3}\chi^{(\gamma)}_{l_1l_2l_3} = \chi^{(\gamma)}_{n_1+l_1,\,n_2+l_2,\,n_3+l_3}. \tag{21}$$

Furthermore, for these one-dimensional representations the character of the first class (the identity) is equal to unity, thus,

$$\chi^{(\gamma)}_{000} = \chi^{(\gamma)}_{\mathcal{N}_100} = \chi^{(\gamma)}_{00\mathcal{N}_3} = \dots = \chi^{(\gamma)}_{\mathcal{N}_1\mathcal{N}_2\mathcal{N}_3} = 1. \tag{22}$$

These two conditions are sufficient to determine the form of the representations,

$$\chi_{\tau_n}^{(k)} = \exp\,[2\pi i\, \mathbf{k} \cdot \boldsymbol{\tau}_n], \tag{23}$$

where $\boldsymbol{\tau}_n$ is the primitive translation vector and $\mathbf{k}$ is the wave vector. The wave vector can be written in the form

$$\mathbf{k} = \frac{s_1}{\mathcal{N}_1}\mathbf{b}_1 + \frac{s_2}{\mathcal{N}_2}\mathbf{b}_2 + \frac{s_3}{\mathcal{N}_3}\mathbf{b}_3, \tag{24}$$

where the $\mathbf{b}_i$'s are the reciprocal lattice vectors, and s_1, s_2, and s_3 are integers running from 0 to $\mathcal{N}_1 - 1$, $\mathcal{N}_2 - 1$, and $\mathcal{N}_3 - 1$, respectively. Hence, Eqn (23) represents a set of $(\mathcal{N}_1\mathcal{N}_2\mathcal{N}_3)^2$ numbers which are the entries in the character table for the group $\mathfrak{T}$.

The group $\mathfrak{T}_1$ is the translation group of order $\mathcal{N}_1$, for which there are $\mathcal{N}_1{}^2$ characters. From Eqns (23) and (24) the characters are given by

$$\chi_{n_1}^{(s_1)} = \mathrm{e}^{\,2\pi i s_1 n_1/\mathcal{N}_1}, \qquad s_1, n_1 = 0, 1, \dots \mathcal{N}_1 - 1 \tag{25}$$

as shown in Table 5, where the more conventional notation

$$\chi_{j_1}^{(\gamma_1)} = \mathrm{e}^{\,2\pi i(\gamma_1 - 1)(j_1 - 1)/\mathcal{N}_1} \qquad \gamma_1, j_1 = 1, 2, \dots \mathcal{N}_1 \tag{26}$$

has been introduced. Note that $\gamma_1 = s_1 + 1$ and $j_1 = n_1 + 1$; hence, $s_1 = 0$ corresponds to the totally symmetric species and $j_1 = 1$ to the class of the identity.

Consider now a finite crystal consisting of $\mathcal{N}_1\mathcal{N}_2\mathcal{N}_3$ primitive cells, each of which contains m atoms. The total number of normal vibrations of the crystal is equal to $3m\,\mathcal{N}_1\mathcal{N}_2\mathcal{N}_3$. In order to determine the number of vibrations, $n^{(\gamma)}$, corresponding to each irreducible representation of the space group, it is first necessary to consider the subgroup of translations, $\mathfrak{T}$. The values of the characters of the irreducible representations are given by Eqn (23). The characters of the reducible representations can be easily found by considering the usual transformation matrix $\mathbf{R}$ of Eqn (2) for each atom which is transformed into itself by the corresponding symmetry operation. It is evident that

$$\chi_j = \begin{cases} 3m\,\mathcal{N}_1\mathcal{N}_2\mathcal{N}_3 & \text{if } \mathbf{R} = \mathbf{E} \\ 0 & \text{if } \mathbf{R} \neq \mathbf{E} \end{cases} \tag{27}$$

for the class j to which the operation $\mathbf{R}$ belongs. The first part of Eqn (27) is obvious, as the order of the unit matrix, $\mathbf{E}$, is $3m\,\mathcal{N}_1\mathcal{N}_2\mathcal{N}_3$. The second part follows immediately from the fact that no atoms are invariant

TABLE 5

	$\{E\|\mathbf{t}_1\}^{j-1}$								
$\Gamma^{(\gamma)}$	$\{E\|\mathbf{t}_1\}^0$	$\{E\|\mathbf{t}_1\}^1$	$\{E\|\mathbf{t}_1\}^2$	$\dots$	$\{E\|\mathbf{t}_1\}^{\mathcal{N}_1-1}$	$n_{tot}^{(\gamma)}$	$n_{trans}^{(\gamma)}$	$n_{rot}^{(\gamma)}$	Activity
$\Gamma^{(1)}$	1	1	1		1	$3m$	3	1	*IR, R*
$\Gamma^{(2)}$	1	$e^{2\pi i/\mathcal{N}_1}$	$e^{2\pi i2/\mathcal{N}_1}$		$e^{2\pi i(\mathcal{N}_1-1)/\mathcal{N}_1}$	$3m$	0	0	—
$\Gamma^{(3)}$	1	$e^{2\pi i2/\mathcal{N}_1}$	$e^{2\pi i4/\mathcal{N}_1}$		$e^{2\pi i2(\mathcal{N}_1-1)/\mathcal{N}_1}$	$3m$	0	0	—
$\vdots$									
$\Gamma^{(\mathcal{N}_1)}$	1	$e^{2\pi i(\mathcal{N}_1-1)/\mathcal{N}_1}$	$e^{2\pi i2(\mathcal{N}_1-1)/\mathcal{N}_1}$		$e^{2\pi i(\mathcal{N}_1-1)^2/\mathcal{N}_1}$	$3m$	0	0	—

under the translation operations other than the trivial one, the identity. Substitution of the values of χ_j given in Eqn (27) into the "magic formula",

$$n^{(\gamma)} = \frac{1}{g}\sum_j g_j \chi_j^{(\gamma)*} \chi_j, \tag{28}$$

yields the result

$$n^{(\gamma)} = \frac{1}{\mathcal{N}_1 \mathcal{N}_2 \mathcal{N}_3} \chi_E^{(\gamma)} \chi_E = \frac{1}{\mathcal{N}_1 \mathcal{N}_2 \mathcal{N}_3} 3m \mathcal{N}_1 \mathcal{N}_2 \mathcal{N}_3 = 3m. \tag{29}$$

As $n^{(\gamma)} = 3m$ is independent of γ, every irreducible representation of the translation group contains $3m$ normal vibrations.

The character of the representation corresponding to the acoustic modes is found from the expression

$$\chi_j\,(\text{acous}) = \chi_j\,(\text{trans}) = 2\cos\phi_j \pm 1. \tag{30}$$

Since the acoustic vibrations represent simple translational motions of successive unit cells, they transform in the same way as the translation vector [Eqn (30)]. Furthermore, as no rotation is involved, $\phi_j = 0$ and $\chi_j\,(\text{acous}) = 3$. Then,

$$n_{\text{acous}}^{(\gamma)} = \frac{1}{\mathcal{N}_1 \mathcal{N}_2 \mathcal{N}_3} \Sigma_j\, 3\chi_j^{(\gamma)} = \begin{cases} 3 & \text{if} \quad \gamma = 1 \\ 0 & \text{if} \quad \gamma \neq 1, \end{cases} \tag{31}$$

and all three acoustic modes belong to the totally symmetric species, $\Gamma^{(1)}$.

As pointed out in Chapter 2, selection rules for infrared absorption and the Raman effect are determined by the species of the components of the dipole moment vector and the polarizability tensor, respectively. Since the dipole moment vector transforms in the same way as the translation vector, the result of Eqn (31) applies to $\mathfrak{m}_x$, $\mathfrak{m}_y$, and $\mathfrak{m}_z$. Hence, only vibrational fundamentals of the totally symmetric species, for which $\mathbf{k} = \mathbf{0}$, have the possibility of being infrared–active.

The characters of transformations of the polarizability tensor are given by Eqn (70), Chap. 1,

$$\chi_j(\alpha) = 2\cos\phi_j\,(2\cos\phi_j \pm 1). \tag{32}$$

Since $\phi_j = 0$, $\chi_j(\alpha) = 6$ for all operations of the translation group. Equation (28) then yields the result

$$n^{(\gamma)}\,(\alpha) = \begin{cases} 6 & \text{if} \quad \gamma = 1 \\ 0 & \text{if} \quad \gamma \neq 1. \end{cases} \tag{33}$$

That is, all six components of the symmetric polarizability tensor fall in the totally symmetric species. Hence, in the Raman effect, as in infrared absorption, only fundamentals in species $\Gamma^{(1)}$ of the translation group have the possibility of activity. However, the activity of these fundamentals may be further limited by the symmetry of the factor group.

VII. Factor-Group Analysis

In the development of the properties of the translation group, the lattice was constructed entirely of points, the internal structure of individual primitive cells having been ignored. However, the symmetry of a lattice with a *basis*, that is, one whose primitive cell contains, in general, a number of atoms, is further specified by the operations of the factor group. As shown above, potentially infrared- or Raman-active vibrations correspond to the totally symmetric representation of the translation group, for which $\mathbf{k} = \mathbf{0}$. Hence, for spectroscopic purposes only those representations of the space group which are derived from the irreducible representations of the factor group need be considered.

The distribution of the vibrations among the irreducible representations of the factor group can be determined by the method of Winston and Halford (**8**). As in the molecular problem, the calculation is essentially an application of the magic formula. In this case the order of the translation group is $\mathcal{N}_1 \mathcal{N}_2 \mathcal{N}_3$ and, if there are h operations in the factor group, $\mathfrak{U}$, the number of vibrations of species γ in $\mathfrak{U}$ is given by

$$n^{(\gamma)} = \frac{1}{\mathcal{N}_1 \mathcal{N}_2 \mathcal{N}_3 h} \sum_{R,n} \chi_R{}^{(\gamma)*} \chi_{R,n}, \tag{34}$$

where the summation extends over all of the operations of both the factor group and the translation group. Equation (34) can be written in the form

$$n^{(\gamma)} = \frac{1}{\mathcal{N}_1 \mathcal{N}_2 \mathcal{N}_3 h} \sum_R \sum_n \chi_R{}^{(\gamma)*} \chi_{R,n}$$

$$= \frac{1}{\mathcal{N}_1 \mathcal{N}_2 \mathcal{N}_3 h} \sum_R \chi_R{}^{(\gamma)*} \sum_n \chi_{R,n}, \tag{35}$$

as $\chi_R{}^{(\gamma)}$ is independent of translation-group operations. The second summation in Eqn (35) is simply the character of a reducible representation, χ_R, of the factor group summed over the $\mathcal{N}_1 \mathcal{N}_2 \mathcal{N}_3$ primitive cells of the cyclic unit. Thus, as the value of χ_R is independent of the cell chosen,

$$\sum_n \chi_{R,n} = \mathcal{N}_1 \mathcal{N}_2 \mathcal{N}_3 \chi_R, \tag{36}$$

and Eqn (35) becomes

$$n^{(\gamma)} = \frac{1}{h} \sum_R \chi_R^{(\gamma)*} \chi_R. \tag{37}$$

This result is seen to be independent of the size of the cyclic unit. If the h operations of $\mathfrak{U}$ are grouped into classes, Eqn (37) becomes

$$n^{(\gamma)} = \frac{1}{h} \sum_j h_j \chi_j^{(\gamma)*} \chi_j, \tag{38}$$

where h_j is the number of symmetry operations in the jth class.

The result of this argument is that the vibrations of a crystal can be classified according to the irreducible representations of the factor group, $\mathfrak{U}$. As Eqn (38) is equivalent to Eqn (28), the factor-group method consists of treating a primitive cell of a crystal as if it were a single polyatomic molecule. The method developed above is exactly that originally proposed by Bhagavantam and Venkatarayudu (**9–10**).

The problem of determining χ_j becomes one of counting the number of atoms which are transformed into themselves by an operation R of class j of the factor group. However, in this case, if a given atom is transformed into an atom in a corresponding position in another unit cell, it is considered to be unshifted. This modification of the concept of an unshifted atom automatically takes into account the translational symmetry of the vibrating lattice when $\mathbf{k} = \mathbf{0}$, that is, when corresponding atoms in each unit cell move in phase.

Proceeding as in the molecular case, the characters of the reducible representations of the factor group are given by

$$\chi_j = \mathscr{m}_j (2 \cos \phi_j \pm 1), \tag{39}$$

where $\mathscr{m}_j$ is the number of atoms which remain fixed or are transformed into translationally equivalent atoms by an operation of class j.

As a simple example, consider now the unit cell of cesium chloride, shown in Fig. 8. The space group is $Pm3m \equiv \mathscr{O}_h^1$, and the factor group is isomorphic with the point group $\mathscr{O}_h$. The cell of Fig. 8 is primitive and contains one CsCl formula unit, that is, one Cs^+ and one Cl^-. Although there are two ions in the cell, they cannot be located on symmetrically equivalent points, as they are different atomic species. The ion in the center of the cube is clearly invariant under all 48 operations of the factor group. The chloride ions on the corners of the cube are, in general, interchanged by the various symmetry operations. However, as the corners are equivalent under primitive translations of the cube, the chloride ion is also effectively invariant. Thus, $m_j = 2$ for all operations of the group. The resulting values of χ_j calculated from Eqn (39) are given in the last line of Table 6. The column headed $n_{\text{tot}}^{(\gamma)}$ gives the distribution of the six degrees of freedom over the symmetry species, as calculated from the character using the magic formula. It is seen that the six vibrations correspond to two triply degenerate frequencies of species F_{1u}. However, application of Eqn (30) shows that three of these degrees of freedom result in acoustic modes. Hence, there remains but one triply degenerate optical frequency, as shown in the last column. As the components of the translation vector, and, hence, the dipole–moment vector, are found in the same species, the single optical frequency should be infrared-active. It will of course be Raman inactive, as no components of α fall in species F_{1u}.

A second example is sodium chloride, whose crystallographic unit cell contains four formula units, as shown in Fig. 9. The space group is $Fm\,3m = \mathcal{O}_h^5$. Thus the crystallographic unit cell is not a primitive cell. Furthermore, the rhombohedral primitive cell constructed in the usual way, as shown by the dashed lines in Fig. 9, does not have cubic symmetry. The symmetrically correct primitive cell for this lattice is the Wigner–Seitz cell shown in Fig. 10. This cell contains only a single NaCl formula unit and, furthermore, has the desired symmetry, $\mathcal{O}_h$. Once again, both ions are invariant under all 48 factor-group operations. Hence, the results of the factor-group analysis are identical to those for cesium chloride.

In the two examples considered above, the crystal structures can be described by two interpenetrating lattices, one associated with the cation,

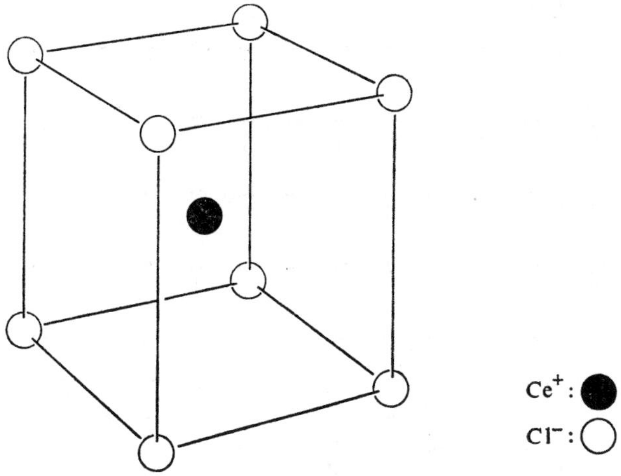

FIG. 8. Unit cell of caesium chloride.

TABLE 6

$\mathcal{O}_h$	E	$8C_3$	$3C_2$	$6C_4$	$6C_2'$	i	$8S_6$	$3\sigma_h$	$6S_4$	$6\sigma_d$		$n_{tot}^{(\gamma)}$	$n_{acous}^{(\gamma)}$	$n_{opt}^{(\gamma)}$
A_{1g}	1	1	1	1	1	1	1	1	1	1	$\alpha_{xx} + \alpha_{yy} + \alpha_{zz}$	0	0	0
A_{2g}	1	1	1	−1	−1	1	1	1	−1	−1		0	0	0
E_g	2	−1	2	0	0	2	−1	2	0	0	$(2\alpha_{zz} - \alpha_{xx} - \alpha_{yy}, \alpha_{xx} - \alpha_{yy})$	0	0	0
$F_{1g} \equiv T_{1g}$	3	0	−1	1	−1	3	0	−1	1	−1	**R**	0	0	0
$F_{2g} = T_{2g}$	3	0	−1	−1	1	3	0	−1	−1	1	$(\alpha_{xy}, \alpha_{yz}, \alpha_{zx})$	0	0	0
A_{1u}	1	1	1	1	1	−1	−1	−1	−1	−1		0	0	0
A_{2u}	1	1	1	−1	−1	−1	−1	−1	1	1		0	0	0
E_u	2	−1	2	0	0	−2	1	−2	0	0		0	0	0
$F_{1u} \equiv T_{1u}$	3	0	−1	1	−1	−3	0	1	−1	1	**T**	2	1	1
$F_{2u} \equiv T_{2u}$	3	0	−1	−1	1	−3	0	1	1	−1		0	0	0
m_j	2	2	2	2	2	2	2	2	2	2				
χ_j	6	0	−2	2	−2	−6	0	2	−2	2				

the other with the anion. The primitive cell chosen for CsCl was just the simple cubic Wigner–Seitz cell for either ionic species. In the case of NaCl the primitive cell was the dodecahedral Wigner–Seitz cell for the face-centered Bravais lattice of either ion.

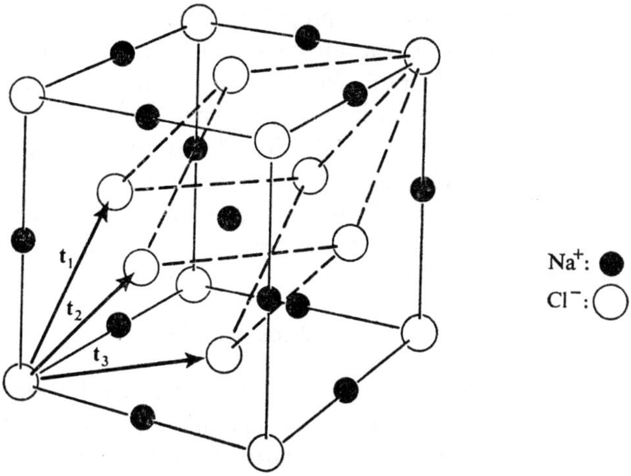

FIG. 9. Crystallographic unit cell of sodium chloride.

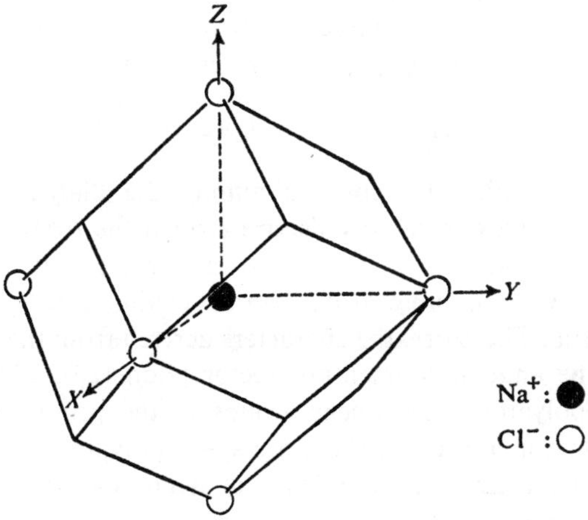

FIG. 10. Wigner–Seitz cell for sodium chloride.

VIII. Molecular and Complex Ionic Crystals

Crystals composed of polyatomic ions or molecules can be represented in general by a set of interpenetrating lattices. However, when interatomic forces in the polyatomic groups are much stronger than the forces between groups, it is convenient to distinguish internal and external group motions, as pointed out in Chapter 3. The external motions arise from the translational and rotational degrees of freedom of the N-atomic groups, while the internal motions are derived from the $3N - 6$ vibrations of each group.

As the method developed in the preceding section is completely general, it can be applied directly to the analysis of molecular or complex ionic crystals. The quantity m_j in Eqn (39) is again the number of atoms which remain fixed or are transformed into translationally equivalent atoms by an operation of class j. In order to classify the external modes which are translational in origin, each group of atoms, i.e. ion or molecule, is treated as if it were a single particle. Thus, if there is a total of s groups in the primitive cell, the characters of the representations associated with these vibrations are given by

$$\chi_j' = m_j(s)(2\cos\phi_j \pm 1), \tag{40}$$

where $m_j(s)$ is the number of groups of atoms which transform into themselves or their equivalents under an operation of the jth class. Three of these translational external motions correspond to acoustical vibrations. Therefore, as only the optical vibrations are of spectroscopic interest, the characters of Eqn (40) should be reduced by the acoustical contribution given by Eqn (30). Thus, the characters for optical vibrations resulting from translational motion of atomic groups are given by

$$\chi_j(\text{trans}) = [m_j(s) - 1](2\cos\phi_j \pm 1). \tag{41}$$

These vibrations, which are counted among the "lattice vibrations" of molecular and complex ionic crystals, are termed the *translatory vibrations*.

Other external vibrations arise from overall rotational motion of polyatomic groups. A monatomic group cannot exhibit rotation, as it has no moment of inertia. Therefore, the characters derived from the transformation properties of the angular momentum vector [Eqn (26), Chap. 2] must be added for all polyatomic nonlinear groups in the primitive cell. Thus, if there are v monatomic groups, there are $s - v$ polyatomic groups, and the corresponding characters for the $3(s - v)$ degrees of rotational freedom are given by

$$\chi_j(\text{rot}) = m_j(s - v)(\pm 2\cos\phi_j + 1), \tag{42}$$

where $m_j(s - v)$ is the number of nonlinear polyatomic groups which are invariant under an operation of class j. The motions considered here, which are all optical vibrations, are referred to as the *rotatory vibrations*. The ensemble of translatory and rotatory vibrations of a crystal constitute the "lattice modes", which are normally observed in the far-infrared region of the absorption spectrum and in the Raman effect.

The internal vibrational modes can be classified by difference using the above results. Thus, Eqn (39) for the total characters, less Eqns (40) and (42) for the characters associated with external motions, leads to the expression

$$\chi_j \text{(vib)} = [m_j - m_j(s)](2\cos\phi_j \pm 1) - m_j(s-v)(\pm 2\cos\phi_j \pm 1) \quad (43)$$

for the characters of the representations of the internal vibrations of the polyatomic species present in the crystal.

From the above discussion it is seen that the optical vibrations of crystals can be classified into translatory and rotatory lattice vibrations and internal vibrations. Furthermore, the symmetries of each of these types of vibrations can be described by the irreducible representations of the factor group with the aid of its character table, which is the same as that of the isomorphic point group.

IX. Factor-Group Analysis of Molecular Crystals: Examples

Naphthalene crystallizes into a monoclinic lattice of space group $P2_2/b \equiv \mathscr{C}_{2h}^5$, whose factor-group operations have already been demonstrated (Fig. 3). The crystallographic unit cell, which is also a primitive cell, contains two naphthalene molecules, which are numbered 1 and 2 in Fig. 11. The long axis of each molecule is approximately parallel to the c axis, while the molecular planes are tilted by about 65° with respect to the b, c plane of the unit cell.

The two molecules per unit cell must occupy a set of sites having a multiplicity of two. One of the four sets represented by $\mathscr{C}_i(2)$ in Appendix E is, therefore, dictated and the centers of inversion of the two molecules must be coincident with the centers of inversion of the unit cell (See Fig. 3).

An isolated naphthalene molecule has point symmetry $\mathscr{D}_{2h}$. Thus, in addition to its center of inversion, it contains three planes and three two-fold axes of symmetry. Its 18 atoms result in $3 \times 18 - 6 = 48$ degrees of vibrational freedom represented by

$$\Gamma_{\text{vib}} = 9A_g + 3B_{1g} + 4B_{2g} + 8B_{3g} + 4A_u + 8B_{1u} + 8B_{2u} + 4B_{3u}, \quad (44)$$

using the method of Chapter 2, Section X. All *gerade* species are Raman-active, while those of species B_{1u}, B_{2u}, and B_{3u} are active in the infrared spectrum.

In the primitive cell containing two molecules there are $2 \times 48 = 96$ internal vibrational frequencies, which would correspond to 48 pairs of doubly degenerate vibrations if there were no interaction between the two molecules. In addition, there are $2 \times 6 - 3 = 9$ external optical vibrations or "lattice modes".

The factor-group analysis of the unit cell of naphthalene is carried out using the character table for the point group $\mathscr{C}_{2h}$, which is isomorphic with the factor group. The analysis of napthalene is summarized in Table 7. The values of m_j follow immediately from the observation that only the identity operation leaves any atoms invariant. The resulting values of $n_{tot}^{(\gamma)}$ describe the distribution of the 108 total degrees of freedom of the system.

Further approximate classification can be made in this case, because the intramolecular forces are much stronger than those between the two molecules. Thus, the unit cell is considered to be made up to two groups ($s = 2$) and no monatomic species ($v = 0$). Application of Eqns (41) and (42) then allows the numbers of translatory and rotatory lattice vibrations to be calculated, as given in Table 7.

Since the site group is a subgroup of both the factor group and the molecular point group of the free molecule, it can be used to establish the

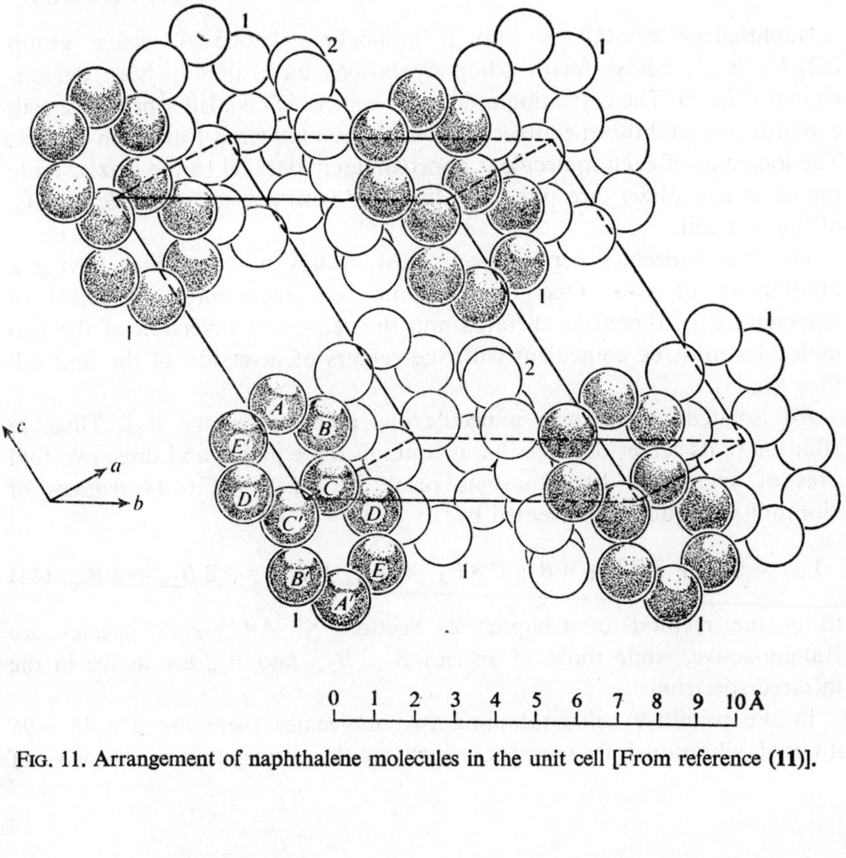

FIG. 11. Arrangement of naphthalene molecules in the unit cell [From reference (11)].

TABLE 7

$\mathscr{C}_{2h}^{\ 5}$	E	C_2^s	i	σ^g	$n_{\text{tot}}^{(\gamma)}$	$n_{\text{acous}}^{(\gamma)}$	$n_{\text{trans}}^{(\gamma)}$	$n_{\text{rot}}^{(\gamma)}$	$n_{\text{vib}}^{(\gamma)}$	Activity
A_g	1	1	1	1	27	0	0	3	24	*R*
B_g	1	−1	1	−1	27	0	0	3	24	*R*
A_u	1	1	−1	−1	27	1	2	0	24	*IR*
B_u	1	−1	−1	1	27	2	1	0	24	*IR*
χ_j (tot)	108	0	0	0						
χ_j (trans)	3	1	−3	−1						
χ_j (rot)	6	0	6	0						

correlation of these two groups following the method outlined in Sec. XIV, Chap. 2. Inspection of the characters given in the table for $\mathscr{C}_i$ in Appendix C with the entries in the columns headed E and i in the table for $\mathscr{C}_{2h}$ shows that A_g (in $\mathscr{C}_i$) $\Rightarrow A_u + B_u$ (in $\mathscr{C}_{2h}$). Similar considerations apply to the mapping of $\mathscr{C}_i$ on $\mathscr{C}_{2h}$. The resulting correlation diagram is shown as Table 8. It is immediately clear that the effect of crystal symmetry on the selection rules governing molecular vibrations can be decomposed into two parts. First, there is the possible effect of site symmetry, which in this case results in lowering of the effective molecular symmetry from $\mathscr{D}_{2h}$ to $\mathscr{C}_i$. As a result, the four vibrations of species A_u become allowed in the infrared spectrum. Hence, under site-group selection rules all 24 vibrational fundamentals of *ungerade* species are infrared-active. Raman selection rules are unaffected, and, as the center of inversion is preserved, the rule of mutual exclusion is still applicable (See Appendix G).

TABLE 8

Isolated molecule $\mathscr{D}_{2h}$	Site $\mathscr{C}_i$	Unit cell $\mathscr{C}_{2h}^5$
$9\ A_g\ (R)$		$(24 + 3r)\ A_g\ (R)$
$3\ B_{1g}\ (R)$	$24\ A_g\ (R)$	
$4\ B_{2g}\ (R)$		$(24 + 3r)\ B_g\ (R)$
$8\ B_{3g}\ (R)$		
$4\ A_u$ (inactive)		$(24 + 2t)\ A_u\ (IR)$
$8\ B_{1u}\ (IR)$	$24\ A_u\ (IR)$	
$8\ B_{2u}\ (IR)$		$(24 + t)\ B_u\ (IR)$
$4\ B_{3u}\ (IR)$		

The second effect of crystal structure results from the possibility of vibrational coupling between the two molecules in the unit cell. In the present example each of the 24 Raman–active vibrations of species A_g in the site group is split into a doublet with one component each of species A_g and B_g in the factor group. Similarly, each of the A_u-species vibrations is split into an infrared-active doublet $(A_u + B_u)$ as a result of coupling. From

Table 7 it is seen that the two vibrations corresponding to the components of a given doublet are distinguished by their symmetries with respect to the screw-rotation operation of the factor group. This type of splitting, which is known as *correlation splitting* (or *Davydov splitting*), cannot be explained by an analysis based on the site symmetry alone.

An additional advantage of the factor-group method is that it provides a basis for the prediction of the infrared and Raman spectra in the lattice-mode region. From Table 7 it is found that the three translatory lattice modes should be infrared-active, while all six of the rotatory modes would be expected to be found in the Raman spectrum.

The lowering of symmetry which a molecule suffers when it is placed at a site in a crystal is the direct result of the surrounding crystalline field. Although the geometrical distortion of the molecule is often quite small, it has been observed by X-ray methods in numerous cases. The results for naphthalene (**11**) are shown in Fig. 12, where it is clear that the only remaining symmetry operations are those of the site group $\mathscr{C}_i$.

Numerous spectroscopic studies have been made on crystalline naphthalene (**12–15**). Although the experimental results are, in general, in agreement with the factor-group selection rules developed here, correlation splitting of the internal vibrations has not been resolved. Thus, although a factor-group analysis of the entire primitive cell is necessary in order to account for the observed lattice vibrations, the internal molecular vibrations are in this case completely explained by the site-group method. The lattice vibrations of naphthalene are discussed in Chapter 5, where the subject of the polarized Raman spectra of crystals is treated, and in Chapter 6 in connection with the calculation of intermolecular force fields.

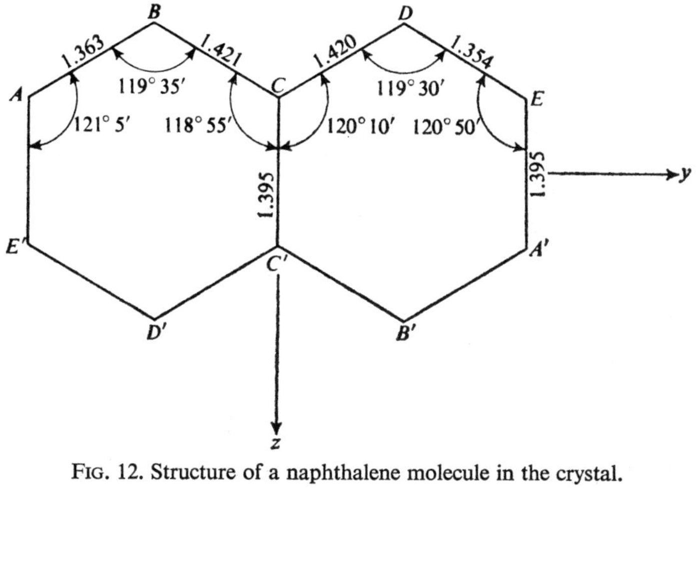

FIG. 12. Structure of a naphthalene molecule in the crystal.

As a second example of the factor-group analysis of an organic crystal, consider the unit cell of cyanuric triazide, shown in Fig. 13. This crystal is of the hexagonal class with space group $\mathscr{C}_{6h}{}^2 \equiv P\,6_3/m$ and two molecules per primitive unit cell (**16**). From Appendix E it is found that this space group contains only two types of sites of multiplicity two. One of these types is of point group $\mathscr{S}_6 \equiv \mathscr{C}_3 \times \mathscr{C}_i$, which includes a center of symmetry, and which cannot, therefore, accomodate a cyanuric triazide molecule. The other is of symmetry $\mathscr{C}_{3h}$ of which there are three sets. One of these latter sets is, therefore, occupied by the two molecules, and the site symmetry is $\mathscr{C}_{3h}$. This point group is also very probably the molecular symmetry group of an isolated molecule.

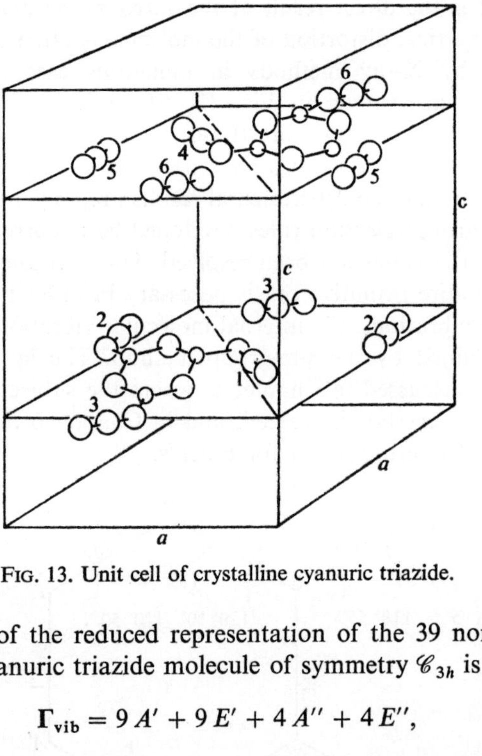

FIG. 13. Unit cell of crystalline cyanuric triazide.

The structure of the reduced representation of the 39 normal modes of vibration of a cyanuric triazide molecule of symmetry $\mathscr{C}_{3h}$ is found to be

$$\Gamma_{\text{vib}} = 9\,A' + 9\,E' + 4\,A'' + 4\,E'', \tag{45}$$

where the vibrations of species E' and A'' are infrared–active. Vibrational fundamentals of species A' are Raman-active. This analysis is appropriate for a single molecule at a $\mathscr{C}_{3h}$ site, and presumably, for an isolated molecule in the gas phase or in solution.

For the unit cell containing two molecules, consideration of the operations involving pure translation in the four crystallographic directions allows the atoms to be classified as equivalent or nonequivalent. The azide groups can be similarly classified, as shown in Fig. 13, in that groups marked with the same number can be carried into each other by one or more translation operations. By successive application of the symmetry operations of the factor group $\mathscr{C}_{6h}^2$ (See Fig. 14) it is found that only the operations E and σ_h carry atoms or groups into themselves or translationally-equivalent atoms or groups. Therefore, $\chi_E = 90$ and $\chi_{\sigma_h} = 30$, and the number of vibrations of each symmetry species is easily calculated, as given in Table 9. After removal of the acoustic modes, the optical modes can be approximately classified as translatory, rotatory, and internal or vibrational, as shown. These results can be correlated with the vibrations of a single molecule using the diagram of Table 10. It is seen that in this example, correlation coupling of the two molecules in the unit cell does not result in splitting of the vibrational fundamentals.

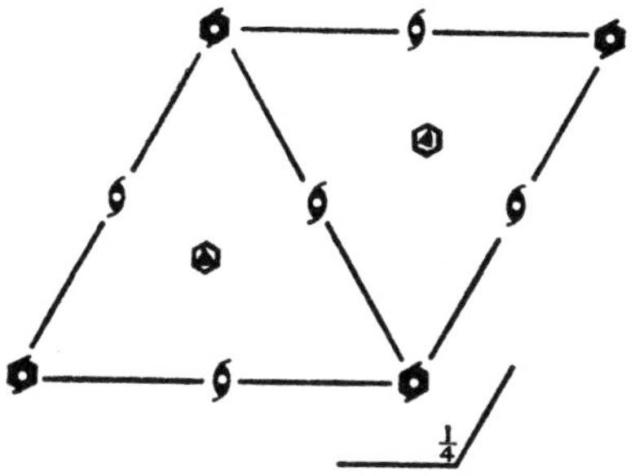

FIG. 14. Symmetry elements for space group $P6_3/m \equiv \mathscr{C}_{6h}^2$ [From reference (3)].

In the isolated molecule there are nine fundamentals of species E', which are active in both the infrared spectrum and Raman effect. From Table 10 it is apparent that in the crystal the coincidence of these nine frequencies in the infrared and Raman spectra is removed by the presence of a center of symmetry.

Table 9 shows that no lattice modes are infrared–active for this crystal. However, two rotatory vibrations (species A_g and E_{1g}) and one translatory vibration (species E_{2g}) should be observable in the low-frequency Raman spectrum.

The infrared and Raman spectra of crystalline cyanuric triazide have recently been reported (**17**). Of the 26 internal vibrations, 23 were assigned. The Raman-active fundamentals of species E' were found at frequencies which were generally 5 to 20 cm^{-1} higher than the corresponding infrared fundamentals, in agreement with the rule of mutual exclusion.

Two Raman bands (77 cm^{-1} and 100 cm^{-1}) were assigned to lattice vibrations. However, the far-infrared spectrum of this compound was not investigated. The low-frequency Raman frequencies undoubtedly correspond

TABLE 9

$\mathscr{C}_{6h}{}^2$	$n_{tot}^{(\gamma)}$	$n_{acous}^{(\gamma)}$	$n_{trans}^{(\gamma)}$	$n_{rot}^{(\gamma)}$	$n_{vib}^{(\gamma)}$	Activity
A_g	10	0	0	1	9	*R*
B_g	5	0	1	0	4	
E_{1g}	5	0	0	1	4	*R*
E_{2g}	10	0	1	0	9	*R*
A_u	5	1	0	0	4	*IR*
B_u	10	0	0	1	9	
E_{1u}	10	1	0	0	9	*IR*
E_{2u}	5	0	0	1	4	
degrees of freedom	90	3	3	6	78	

TABLE 10

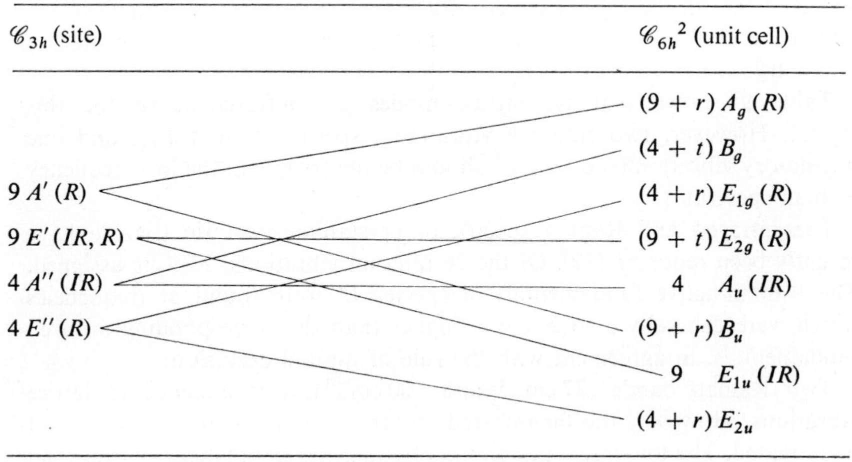

to the two rotatory vibrations predicted in Table 10. The third lattice vibration which is translatory in nature, is probably much weaker. This result is consistent with the general principle [(**9**), p. 148] that the in Raman effect the intensities of rotatory fundamentals are always much stronger than those of translatory motions. The inverse might be expected to hold for infrared spectra.

X. Factor-Group Analysis of Complex Ionic Crystals: Examples

If a crystal contains a complex ion, the equations derived above for the characters of the representations of translatory, rotatory, and vibrational degrees of freedom are still applicable. As an illustration, consider sodium metathioborate, $Na_3B_3S_6$. The anion, $B_3S_6^{3-}$, has the planar structure (**18, 19**) shown in Fig. 15, which is of point group $\mathscr{D}_{3h}$. Using the methods of Chapter 2, the structure of the reduced representation of the $3N - 6 = 21$ modes of vibration is found to be

$$\Gamma_{\text{vib}} = 3\,A_1' + 2\,A_2' + 5\,E' + 2\,A_2'' + 2\,E''. \tag{46}$$

Vibrations of species E' and A_2'' are infrared–active, while those of species A_1', E', and E'' are active in the Raman effect.

In the crystalline solid the sodium salt of $B_3S_6^{3-}$ has six formula units, $Na_3B_3S_6$, per hexagonal unit cell or two such units in the rhombohedral cell (See Fig. 16). The space group is $R\,\bar{3}c \equiv \mathscr{D}_{3d}^6$. Hence, the factor-group analysis of this compound is somewhat analogous to that of the classic example, calcite [(**9**), p. 144; (**20**), p. 460]. From the International Tables it is found that the anions occupy the two equivalent sites of

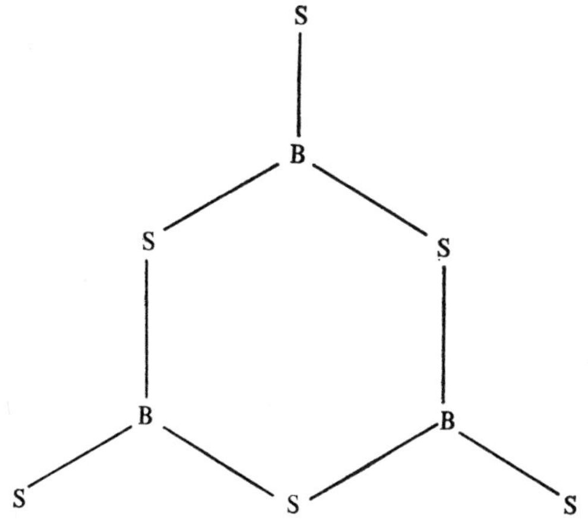

Fig. 15. Structure of the metathioborate ion, $B_3S_6^{3-}$.

symmetry $\mathscr{D}_3 \equiv 32$, whose coordinates are $0, 0, \frac{1}{4}$ and $0, 0, \frac{3}{4}$. The cations are found at six equivalent sites of symmetry $\mathscr{C}_2 \equiv 2$. Three cations are coplanar with each anion and are located along the extensions of the >B—S side chains as shown in Fig. 17.

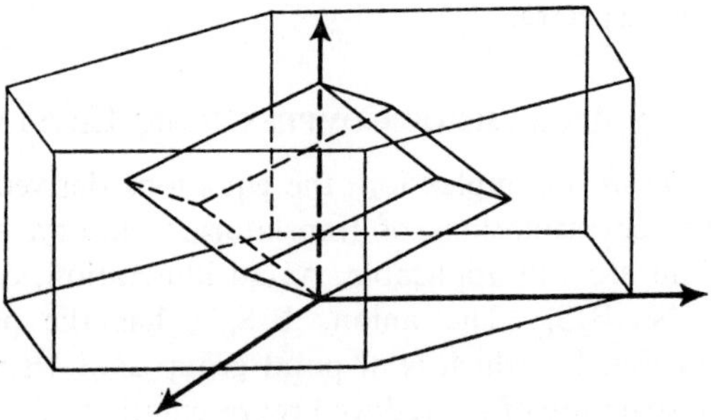

FIG. 16. Rhombohedral unit cell of sodium metathioborate, $Na_3B_3S_6$.

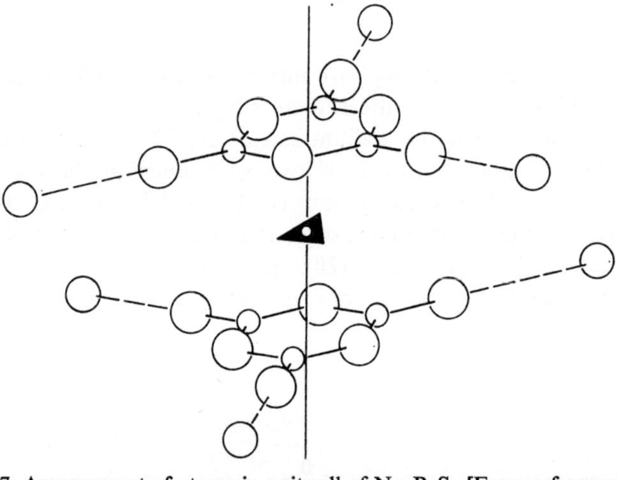

FIG. 17. Arrangement of atoms in unit cell of $Na_3B_3S_6$ [From reference (**19**)].

Consideration of the effect of the factor-group operations shows that atoms are unshifted only under operations E and C_2 (See Fig. 18). The latter operation is a 180° rotation around the axis of a >B—S side chain, which leaves four atoms (S⟨⟩B—S ··· Na) invariant and moves another such set of four atoms into a translationally equivalent set in an adjacent unit cell. Thus, $m_{C_2} = 8$, and Eqn (39) then yields the values of χ_j shown in Table 11, which lead in turn to the entries under $n_{\text{tot}}^{(\gamma)}$.

As the unit cell of this compound contains two complex anions and six monatomic cations, $s = 8$ and $v = 6$ in Eqns (41) to (43). The effect of

TABLE 11

$\mathscr{D}_{3d}{}^6$	E	$2C_3$	$3C_2$	i	$2S_6$	$3\sigma_d{}^g$		$n_{\text{tot}}^{(\gamma)}$	$n_{\text{trans}}^{(\gamma)}$	$n_{\text{rot}}^{(\gamma)}$	$n_{\text{acous}}^{(\gamma)}$	$n_{\text{vib}}^{(\gamma)}$
A_{1g}	1	1	1	1	1	1	$\alpha_{xx}+\alpha_{yy}, \alpha_{zz}$	4	1	0	0	3
A_{2g}	1	1	−1	1	1	−1	R_z	8	3	1	0	4
E_g	2	−1	0	2	−1	0	$(R_x, R_y)\ (\alpha_{xx}-\alpha_{yy}, \alpha_{xy})\ (\alpha_{yz}, \alpha_{zx})$	12	4	1	0	7
A_{1u}	1	1	1	−1	−1	−1		4	1	0	0	3
A_{2u}	1	1	−1	−1	−1	1	T_z	8	2	1	1	4
E_u	2	−1	0	−2	1	0	(T_x, T_y)	12	3	1	1	7
m_j	24	0	8	0	0	0						
χ_j	72	0	−8	0	0	0						
$m_j(s)$	8	2	4	0	0	0						
χ_j (trans)	21	0	−3	3	0	−1						
$m_j(s-v)$	2	2	2	0	0	0						
χ_j (rot)	6	0	−2	0	0	0						
χ_j (vib)	42	0	−2	0	0	0						

the symmetry operations on the six cations and the centers of mass of the two anions is represented by $m_j(s)$ in Table 11. Application of Eqn (41) then gives the characters χ_j (trans) and the magic formula yields the values of $n^{(\gamma)}_{\text{trans}}$ shown. Only the two anions contribute to χ_j (rot), as the cations

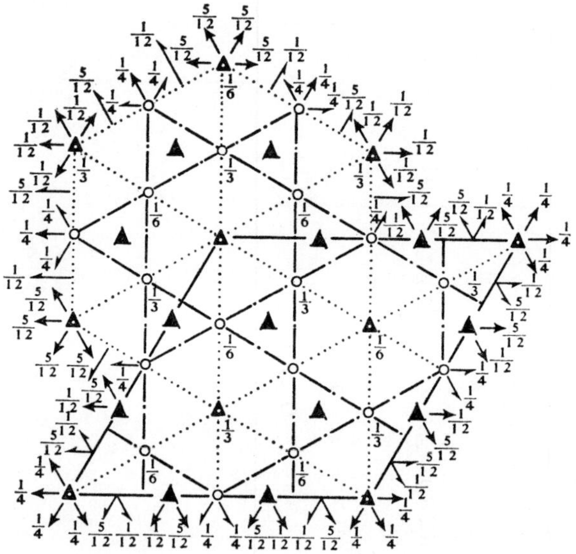

FIG. 18. Symmetry elements of space group $R\bar{3}c \equiv \mathscr{D}^6_{3d}$ [From reference (3)].

have no moments of inertia ($s - v = 2$) and the results given in Table 11 are obtained. The three acoustical degrees of freedom can be classified using Eqn (30) or by inspection of the character table for point group $\mathscr{D}_{3d}$ in Appendix C.

Finally, the distribution of the vibrational modes over the symmetry species can be found by difference from the right-hand part of Table 11 (column 1 minus the sum of columns 2, 3, and 4) or directly using Eqn (43) to obtain the characters shown in the last line of the table.

The correlation of the above results with Eqn (46) for the free ion is made *via* the site group, $\mathscr{D}_3$, as shown in Table 12. Several effects of the crystal symmetry can be noted. First, the presence of a center of symmetry in the unit cell removes the requirement of infrared and Raman coincidence of the five vibrations of species E' of the free ion. Secondly, the two vibrations of species E'', which are only Raman-active for the free ion, become infrared–active in the crystal. No vibrational fundamentals would be expected to be split. Finally, the far–infrared spectrum would exhibit five translatory ($2\,A_{2u} + 3\,E_u$) and two rotatory ($A_{2u} + E_u$) lattice vibrations.

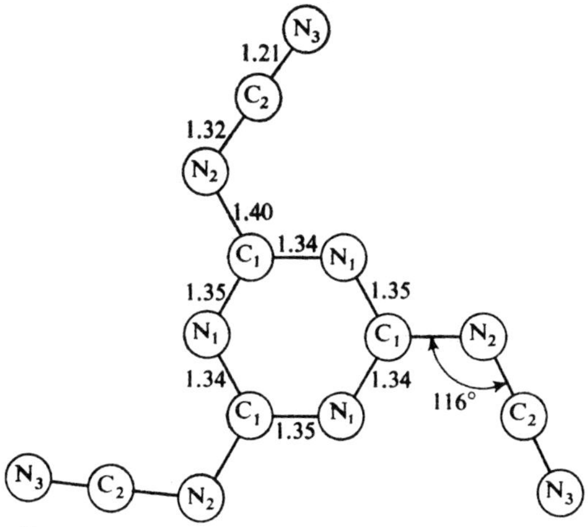

FIG. 19. Structure of the cyanuric tricyanamide ion.

In an infrared study of $Na_3B_3S_6$ (**21**), all five E'-species fundamentals as well as the five translatory lattice modes, were assigned. The Raman spectrum of this compound has not yet been investigated.

A rather more complex ionic crystal is that of sodium cyanuric tricyanamide trihydrate, $Na_3N_3C_3(NCN)_3 . 3H_2O$. The anion is isoelectric with cyanuric triazide and has the structure shown in Fig. 19 (**22**). The vibrational frequencies of this ion can, therefore, be classified by Eqn (45).

TABLE 12

$\mathscr{D}_{3h}$ (free ion)	$\mathscr{D}_3$ (site)	$\mathscr{D}_{3d}^6$ (unit cell)	
$3\,A_1'$ (R)		$A_{1g}\,(3 + t)$	(R)
$2\,A_2'$	A_1	$A_{2g}\,(4 + 3t + r)$	
$5\,E'$ (IR, R)	A_2	$E_g\,(7 + 4t + r)$	(R)
$0\,A_1''$	E	$A_{1u}\,(3 + t)$	
$2\,A_2''$ (IR)		$A_{2u}\,(4 + 2t + r)$	(IR)
$2\,E''$ (R)		$E_u\,(7 + 3t + r)$	(IR)

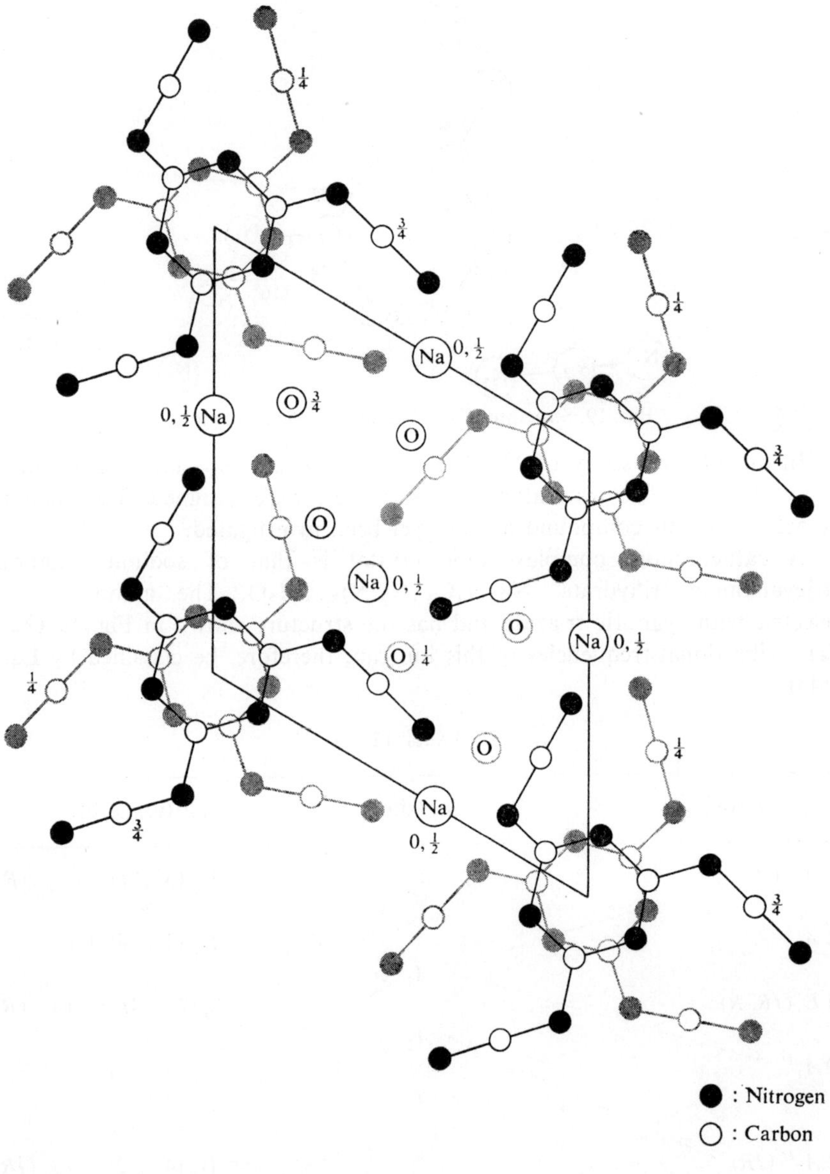

FIG. 20. Structure of the unit cell of sodium cyanuric tricyanamide trihydrate. Hydrogen atoms (not shown) form hydrogen bonds between pairs of oxygen atoms.

The unit cell of the crystal is of space group $P\bar{6}2c \equiv \mathscr{D}_{3h}{}^4$ with two anions, six cations, and six molecules of water of hydration per primitive unit cell. The structure is shown in Fig. 20. while the factor–group operations can be identified using Fig. 21. The anions are found to occupy sites of $\mathscr{C}_{3h}$ symmetry located at 0, 0, 0 and 0, 0, $\frac{1}{2}$.

As this structure contains two different polyatomic groups, the formulas

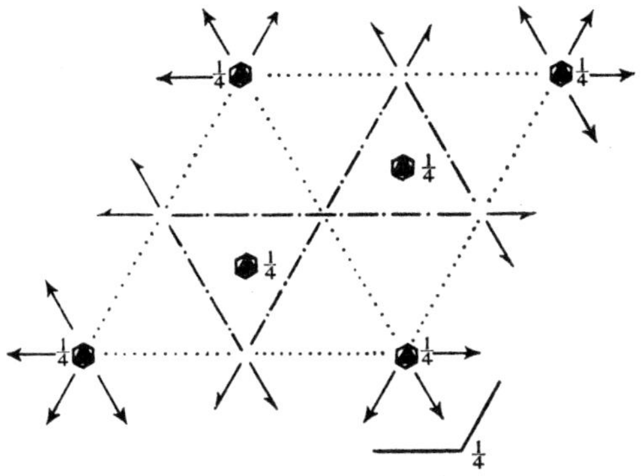

FIG. 21. Symmetry elements for space group $P\bar{6}2c \equiv \mathscr{D}^4_{3h}$. [From ref. (3)].

TABLE 13

$\mathscr{D}^4_{3h}$		E	$2C_3$	$3C_2$	σ_h	$2S_3$	$3\sigma_v$
anion	χ_j (tot)	90	0	0	30	0	0
	χ_j (trans)	6	0	0	2	−4	0
	χ_j (rot)	6	0	0	−2	4	0
cation	χ_j (trans)	18	0	−2	0	0	0
water	χ_j (tot)	54	0	0	18	0	0
	χ_j (trans)	18	0	0	6	0	0
	χ_j (rot)	18	0	0	−6	0	0

developed earlier for the characters [Eqns (41) to (43)] must be reconsidered, if, for example, the internal vibrations of the cation and water are to be distinguished. Following Hass and Sutherland (**23**) the correct result will be obtained if Eqns (41) to (43) are first applied to the ionic lattice, neglecting the water molecules. In the second step the lattice of water molecules is treated using the same set of formulas. The total of the two results gives the analysis of the degrees of freedom of the crystal. Alternatively, each group can be considered separately, as shown in Table 13. For the external vibrations the normal modes will, in general, be linear combinations of the translatory and rotatory coordinates of the various groups present in the unit cell.

The results of the factor-group analysis of $Na_3N_3C_3(NCN)_3 . 3\,H_2O$ are summarized in Table 14. The correlation diagram relating the factor group to the site group is shown in Table 15. There it is seen that the presence of two molecules in the unit cell can result in doubling of the E'-species fundamental in the infrared spectrum and both E' and E'' species in the Raman. Coincidences of the 18 fundamentals of species E' in the infrared and Raman are still expected. Three translatory lattice modes ($2\,E' + A_2''$) are predicted in the far-infrared spectrum, while three rotatory ($A' + 2\,E''$) and two translatory ($2\,E'$) vibrations should be Raman-active.

In a recent study of the infrared and Raman spectrum of $Na_3N_3C_3(NCN)_3 . 3H_2O$, all of the internal vibrations were assigned (**24**). Most of the fundamentals of species E'' were split into doublets as, predicted, and all three of the translatory lattice vibrations were assigned in the far–infrared region. The Raman spectrum was of poor quality due to strong flourescence of the sample. However, E'-species bands observed in the infrared and Raman were coincident to within a few wavenumbers. The lattice-mode region was obscured by the sample flourescence, making assignment of additional lattice modes impossible.

As a final example, the factor-group analysis of the ionic crystal potassium azide will be given. The crystallographic unit cell of this compound is of space group $I4/mcm \equiv \mathscr{D}_{4h}^{18}$ and contains four formula units of KN_3 (**25**). The linear anions, N_3^-, occupy sites of $\mathscr{D}_{2h}$ symmetry located at positions $0, \frac{1}{2}, 0$ and $\frac{1}{2}, 0, 0$, while the cations are at $0, 0, \frac{1}{4}$ and $0, 0, \frac{3}{4}$. The crystal structure is shown in Fig. 22. As it is a body-centered, tetragonal Bravais lattice, the crystallographic unit cell has twice the volume of a primitive cell. The latter cell contains eight atoms and has, therefore, 24 degrees of freedom, which can be classified under the irreducible representations of the factor group $\mathscr{D}_{4h}^{18}$.

In this example the anion has a linear structure. Thus, although Eqns (39) and (41), which give the characters χ_j (tot) and χ_j (trans), respectively,

TABLE 14

$\mathscr{D}_{3h}^{4}$	Anion				Cation	Water				Activity
	$n_{tot}^{(\gamma)}$	$n_{trans}^{(\gamma)}$	$n_{rot}^{(\gamma)}$	$n_{vib}^{(\gamma)}$	$n_{trans}^{(\gamma)}$	$n_{tot}^{(\gamma)}$	$n_{trans}^{(\gamma)}$	$n_{rot}^{(\gamma)}$	$n_{vib}^{(\gamma)}$	
A_1'	10	0	1	9	1	6	2	1	3	*R*
A_2'	10	0	1	9	2	6	2	1	3	
E'	20	2	0	18	3	12	4	2	6	*IR*, *R*
A_1''	5	1	0	4	1	3	1	2	0	
A_2''	5	1	0	4	2	3	1	2	0	*IR*
E''	10	0	2	8	3	6	2	4	0	*R*
degrees of freedom	90	6	6	78	18	54	18	18	18	

are still valid, Eqn (42) for the characters of the representations of the rotatory motions must be modified. This problem has been treated by Mitra in the following form (**26**).

Let the z axis be coincident with the infinite rotation axis of the linear group (N_3^-, in the present example). The two components of angular

TABLE 15

$\mathscr{C}_{3h}$ (site)	$\mathscr{D}_{3h}$ (unit cell)
$9\,A'\,(R)$	$A_1'\,(9 + r)\quad (R)$
	$A_2'\,(9 + r)$
$9\,E'\,(IR, R)$	$E'\,(18 + 2t)\quad (IR, R)$
$4\,A''\,(IR)$	$A_1''\,(4 + t)$
	$A_2''\,(4 + t)\quad (IR)$
$4\,E''\,(R)$	$E''\,(8 + 2r)\quad (R)$

momentum transform as

$$\begin{pmatrix} P_x' \\ P_y' \end{pmatrix} = \begin{pmatrix} \pm\cos\phi & \pm\sin\phi \\ \mp\sin\phi & \pm\cos\phi \end{pmatrix} \begin{pmatrix} P_x \\ P_y \end{pmatrix}, \tag{47}$$

under symmetry operations $C_n^k(z)$ and $S_n^k(z)$, for which the z axis is the symmetry element. For operations such as σ_v, or C_2 with respect to an axis perpendicular to z, the corresponding transformation is given by

$$\begin{pmatrix} P_x' \\ P_y' \end{pmatrix} = \begin{pmatrix} \cos 2\theta & \sin 2\theta \\ \sin 2\theta & -\cos 2\theta \end{pmatrix} \begin{pmatrix} P_x \\ P_y \end{pmatrix}, \tag{48}$$

where $\theta = 2\pi n$ for the latter operations.

Thus, in the case of linear groups the expression for the character for rotatory motion becomes

$$\chi_j\,(\text{rot}) = [\mathscr{m}_j\,(s - v)]\,P_j, \tag{49}$$

$$\text{where}\quad P_j = \begin{cases} \pm 2\cos\phi_j & \text{for classes containing } C_n^k(z) \text{ and } S_n^k(z) \\ 0 & \text{for classes containing } C_2(\perp z) \text{ and } \sigma_v. \end{cases} \tag{50}$$

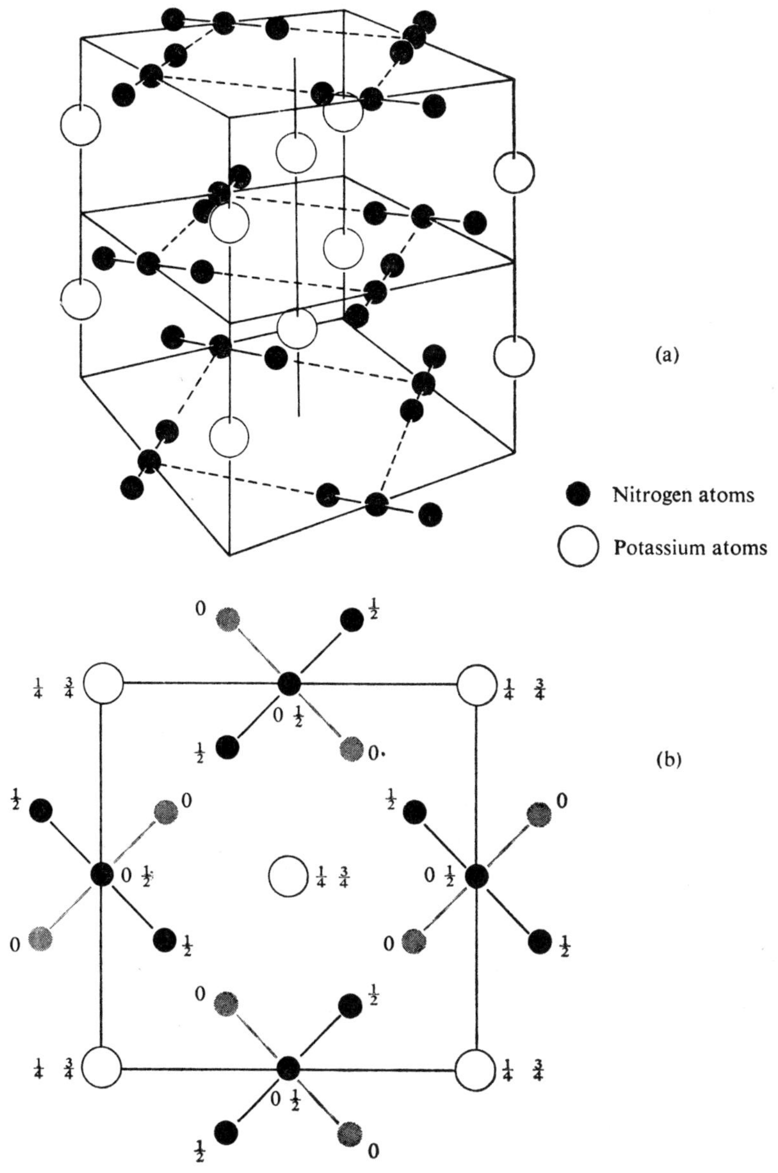

FIG. 22. Unit cell of potassium azide (a) Perspective view, (b) Projection on the 001 plane [From reference (**27**)].

TABLE 16

$\mathscr{D}_{4h}^{18}$	E	$2C_4$	C_2	$2C_2'$	$2C_2''$	i	$2S_4$	σ_h	$2\sigma_v^s$	$2\sigma_d$
m_j	8	2	4	2	6	2	0	6	0	4
$m_j\,(s)$	4	2	4	2	4	2	0	2	0	2
$m_j\,(s-v)$	2	0	2	0	2	2	0	2	0	2
P_j	2	1	0	−1	−1	2	1	0	−1	−1
χ_j (tot)	24	2	−4	−2	−6	−6	0	6	0	4
χ_j (trans)	9	1	−3	−1	−3	−3	1	1	−1	1
χ_j (rot)	4	0	0	0	−2	4	0	0	0	−2
χ_j (vib)	8	0	0	0	0	−4	0	4	0	4

For linear groups, then, Eqns (49) and (50) must be substituted for Eqn (42), and Eqn (43) must be modified accordingly.

Returning to the example of KN_3, the effect of each symmetry operation of the factor group is represented by the values of m_j, $m_j(s)$, and $m_j\,(s - v)$ listed in Table 16. These results are obtained by consideration of the symmetry elements shown in Fig. 23. The classification of the various types of motion is summarized in Table 17. These results, which were obtained from Table 16 using the magic formula, can be correlated with the

TABLE 17

$\mathscr{D}_{4h}^{18}$	$n_{\text{tot}}^{(\gamma)}$	$n_{\text{trans}}^{(\gamma)}$	$n_{\text{rot}}^{(\gamma)}$	$n_{\text{acous}}^{(\gamma)}$	$n_{\text{vib}}^{(\gamma)}$	Activity
A_{1g}	1	0	0	0	1	*R*
A_{2g}	2	1	1	0	0	
B_{1g}	1	0	1	0	0	*R*
B_{2g}	1	0	0	0	1	*R*
E_g	2	1	1	0	0	*R*
A_{1u}	0	0	0	0	0	
A_{2u}	3	1	0	1	1	*IR*
B_{1u}	2	1	0	0	1	
B_{2u}	0	0	0	0	0	
E_u	5	2	0	1	2	*IR*
degrees of freedom	24	9	4	3	8	

TABLE 18

$\mathscr{D}_{\infty h}$ (free ion)	$\mathscr{D}_{2h}$ (site)	$\mathscr{D}_{4h}^{18}$ (unit cell)	
ν_1 Σ_g^+	A_g	A_{1g} (1)	R
		A_{2g} $(t + r)$	
Σ_g^-	B_{1g}		
		B_{1g} (r)	R
	B_{2g}	B_{2g} (1)	R
Π_g			
⋮	B_{3g}	E_g $(t + r)$	R
ν_3 Σ_u^+	A_u	A_{1u}	
		A_{2u} $(1 + t)$	IR
Σ_u^-	B_{1u}		
		B_{1u} $(1 + t)$	
	B_{2u}	B_{2u}	
ν_2 Π_u			
⋮	B_{3u}	E_u $(2 + 2t)$	IR

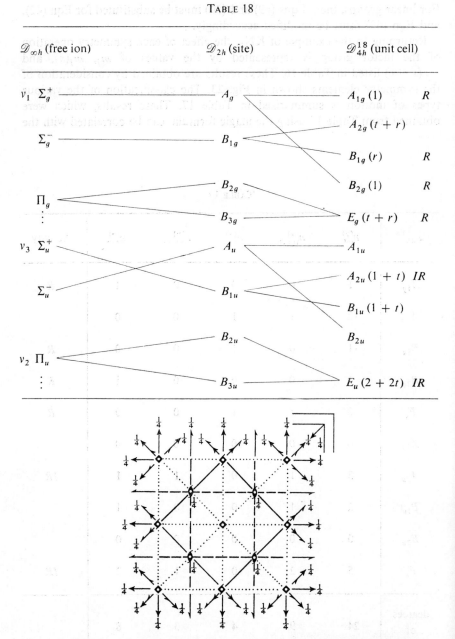

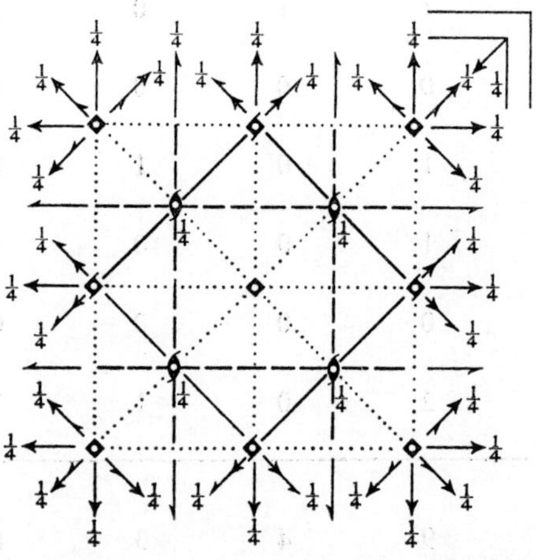

FIG. 23. Symmetry elements for space group $I4/mcm \equiv \mathscr{D}_{4h}{}^{18}$ [From reference (3)].

$\mathscr{C}_{2v}^1 \equiv Pmm2$ $\quad$ $\mathscr{C}_{2v}^4 \equiv Pma2$ $\quad$ $\mathscr{C}_{2v}^2 \equiv Pmc2_1$ $\quad$ $\mathscr{C}_{2v}^7 \equiv Pmn2_1$

symmetry properties of the vibrations of the azide ion in its $\mathscr{D}_{2h}$ site and in the free state by means of Table 18.

The infrared spectrum of crystalline KN_3 has been studied by Bryant (**27**). In agreement with the above predictions, the asymmetric stretching mode of N_3^- (species Σ_u^+) is observed as a single strong feature at 2036 cm^{-1}. The Π_u-species bending mode (ν_2), on the other hand, appears as a doublet centered at approximately 642 cm^{-1}, with a splitting of 11 cm^{-1}. Although neither the far-infrared region nor the Raman spectrum was investigated, the frequencies of a number of lattice modes were inferred from observed combination bands.

REFERENCES

1. Landau, L. D. and Lifshitz, E. M. "Statistical Physics", Pergamon, London (1958).
2. Zhdanov, G. S. "Crystal Physics", Oliver and Boyd, London (1965).
3. "International Tables for X-Ray Crystallography", Vol. I, Kynoch, Birmingham (1965).
4. Seitz, F. *Zeit. Krist.* **88,** 433 (1934); **90,** 289 (1935); **91,** 336 (1935); **94,** 100 (136); *Ann. Math.* **37,** 17 (1936).
5. Lomont, J. S. "Applications of Finite Groups", Academic Press, New York (1959).
6. Wigner, E. and Seitz, F. *Phys. Rev.* **43,** 804 (1933).
7. Ziman, J. M. "Principles of the Theory of Solids", Cambridge (1964).
8. Winston, H. and Halford, R. S. *J. Chem. Phys.* **17,** 607 (1949).
9. Bhagavantam, S. and Venkatarayudu, T. "Theory of Groups and its Application to Physical Problems", Third edition, Andhra University, Waltair (1962).
10. Bhagavantam, S. and Venkatarayudu, T. *Proc. Indian Acad. Sci.* **9A,** 224 (1939).
11. Abrahams, S. C., Robertson, J. M. and White, J. G. *Acta Cryst.* **2,** 238 (1949).
12. McClellan, A. L. and Pimentel, G. C. *J. Chem. Phys.* **23,** 245 (1955).
13. Mitra, S. S. and Bernstein, H. J. *Can. J. Chem.* **37,** 553 (1959).
14. Scully, D. B. and Whiffen, D. H. *Spectrochim. Acta* **16,** 1409 (1960).
15. Scherer, J. R. *J. Chem. Phys.* **36,** 3308 (1962).
16. Knaggs, I. *Proc. Roy. Soc.* (*London*) **A150,** 576 (1935); *J. Chem. Phys.* **3,** 241 (1935).
17. Shearer, S. J., Turrell, G. C., Bryant, J. I. and Brooks, R. L. III, *J. Chem. Phys.* **44,** 1138 (1968).
18. Chopin, F. and Hardy, A. *Compt. Rend.* **260,** 142 (1965).
19. Chopin, F. Thesis, Bordeaux (1966).
20. Mathieu, J. -P. "Spectres de vibration et symétrie des molécules et des cristaux", Hermann, Paris (1945).
21. Chopin, F. and Turrell, G. *J. Mol. Structure* **3,** 57 (1969).
22. Hoard, J. *J. Am. Chem. Soc.* **60,** 1194 (1938).
23. Hass, M. and Sutherland, G. B. B. M. *Proc. Roy. Soc.* (*London*) **A236,** 427 (1956).
24. Shearer-Turrell, S. Diplôme d'Études Supérieures, Paris (1967).
25. Hendricks, S. and Pauling, L. *J. Am. Chem. Soc.* **47,** 2904 (1925).
26. Mitra, S. S. *Zeit. Krist.* **116,** 149 (1961).
27. Bryant, J. I. *J. Chem. Phys.* **38,** 2845 (1963).

Chapter 5

Optical Properties of Crystals

I. Electromagnetic Basis †

An electromagnetic field in a vacuum is described by two vector quantities, the electric field $\mathscr{E}$ and the magnetic field $\mathscr{H}$. In a dielectric medium two additional quantities, the electric displacement $\mathscr{D}$ and the magnetic induction, $\mathscr{B}$, must be specified. The macroscopic electromagnetic properties of the medium are then determined by Maxwell's equations, which take the general form

$$\nabla \cdot \mathscr{D} = 4\pi\rho, \tag{1}$$

$$\nabla \cdot \mathscr{B} = 0, \tag{2}$$

$$\nabla \times \mathscr{H} = 4\pi \mathbf{j} + \frac{1}{c}\dot{\mathscr{D}}, \tag{3}$$

and

$$\nabla \times \mathscr{E} = -\frac{1}{c}\dot{\mathscr{B}}, \tag{4}$$

where ρ is the charge density in the medium, $\mathbf{j}$ is the current density and c is the velocity of electromagnetic waves.‡

In isotropic media $\mathscr{D}$ and $\mathscr{E}$ are related by $\mathscr{D} = \varepsilon\mathscr{E}$, where ε is the scalar *dielectric constant*.§ Furthermore, for homogeneous, electrically neutral media, there is no net charge, that is, $\rho = 0$. And, from Ohm's law the current is given by $\mathbf{j} = \sigma\mathscr{E}/c$, where σ is the electrical conductivity.

† See (**1**), Chap. 1; (**2**), Chap. 19.

‡ Gaussian, c.g.s. units will be used throughout this chapter. The constant c is then the ratio of the e.m.u. to the e.s.u. of charge.

§ The usual term *dielectric constant* is used here, although some authors prefer *"dielectric parameter"*. The latter term is in a sense more logical, as this quantity is not in general constant but, rather, a function of frequency.

Finally, as magnetic phenomena are not of direct interest here, unit permeability will be assumed and $\mathscr{B}$ will be replaced by $\mathscr{H}$. Equations (1) to (4) then become, respectively,

$$\nabla \cdot \mathscr{D} = 0, \tag{5}$$

$$\nabla \cdot \mathscr{H} = 0, \tag{6}$$

$$\nabla \times \mathscr{H} = \frac{4\pi}{c} \sigma \mathscr{E} + \frac{1}{c} \varepsilon \dot{\mathscr{E}} \tag{7}$$

and

$$\nabla \times \mathscr{E} = -\frac{1}{c} \dot{\mathscr{H}}. \tag{8}$$

By taking the curl of Eqn (8) one finds

$$\nabla \times (\nabla \times \mathscr{E}) = -\frac{1}{c} \nabla \times \dot{\mathscr{H}}, \tag{9}$$

which, using Eqn (7) yields

$$\nabla \times (\nabla \times \mathscr{E}) = -\frac{4\pi\sigma}{c^2} \dot{\mathscr{E}} - \frac{\varepsilon}{c^2} \ddot{\mathscr{E}}. \tag{10}$$

Since

$$\nabla \times (\nabla \times \mathscr{E}) = \nabla(\nabla \cdot \mathscr{E}) - \nabla^2 \mathscr{E}, \tag{11}$$

the differential equation for the electric field becomes

$$\nabla^2 \mathscr{E} - \nabla(\nabla \cdot \mathscr{E}) = \frac{1}{c^2} (\varepsilon \ddot{\mathscr{E}} + 4\pi\sigma \dot{\mathscr{E}}). \tag{12}$$

Plane–wave solutions to Eqn (12) are of the form

$$\mathscr{E} = \mathscr{E}^0 \, \mathrm{e}^{-2\pi i(\mathbf{K}\cdot\mathbf{r} - \nu t)} \tag{13}$$

for monochromatic waves of frequency ν propagating in the direction of $\mathbf{r}$. The propagation vector associated with the wave is represented by $\mathbf{K}$. From Eqn (13) the following relations are easily obtained

$$\dot{\mathscr{E}} = 2\pi i \nu \mathscr{E}, \tag{14}$$

$$\nabla \cdot \mathscr{E} = -2\pi i \, \mathbf{K} \cdot \mathscr{E} \tag{15}$$

and

$$\nabla \times \mathscr{E} = -2\pi i \, \mathbf{K} \times \mathscr{E}. \tag{16}$$

Substitution of Eqns (14) to (16) into Eqn (12) yields the relation

$$-(\mathbf{K}\cdot\mathscr{E})\mathbf{K} + (\mathbf{K}\cdot\mathbf{K})\mathscr{E} = \frac{v^2}{c^2}\left(\varepsilon - \frac{2\sigma i}{v}\right)\mathscr{E} = \frac{v^2}{c^2}N^2\mathscr{E}, \tag{17}$$

where, by definition, $N^2 = \varepsilon - (2\sigma i/v)$ is the square of the refractive index of the medium. By taking the scalar product of **K** with Eqn (17) it is found that

$$\frac{v^2 N^2}{c^2}\mathbf{K}\cdot\mathscr{E} = 0, \tag{18}$$

so that either $N = 0$ or $\mathbf{K}\cdot\mathscr{E} = 0$. Ordinarily $N \neq 0$, and the resulting condition, $\mathbf{K}\cdot\mathscr{E} = 0$, describes a transverse wave, that is, a wave in which the electric field is perpendicular to the propagation direction. With $\mathbf{K}\cdot\mathscr{E} = 0$, Eqn (17) becomes

$$\left(\frac{c}{v}\mathbf{K}\right)\cdot\left(\frac{c}{v}\mathbf{K}\right) = N^2 = \varepsilon - \frac{2\sigma i}{v}, \tag{19}$$

or

$$N = \left(\varepsilon - \frac{2\sigma i}{v}\right)^{1/2} = n - i\kappa. \tag{20}$$

This complex refractive index completely characterizes the optical properties of the medium. In an anisotropic crystal, for example, both ε and σ are tensor quantities which depend on the propagation and polarization directions. Hence, a crystal may in general, exhibit anisotropy with respect to both real and imaginary parts of N.

The condition represented by Eqn (18) can be satisfied in an alternative way, namely, $N = 0$. In this case $\mathscr{E}$ is not necessarily perpendicular to **K** and the wave has a longitudinal component. Waves of this type will be discussed later in this chapter.

II. Absorption and Reflection of Radiation (3)

For the case of homogeneous transverse waves for which the real and imaginary parts of **K** are parallel, Eqns (19) and (20) yield

$$K = \frac{v}{c}N = \frac{vn}{c} - i\frac{v\kappa}{c} \tag{21}$$

and the wave described by Eqn (13) can be represented by the expression

$$\mathscr{E} = \mathscr{E}^0\,[\mathrm{e}^{-2\pi v\kappa Z/c}]\,\mathrm{e}^{-2\pi i v[(nZ/c)-t]}, \tag{22}$$

for a wave propagating in the Z direction. Hence, the velocity becomes c/n and the wave is exponentially damped by the factor in brackets. The damping arises directly from losses, that is, from the absorption of electromagnetic energy by the medium.

The intensity, $\mathscr{I}$, of the wave described by Eqn (22) is proportional to

$$\mathscr{E} \cdot \mathscr{E}^* = \mathscr{E}^{02}\, e^{-4\pi\nu\kappa Z/c}. \tag{23}$$

Therefore,

$$\frac{\mathscr{I}}{\mathscr{I}_0} = \frac{\mathscr{E} \cdot \mathscr{E}^*}{\mathscr{E}^{02}} = e^{-4\pi\nu\kappa Z/c}, \tag{24}$$

and by comparison with the integral of Eqn (71), Chap. 1, the absorption coefficient becomes

$$\alpha = 4\pi\nu\kappa/c, \tag{25}$$

where the thickness, l, is measured in the Z direction. From Eqns (19) and (20),

$$\varepsilon - 2\sigma i/\nu = n^2 - 2n\,\kappa i - \kappa^2, \tag{26}$$

and, separating real and imaginary parts,

$$\varepsilon = n^2 - \kappa^2 \tag{27}$$

and

$$\sigma = \nu n \kappa. \tag{28}$$

By combining Eqns (25) and (28) it becomes evident that the absorption coefficient is directly proportional to the electrical conductivity of the medium. This result accounts for the relative opaqueness of metals compared with nonconducting crystals. It should be noted, however, that even in the case of real, nonconducting crystals, losses are present and the refractive index contains an imaginary part.

Thus, for a nonconducting but lossy medium, one writes Eqn (26) in the form

$$\varepsilon = \varepsilon' - i\,\varepsilon'' = n^2 - 2n\,\kappa i - \kappa^2 \tag{29}$$

which leads to

$$\varepsilon' = n^2 - \kappa^2 \tag{30}$$

and

$$\varepsilon'' = 2n\kappa. \tag{31}$$

Equation (31) describes the dielectric loss of the medium. If the losses are low, κ^2 can be neglected in Eqn (30).

When a monochromatic electromagnetic wave is incident on a plane surface of a crystal, a portion of the energy will be absorbed following Eqn (23), while the remainder will be associated with a reflected wave. In the simple case of normal incidence, this result is represented in Fig. 1. According to Eqn (13) the wave inside the medium, where $Z > 0$, is described by

$$\mathscr{E}_{Z>0} = \mathscr{E}^0_{\mathrm{abs}}\, e^{2\pi i\nu[t-(NZ/c)]}, \tag{32}$$

where N is defined by Eqn (21). In free space the electric field is given by

$$\mathscr{E}_{Z<0} = \mathscr{E}^0_{\mathrm{inc}}\, e^{2\pi i\nu[t-(Z/c)]} + \mathscr{E}^0_{\mathrm{ref}}\, e^{2\pi i\nu[t+(Z/c)]}. \tag{33}$$

At the boundary, where $Z = 0$, the tangential component of the electric field must be continuous; hence,

$$\mathscr{E}^0_{\mathrm{abs}} = \mathscr{E}^0_{\mathrm{inc}} + \mathscr{E}^0_{\mathrm{ref}}. \tag{34}$$

The tangential component of the magnetic field must obey similar boundary conditions, viz.

$$\mathscr{H}^0_{\mathrm{abs}} = \mathscr{H}^0_{\mathrm{inc}} + \mathscr{H}^0_{\mathrm{ref}}, \tag{35}$$

and, by applying Eqn (8) to the waves of Eqns (32) and (33), the latter condition is found to be equivalent to

$$N\,\mathscr{E}^0_{\mathrm{abs}} = \mathscr{E}^0_{\mathrm{inc}} - \mathscr{E}^0_{\mathrm{ref}}. \tag{36}$$

The combination of Eqns (34) and (36) leads to the relation

$$\frac{\mathscr{E}^0_{\mathrm{ref}}}{\mathscr{E}^0_{\mathrm{inc}}} = \frac{1-N}{1+N}. \tag{37}$$

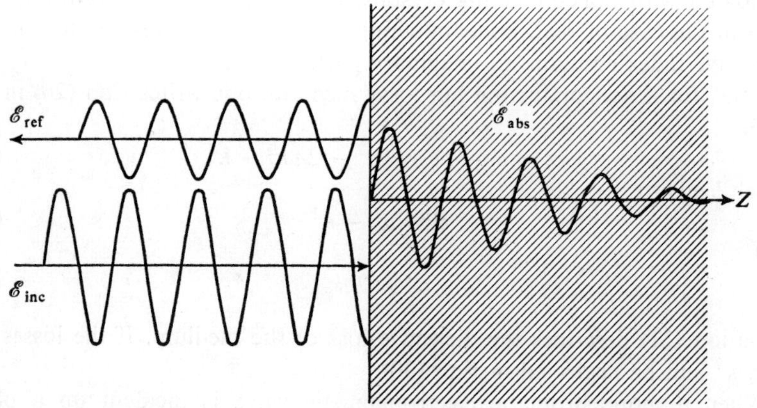

FIG. 1. Schematic representation of absorption and reflection of radiation by a dielectric medium.

The reflection coefficient, R, is defined as the ratio of reflected to incident energy. Thus,

$$R = \frac{\mathscr{E}^0_{ref} \cdot \mathscr{E}^{0*}_{ref}}{\mathscr{E}^0_{inc} \cdot \mathscr{E}^{0*}_{inc}} = \frac{1 - N}{1 + N} \cdot \frac{1 - N^*}{1 + N^*}, \tag{38}$$

and, with $N = n - i\kappa$,

$$R = \frac{(n - 1)^2 + \kappa^2}{(n + 1)^2 + \kappa^2}. \tag{39}$$

From the above results it is apparent that independent measurements of the absorption and reflection coefficients of a crystal yield values of both n and κ.† In practice, however, even isotropic crystals may offer experimental difficulties. For example, it is not always convenient to achieve normal incidence in reflection measurements, while corrections must be introduced if absorption studies are made of very thin films. In addition, in the case of anisotropic crystals both absorption and reflection depend on the orientation of the sample and the polarization of the incident radiation.

III. Dielectric Dispersion

In order to describe the behavior of a dielectric medium under the influence of an electric field, it is convenient to introduce an additional variable, $\mathscr{P}$, the polarization. This quantity, which represents the electric dipole moment per unit volume, is defined by the relation

$$\mathscr{D} = \mathscr{E} + 4\pi \mathscr{P}. \tag{40}$$

In the special case of an isotropic medium the polarization is parallel to the electric field. Furthermore, in this case

$$\mathscr{D} = \varepsilon \mathscr{E}, \tag{41}$$

where ε is the scalar dielectric constant of Eqn (29), and from Eqn (40) one obtains

$$\mathscr{P} = \frac{\varepsilon - 1}{4\pi} \mathscr{E}. \tag{42}$$

The complex factor $(\varepsilon - 1)/4\pi = \chi$, which is known as the *electric susceptibility*, is usually decomposed into electronic and atomic contributions. For an isotropic crystal, for example, the dielectric constant becomes

$$\varepsilon = 1 + 4\pi\chi_{elec} + 4\pi\chi_{ion}, \tag{43}$$

† For a detailed treatment of optical reflection by crystals the reader is referred to the article by E. E. Bell (3).

where the susceptibility χ_{elec} is due to distortion of the electron distribution by the electric field and χ_{ion} arises from ionic displacements during lattice vibrations. It is usual to define a high-frequency, or "optical" dielectric constant by the relation

$$\varepsilon(\infty) = 1 + 4\pi\chi_{\text{elec}}. \tag{44}$$

The dielectric constant takes on this value at frequencies which are high compared to molecular and ionic vibrational frequencies. At lower frequencies $\varepsilon(\infty)$ is essentially constant and the variation in dielectric constant with frequency is due to the third term on the right-hand side of Eqn (43).

It is of interest to develop a specific expression for the low-frequency contribution to the dielectric constant, using a simple model of an ionic lattice. Consider the linear diatomic chain treated in Chapter 3 [See Fig. 5. Chap. 3]. With $f = f'$ this system becomes analogous to an ionic lattice, following Born's analysis. The effect of a radiation field on such a lattice can be included by the addition of a forcing function to the equations of motion. From Eqns (37) and (38), Chap. 3,

$$F_{2n} = f(\rho_{2n-1} + \rho_{2n+1} - 2\rho_{2n}) + e\mathscr{E}^0 \mathrm{e}^{-2\pi i(\nu t - nKd)} = M_1 \ddot{\rho}_{2n} \tag{45}$$

and

$$F_{2n+1} = f(\rho_{2n} + \rho_{2n+2} - 2\rho_{2n+1}) - e\mathscr{E}^0 \mathrm{e}^{-2\pi i[\nu t - (n+\frac{1}{2})Kd]} = M_2 \ddot{\rho}_{2n+1}, \tag{46}$$

where the terms containing exponential functions account for the forces on the ions due to the incident electric field parallel to the Z axis. Here, ions of type 1 are assumed to be the cations with charge $+e$ and those of type 2 the anions with charge $-e$. The frequency of the incident monochromatic radiation is given by ν and its wave number by K.

The velocity of an electromagnetic wave is much greater than those of the lattice waves. Thus, K is, in general, very small compared to the wave number of a lattice, and for all ordinary optical phenomena, K is essentially zero.† Furthermore, as shown in Chapter 3, the optical vibration of the diatomic chain can be active only when the wave number of the lattice wave is also zero. Thus,

$$K = k = 0, \tag{47}$$

and from Eqn (44), Chap 3, the frequency of the optical lattice mode is given by

$$4\pi^2 \nu_0^2 = \frac{2f}{\mu}. \tag{48}$$

† See, however, the last section of this chapter, where the limitations of this approximation are discussed.

As shown in Fig. 6, Chap. 3, with $k = 0$ the displacements are in phase in all unit cells. Thus, n can be set equal to any integer. Using this approximation, then, Eqns (45) and (46) can be simplified by putting $n = 0$, $\rho_{-1} = \rho_1$, and $\rho_2 = \rho_0$, yielding

$$-2f(\rho_0 - \rho_1) + e\mathscr{E}^0 \mathrm{e}^{-2\pi i\nu t} = M\ddot{\rho}_0 \tag{49}$$

and

$$2f(\rho_0 - \rho_1) - e\mathscr{E}^0 \mathrm{e}^{-2\pi i\nu t} = M_2\ddot{\rho}_1. \tag{50}$$

Furthermore, by substituting the relative displacement, $w \equiv \rho_0 - \rho_1$ and eliminating ρ_1 between Eqns (49) and (50), one finds the equation of motion

$$\ddot{w} + \frac{2f}{\mu} w = \frac{e\mathscr{E}^0}{\mu} \mathrm{e}^{-2\pi i\nu t}, \tag{51}$$

which, by substituting Eqn (48), becomes

$$\ddot{w} + 4\pi^2 \nu_0{}^2 w = \frac{e\mathscr{E}^0}{\mu} \mathrm{e}^{-2\pi i\nu t}. \tag{52}$$

Solutions to Eqn (52) are of the form

$$w = A\,\mathrm{e}^{-2\pi i\nu t}, \tag{53}$$

where A, the vibrational amplitude, is found by substitution of these solutions into Eqn (52). Thus, the desired solution is given by

$$w = \frac{e\mathscr{E}^0 \mathrm{e}^{-2\pi i\nu t}}{4\pi^2 \mu(\nu_0{}^2 - \nu^2)}. \tag{54}$$

The dipole moment of the lattice at any time, t, is given $\mathfrak{m}(t) = e\,w$. Hence, the Z component of the polarization becomes $\mathscr{P} = e\,w/V_c$, where V_c is the volume of the unit cell. From Eqn (54)

$$\mathscr{P} = \frac{e^2 \mathscr{E}^0 \mathrm{e}^{-2\pi\nu t}}{4\pi^2 \mu V_c(\nu_0{}^2 - \nu^2)} = \left[\frac{e^2}{4\pi^2 \mu V_c(\nu_0{}^2 - \nu^2)}\right] \mathscr{E}, \tag{55}$$

and by comparison with Eqn (42) it becomes clear that the quantity in brackets is the ionic contribution to the electric susceptibility of this system. Substitution into Eqn (43) yields the following expression for the dielectric constant,

$$\varepsilon(\nu) = \varepsilon(\infty) + \frac{e^2}{\pi\mu V_c(\nu_0{}^2 - \nu^2)}, \tag{56}$$

which is plotted in Fig. 2. In this model all losses have been neglected. Hence, the dielectric constant given by Eqn (56) is equal to the real quantity N^2, the square of the refractive index. At zero frequency Eqn (56) becomes

$$\varepsilon(0) = \varepsilon(\infty) + \frac{e^2}{\pi\mu\, V_c \nu_0^{\,2}}, \tag{57}$$

the static dielectric constant of the medium.

As pointed out earlier, longitudinal waves are possible when $N = 0$. Hence, with $N^2 = \varepsilon(\nu) = 0$, $\nu = \nu_l$, and Eqn (56) reduces to

$$\varepsilon(\nu) = 0 = \varepsilon(\infty) + \frac{e^2}{\pi\mu\, V_c(\nu_0^{\,2} - \nu_l^{\,2})}. \tag{58}$$

The longitudinal frequency can then be found from the relation

$$\nu_l^{\,2} = \nu_0^{\,2} + \frac{e^2}{\pi\mu\, V_c\, \varepsilon(\infty)}. \tag{59}$$

The combination of Eqns (57) and (59) leads to the expression

$$\frac{\nu_l^{\,2}}{\nu_0^{\,2}} = \frac{\varepsilon(0)}{\varepsilon(\infty)}. \tag{60}$$

This result, which is usually referred to as the Lyddane–Sachs–Teller relation (**4**), was derived here on the basis of a specific model. However, it is a completely general relation which has been extended to include the

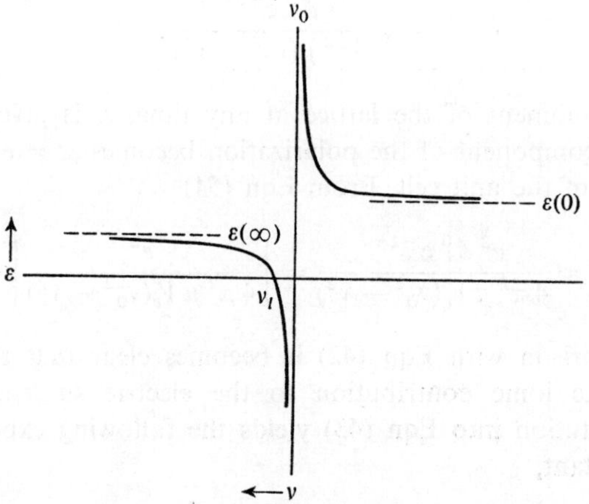

FIG. 2. Frequency dependence of the dielectric constant of a solid in the region of a resonance.

case of primitive cells having a number of optical vibrations (5). The dielectric constant given by Eqn (56) is equal to N^2. For $\nu < \nu_0$ or $\nu > \nu_l$, N is real, while for $\nu_0 < \nu < \nu_l$, it is imaginary. From Eqn (39) the reflection coefficient at normal incidence is found to have the form shown in Fig. 3. For frequencies between ν_0 and ν_l the crystal is totally reflecting. This phenomenon is known as the *Reststrahlen effect.* Since N is purely imaginary in this frequency range, the wave cannot propagate in the crystal. Here, the medium has been assumed to be lossless, hence, no absorption spectrum is predicted.

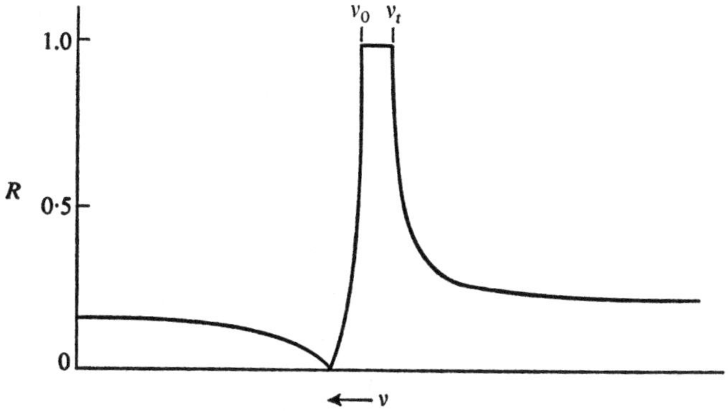

FIG. 3. Reflection coefficient of a crystal as a function of frequency near a resonance. The minimum on the high-frequency side of the Reststrahlen region is known as the Christiansen effect.

As pointed out earlier, even in nonconducting crystals some losses occur due to lattice imperfections, as well as quantum effects associated with the finite lifetimes of excited states. The average value of these losses can be represented mathematically by the inclusion of a first-order damping term in the differential equation of motion, Eqn (52). Then, one finds that

$$\ddot{w} + 2\pi\gamma\,\dot{w} + 4\pi^2\nu_0^{\,2}\,w = \frac{e\,\mathscr{E}^0}{\mu}\,\mathrm{e}^{-2\pi i\nu t}, \tag{61}$$

and that Eqn (54) is now replaced by

$$w = \frac{e\,\mathscr{E}^0\,\mathrm{e}^{-2\pi i\nu t}}{4\pi^2\,\mu(\nu_0^{\,2} - \nu^2 - i\nu\gamma)}. \tag{62}$$

This result leads in turn to

$$\varepsilon(\nu) = \varepsilon(\infty) + \frac{e^2}{\pi\mu\,V_c(\nu_0^{\,2} - \nu^2 - i\nu\gamma)} \tag{63}$$

by analogy with Eqn (56). This complex dielectric constant of a lossy, nonconducting medium is equal to the square of its complex refractive index. By equating real and imaginary parts of Eqn (63), with $\varepsilon(\nu) = \varepsilon'(\nu) - i\,\varepsilon''(\nu) = N^2 = n^2 - \kappa^2 - 2\,n\kappa i$, the following results are obtained:

$$\varepsilon' = n^2 - \kappa^2 = \frac{e^2\,({\nu_0}^2 - \nu^2)}{\pi\mu\,V_c[({\nu_0}^2 - \nu^2)^2 + \nu^2\,\gamma^2]} + \varepsilon'(\infty) \tag{64}$$

and

$$\varepsilon'' = 2\,n\,\kappa = \frac{\nu\gamma\,e^2}{\pi\mu\,V_c[({\nu_0}^2 - \nu^2)^2 + \nu^2\,\gamma^2]}\,. \tag{65}$$

The values of ε' and ε'' calculated from Eqns (64) and (65) for a typical case are plotted in Fig. 4 along with the resulting reflection coefficient evaluated from Eqn (39).

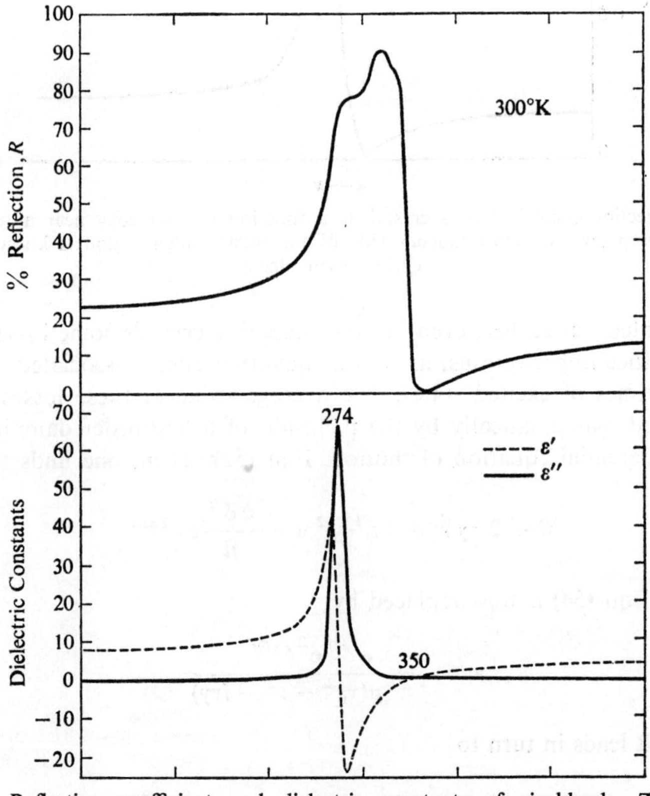

FIG. 4. Reflection coefficient and dielectric constants of zincblende, ZnS [From reference (6)].

If the primitive cell contains more than one infrared–active frequency, Eqns (64) and (65) can be generalized by simply summing over all such frequencies. Furthermore, the ionic charge, e, appearing in these equations can be replaced by an effective charge

$$\sqrt{\mu}\left(\frac{\partial \mathfrak{m}}{\partial Q_{k}}\right)_0, \quad \text{where} \quad \left(\frac{\partial \mathfrak{m}}{\partial Q_{k}}\right)_0$$

is the equilibrium value of the derivative of the dipole moment with respect to the mass-weighted normal coordinate Q_{k}. Equations (64) and (65) then become

$$\varepsilon' = \frac{1}{\pi V_c}\sum_{k}\left(\frac{\partial \mathfrak{m}}{\partial Q_{k}}\right)_0^2 \frac{\nu_{0,k}^2 - \nu^2}{(\nu_{0,k}^2 - \nu^2)^2 + \nu^2 \gamma_{k}^2} + \varepsilon'(\infty) \tag{66}$$

and

$$\varepsilon'' = \frac{1}{\pi V_c}\sum_{k}\left(\frac{\partial \mathfrak{m}}{\partial Q_{k}}\right)_0^2 \frac{\nu \gamma_{k}}{(\nu_{0,k}^2 - \nu^2)^2 + \nu^2 \gamma_{k}^2}, \tag{67}$$

respectively.

In even the most general case the real and imaginary parts of the dielectric constant are related by the *Kramers–Kronig relations*,

$$\varepsilon'(\nu) = \frac{2}{\pi}\int_0^\infty \frac{\nu' \varepsilon''(\nu')}{\nu'^2 - \nu^2}\, d\nu' + \varepsilon'(\infty) \tag{68}$$

and

$$\varepsilon''(\nu) = \frac{2\nu}{\pi}\int_0^\infty \frac{\varepsilon'(\nu') - \varepsilon'(\infty)}{\nu'^2 - \nu^2}\, d\nu', \tag{69}$$

which are derived in Appendix H. The first of these relations is often used to calculate $\varepsilon'(0)$, the static dielectric constant from experimental measurements of $\varepsilon''(\nu)$.

IV. Effective Field and Absorption Intensities

Throughout this discussion it has been assumed that the effective field acting on the vibrating dipoles of a crystal is equal to the applied macroscopic electric field. However, since each dipole occupies a site in the lattice which is surrounded by other atoms, the effect of polarization of these atoms must be considered. Equations (66) and (67) are then modified to become

$$\varepsilon' = \frac{1}{\pi V_c}\sum_{k}\frac{\mathscr{E}_{\text{eff}}^2}{\mathscr{E}^2}\left(\frac{\partial \mathfrak{m}}{\partial Q_{k}}\right)_0^2 \frac{\nu_{0,k}^2 - \nu^2}{(\nu_{0,k}^2 - \nu^2)^2 + \nu^2 \gamma_{k}^2} + \varepsilon'(\infty) \tag{70}$$

and

$$\varepsilon'' = \frac{1}{\pi V_c} \sum_{k} \frac{\mathscr{E}_{\text{eff}}^2}{\mathscr{E}^2} \left(\frac{\partial \mathfrak{m}}{\partial Q_k} \right)_0^2 \frac{\nu \gamma_k}{(\nu_{0,k}^2 - \nu^2)^2 + \nu^2 \gamma_k^2}, \tag{71}$$

respectively. The factor $\mathscr{E}_{\text{eff}}/\mathscr{E}$ is the effective-field correction which arises because the local field acting on a molecule differs from the applied field.

For a molecule in a spherical cavity the effective field is given by

$$\mathscr{E}_{\text{eff}} = \mathscr{E} + \frac{4\pi}{3} \mathscr{P} = \mathscr{E} + \frac{\varepsilon(\infty) - 1}{3} \mathscr{E} = \frac{\varepsilon(\infty) + 2}{3} \mathscr{E}, \tag{72}$$

using Eqn (42). This quantity, which is called the Lorentz field, is equal to the local electric field acting on a molecule having tetrahedral or higher site symmetry (7). The effective-field correction is then

$$\frac{\mathscr{E}_{\text{eff}}}{\mathscr{E}} = \frac{\varepsilon(\infty) + 2}{3}. \tag{73}$$

In the more general case the electric field $\mathscr{E} = \mathscr{E}^0 \, \mathrm{e}^{-2\pi i \nu t}$ is replaced by the effective field $\mathscr{E} + 4\pi \beta \mathscr{P}$. Thus, for example, Eqn (62) can be written in the form

$$\mathbf{w} = \frac{e}{4\pi\mu(\nu_0^2 - \nu^2 - i\nu\gamma)} (\mathscr{E} + 4\pi\beta \mathscr{P}), \tag{74}$$

where the value of β depends on the symmetry of the site involved. Here, for simplicity, the case of a single ionic oscillator per primitive cell has been considered. Substituting Eqns (42) and (63) the polarization becomes

$$\mathscr{P} = \frac{\varepsilon(\nu) - 1}{4\pi} \mathscr{E} = \frac{e\,\mathbf{w}}{V_c} + \chi_{\text{elec}} \mathscr{E}, \tag{75}$$

and, using Eqns (44) and (74), and the definition of the effective field, one finds that

$$\frac{\varepsilon(\nu) - 1}{1 + \beta[\varepsilon(\nu) - 1]} = \frac{e^2}{\pi\mu \, V_c(\nu_0^2 - \nu^2 - i\,\nu\gamma)} + \frac{\varepsilon(\infty) - 1}{1 + \beta[\varepsilon(\infty) - 1]}, \tag{76}$$

where $\varepsilon(\nu)$ has been replaced by $\varepsilon(\infty)$ in the high-frequency limit. More generally,

$$\frac{\varepsilon(\nu) - 1}{1 + \beta[\varepsilon(\nu) - 1]} = \frac{1}{\pi V_c} \left(\frac{\partial \mathfrak{m}}{\partial Q_0} \right)_0^2 \frac{1}{(\nu_0^2 - \nu^2 - i\nu\gamma)} + \frac{\varepsilon(\infty) - 1}{1 + \beta[\varepsilon(\infty) - 1]}, \tag{77}$$

where Q_0 is the normal coordinate of the single vibration of frequency ν_0. For an oscillator on a site of tetrahedral or higher symmetry, $\beta = \frac{1}{3}$ [(**8**), p. 398] and Eqn (77) becomes

$$\frac{\varepsilon(\nu) - 1}{\varepsilon(\nu) + 2} = \frac{1}{3\pi V_c}\left(\frac{\partial \mathfrak{m}}{\partial Q_0}\right)_0^2 \frac{1}{(\nu_0^2 - \nu^2 - i\nu\gamma)} + \frac{\varepsilon(\infty) - 1}{\varepsilon(\infty) + 2}, \tag{78}$$

which is the well-known dispersion relation of Lorentz–Lorenz. Here again, the result is easily generalized for a number of oscillators by summing over normal modes. The field corrections for molecules in sites of lower symmetry have been considered by Born and Kun Huang [(**8**), p. 104] and, more recently by Decius (**9, 10**).

Equation (77) can be rewritten in the form

$$\varepsilon(\nu) = \varepsilon(\infty) + \frac{\rho}{\nu_t^2 - \nu^2 - i\nu\gamma}, \tag{79}$$

where

$$\rho = \frac{1}{\pi V_c}\left(\frac{\partial \mathfrak{m}}{\partial Q_0}\right)_0^2 \{1 + \beta[\varepsilon(\infty) - 1]\}^2 \tag{80}$$

and

$$\nu_t^2 = \nu_0^2 - \frac{\beta}{\pi V_c}\left(\frac{\partial \mathfrak{m}}{\partial Q_0}\right)_0^2 \{1 + \beta[\varepsilon(\infty) - 1]\}. \tag{81}$$

Hence, the true transverse frequency, ν_t, differs from the oscillator frequency, ν_0, due to the effect of the local field. Furthermore, the factor in curly brackets in Eqn (80) is the effective–field correction. Thus,

$$\frac{\mathscr{E}_{\text{eff}}}{\mathscr{E}} = 1 + \beta[\varepsilon(\infty) - 1], \tag{82}$$

and, for molecules on sites of tetrahedral or higher symmetry Eqn (82) becomes

$$\frac{\mathscr{E}_{\text{eff}}}{\mathscr{E}} = \frac{\varepsilon(\infty) + 2}{3}, \tag{83}$$

in agreement with Eqn (72).

If the frequency of a given normal vibration is well separated from all others, its absorption coefficient will be given by [See Eqn (25)]

$$\alpha_k = \frac{4\pi\nu_k}{c} = \frac{2\pi\nu\varepsilon''}{nc} = \frac{2}{ncV_c}\frac{\mathscr{E}_{\text{eff}}^2}{\mathscr{E}^2}\left(\frac{\partial \mathfrak{m}}{\partial Q_k}\right)_0^2 \frac{\nu^2\gamma_k}{(\nu_{t,k}^2 - \nu^2)^2 + \nu^2\gamma_k^2}. \tag{84}$$

The absolute intensity of such an absorption band is then given by

$$\int_{\text{band}} \mathfrak{a}_k \, d\nu = \frac{1}{\bar{n}} \frac{\mathscr{E}_{\text{eff}}^2}{\mathscr{E}^2} \frac{\pi}{c V_c} \left(\frac{\partial \mathfrak{m}}{\partial Q_k} \right)_0^2, \tag{85}$$

where

$$\bar{n} = \int n \, \mathfrak{a} \, d\nu \Big/ \int \mathfrak{a} \, d\nu \tag{86}$$

is the "effective" refractive index through the band. It is customary to replace $\bar{n}$ by $[\varepsilon(\infty)]^{\frac{1}{2}}$, although this approximation is not generally justifiable. For the case of a molecule or complex ion at a site of tetrahedral or higher symmetry, Eqn (85) becomes

$$\int_{\text{band}} \mathfrak{a}_k \, d\nu = \frac{1}{\bar{n}} \left[\frac{\varepsilon(\infty) + 2}{3} \right]^2 \frac{\pi}{c V_c} \left(\frac{\partial \mathfrak{m}}{\partial Q_k} \right)_0^2. \tag{87}$$

Thus, in principle, absolute intensity measurements on crystalline samples can, as in the case of gaseous molecules, yield values of dipole moment derivatives (**11–13**). However, the determination of $\bar{n}$ is often difficult.

V. Wave Propagation in Anisotropic Media

The problem of the interaction of an electromagnetic wave with a dielectric medium has thus far been limited to the case of isotropic media, such as gases, liquids, glasses, or cubic crystals. If the medium being considered is a crystalline solid belonging to any class other than the cubic one, anisotropy of the various electrical properties will have to be included in the development. For example, in an anisotropic medium the electric displacement will become

$$\mathscr{D} = \boldsymbol{\varepsilon} \mathscr{E}, \tag{88}$$

where $\boldsymbol{\varepsilon}$ is now a tensor quantity.† The form of the dielectric tensor, $\boldsymbol{\varepsilon}$, depends directly on the crystal symmetry. However, it is easily shown [(**1**), p. 666] that $\boldsymbol{\varepsilon}$ is symmetric, that is, $\varepsilon_{12} = \varepsilon_{21}$, etc. Hence, $\boldsymbol{\varepsilon}$ will contain, in the general case, six different elements.

Because $\boldsymbol{\varepsilon}$ is symmetric, it can be reduced to diagonal form by a proper choice of axes, the *principal dielectric axes*. In terms of principal axes the dielectric tensor becomes

$$\boldsymbol{\varepsilon} = \begin{pmatrix} \varepsilon_X & 0 & 0 \\ & \varepsilon_Y & 0 \\ \text{sym} & & \varepsilon_Z \end{pmatrix}. \tag{89}$$

† It as assumed here that the electric fields involved are weak enough so that the linear relation of Eqn (88) is valid. In the case of high-field excitation, for example, by a laser beam, nonlinear effects may become important.

For monochromatic waves the elements ε_X, ε_Y, and ε_Z are complex constants which depend only on the medium.

Equation (89) allows three different cases to be distinguished:

(i) Crystals in which three equivalent, mutually orthogonal directions can be chosen. These crystals belong to the cubic system and the quantity $\varepsilon_X = \varepsilon_Y = \varepsilon_Z$ is the scalar dielectric constant used in the first part of this chapter.

(ii) Noncubic crystals in which two or more equivalent directions can be chosen in a plane. These crystals are of the trigonal, tetragonal, or hexagonal systems. One principal dielectric axis must be colinear with the unique axis (3-, 4-, or 6-fold, respectively). If the Z direction defines this axis, $\varepsilon_X = \varepsilon_Y \neq \varepsilon_Z$, and the crystal is said to be *uniaxial.*

(iii) Crystals in which no two equivalent directions can be chosen. These crystals are members of the orthorhombic, monoclinic, or triclinic systems. In this case $\varepsilon_X \neq \varepsilon_Y \neq \varepsilon_Z$, and the principal dielectric axis are not necessarily determined by symmetry.

As a simple example of wave propagation in an anisotropic medium, consider a nonconducting, uniaxial crystal. Equations (3) and (7) are now equivalent to

$$\nabla \times \mathscr{H} = \frac{1}{c}\dot{\mathscr{D}} = \frac{1}{c}\boldsymbol{\varepsilon}\dot{\mathscr{E}}, \tag{90}$$

where $\boldsymbol{\varepsilon}$ is given by Eqn (89) with $\varepsilon_X = \varepsilon_Y \neq \varepsilon_Z$, and Eqn (10) becomes

$$\nabla \times (\nabla \times \mathscr{E}) = -\frac{1}{c^2}\boldsymbol{\varepsilon}\ddot{\mathscr{E}}. \tag{91}$$

Using Eqns (14)–(16), as before, one finds for a transverse wave,

$$-\mathbf{K} \times (\mathbf{K} \times \mathscr{E}) = \frac{v^2}{c^2}\boldsymbol{\varepsilon}\mathscr{E}. \tag{92}$$

Although $\mathscr{D}$ is still perpendicular to $\mathbf{K}$, in general, $\mathscr{E}$ is not.

If $\mathbf{K}$ makes an angle of ϕ with respect to the Z axis, and is chosen to be in the X, Z plane, its components are

$$K_X = K \sin \phi,$$

$$K_Y = 0$$

and

$$K_Z = K \cos \phi.$$

Substitution of these expressions into Eqn (92) yields the set of simultaneous equations

$$K^2 \cos\phi\,(\mathscr{E}_X \cos\phi - \mathscr{E}_Z \sin\phi) = \frac{v^2}{c^2}\,\varepsilon_X \mathscr{E}_X, \tag{94}$$

$$K^2 \mathscr{E}_Y = \frac{v^2}{c^2}\,\varepsilon_X \mathscr{E}_Y, \tag{95}$$

and

$$-K^2 \sin\phi\,(\mathscr{E}_X \cos\phi - \mathscr{E}_Z \sin\phi) = \frac{v^2}{c^2}\,\varepsilon_Z \mathscr{E}_Z. \tag{96}$$

The two nontrivial solutions to these equations are

$$K^2 = \frac{v^2}{c^2}\,\varepsilon_X \quad \text{for} \quad \mathscr{E}_X = 0, \quad \mathscr{E}_Y \neq 0, \quad \mathscr{E}_Z = 0, \tag{97}$$

and

$$K^2 = \frac{v^2}{c^2}\left(\frac{\cos^2\phi}{\varepsilon_X} + \frac{\sin^2\phi}{\varepsilon_Z}\right)^{-1} \quad \text{for} \quad \mathscr{E}_X \neq 0, \quad \mathscr{E}_Y = 0, \quad \mathscr{E}_Z \neq 0. \tag{98}$$

The first solution is associated with a wave whose electric field is parallel to the Y axis and perpendicular to the direction of propagation. Thus, this wave, which is called the *ordinary ray*, is polarized perpendicular to the plane of **K** and the crystallographic axis. For a given frequency the wavelength $\lambda = 1/K$, is independent of the direction of propagation.

The second solution to Eqns (94)–(96) describes a wave in which $\mathscr{E}$ and $\mathscr{D}$ are not parallel. This type of wave, which is known as the *extraordinary ray*, is polarized in the plane of **K** and the crystallographic axis.

The complex refractive indices associated with the ordinary and extraordinary rays can be defined by

$$N_o^2 \equiv \frac{c^2 K_o^2}{v^2} = \varepsilon_X \tag{99}$$

and

$$N_e^2 = \frac{c^2 K_e^2}{v^2} = \left(\frac{\cos^2\phi}{\varepsilon_X} + \frac{\sin^2\phi}{\varepsilon_Z}\right)^{-1}, \tag{100}$$

respectively, using the relation of Eqns (97) and (98). For $\phi = 0°$, $N_e^2 = N_o^2 = \varepsilon_X$, while for $\phi = 90°$, $N_e^2 = \varepsilon_Z$.

VI. Infrared Dichroism†

In the previous section it was shown that in anisotropic media the dielectric constant is a tensor quantity. Furthermore, optical absorption can be accounted for by allowing the elements of the dielectric tensor to take on complex values. Therefore, writing $\boldsymbol{\varepsilon} = \boldsymbol{\varepsilon}' - i\boldsymbol{\varepsilon}''$, Eqn (92) becomes

$$-\mathbf{K}\times(\mathbf{K}\times\boldsymbol{\mathscr{E}}) = \frac{\nu^2}{c^2}(\boldsymbol{\varepsilon}' - i\boldsymbol{\varepsilon}'')\boldsymbol{\mathscr{E}} \tag{101}$$

or

$$\begin{pmatrix} K_Y(K_X\mathscr{E}_Y - K_Y\mathscr{E}_X) - K_Z(K_Z\mathscr{E}_X - K_X\mathscr{E}_Z) \\ K_Z(K_Y\mathscr{E}_Z - K_Z\mathscr{E}_Y) - K_X(K_X\mathscr{E}_Y - K_Y\mathscr{E}_X) \\ K_X(K_Z\mathscr{E}_X - K_X\mathscr{E}_Z) - K_Y(K_Y\mathscr{E}_Z - K_Z\mathscr{E}_Y) \end{pmatrix}$$

$$= -\frac{\nu^2}{c^2}\begin{pmatrix} \varepsilon_{XX}' - i\varepsilon_{XX}'' & \varepsilon_{XY}' - i\varepsilon_{XY}'' & \varepsilon_{XZ}' - i\varepsilon_{XZ}'' \\ \varepsilon_{YX}' - i\varepsilon_{YX}'' & \varepsilon_{YY}' - i\varepsilon_{YY}'' & \varepsilon_{YZ}' - i\varepsilon_{YZ}'' \\ \varepsilon_{ZX}' - i\varepsilon_{ZX}'' & \varepsilon_{ZY}' - i\varepsilon_{ZY}'' & \varepsilon_{ZZ}' - i\varepsilon_{ZZ}'' \end{pmatrix}\begin{pmatrix} \mathscr{E}_X \\ \mathscr{E}_Y \\ \mathscr{E}_Z \end{pmatrix}. \tag{102}$$

In general, if an electromagnetic wave is propagated inside a crystal in the direction **K** and with electric field $\mathscr{E}$, the absorption is determined by the components of $\boldsymbol{\varepsilon}''$.

Equation (102) represents a set of three simultaneous equations for the components of the electric field. Nontrivial solutions are obtained only when the determinant of the coefficients vanishes. This condition leads to a secular equation whose roots are essentially the values of the wave vector **K**. As the tensors $\boldsymbol{\varepsilon}'$ and $\boldsymbol{\varepsilon}''$ are symmetric, each contains, in the general case, six independent elements. However, this number can be reduced by the crystal symmetry, as indicated in the preceding section. → $\mathscr{C}_{2h}$!

Consider the special case of a monoclinic crystal of class $\mathscr{C}_{2v}$. It the b axis is chosen as the two-fold symmetry axis, the plane of symmetry is the a, c plane, as shown in Fig. 5, and $\varepsilon_{12}' = \varepsilon_{23}' = 0$. If principal dielectric axes X, Y, Z are chosen, then $\varepsilon_{13}' = 0$, and the real part of the dielectric tensor takes the diagonal form

$$\boldsymbol{\varepsilon}' = \begin{pmatrix} \varepsilon_{11}' & 0 & 0 \\ 0 & \varepsilon_{22}' & 0 \\ 0 & 0 & \varepsilon_{33}' \end{pmatrix} = \begin{pmatrix} \varepsilon_X' & 0 & 0 \\ 0 & \varepsilon_Y' & 0 \\ 0 & 0 & \varepsilon_Z' \end{pmatrix}. \tag{103}$$

† The term *dichroism* is used here in place of the more general, but less known, *pleochroism*. Strictly speaking, only crystals of type (ii) of the previous section exhibit dichroism, while all optically anisotropic crystals can be pleochroic.

The imaginary part of the dielectric constant depends on the presence of transition dipole moments in the crystal. Since the X, Z plane is a plane of symmetry, the resultant transition moment lies in this plane. Hence,

$$\mathfrak{m} = \begin{pmatrix} \mathfrak{m}_0 \cos \chi \\ 0 \\ \mathfrak{m}_0 \sin \chi \end{pmatrix} \tag{104}$$

where χ is the angle between the transition moment, $\mathfrak{m}_0$, and the X axis. Thus,

$$\varepsilon'' = C' \begin{pmatrix} \mathfrak{m}_X{}^2 & \mathfrak{M}_X\mathfrak{M}_Y & \mathfrak{M}_X\mathfrak{M}_Z \\ \mathfrak{M}_X\mathfrak{M}_Y & \mathfrak{M}_Y{}^2 & \mathfrak{M}_Y\mathfrak{M}_Z \\ \mathfrak{M}_X\mathfrak{M}_Z & \mathfrak{M}_Y\mathfrak{M}_Z & \mathfrak{M}_Z{}^2 \end{pmatrix} \tag{105}$$

$$= C \begin{pmatrix} \cos^2 \chi & 0 & \sin \chi \cos \chi \\ 0 & 0 & 0 \\ \sin \chi \cos \chi & 0 & \sin^2 \chi \end{pmatrix}.$$

where $C = C'\mathfrak{m}_0{}^2$ is a constant.

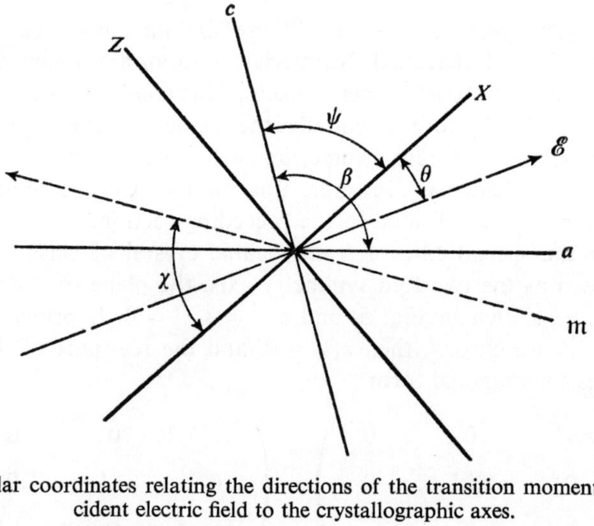

FIG. 5. Angular coordinates relating the directions of the transition moment and the incident electric field to the crystallographic axes.

If a parallel beam of infrared radiation enters the crystal in the direction parallel to the Y axis (the b axis of the crystal), it will not be refracted. Hence, $K_X = K_Z = 0$ inside the crystal. For this special case, then, Eqn (102) becomes

$$- K^2 \mathscr{E}_X = - \frac{\nu^2}{c^2} [(\varepsilon'_X - iC \cos^2 \chi) \mathscr{E}_X - (iC \sin \chi \cos \chi) \mathscr{E}_Z], \tag{106}$$

and

$$- K^2 \mathscr{E}_Z = - \frac{\nu^2}{c^2} [- (iC \sin \chi \cos \chi) \mathscr{E}_X + (\varepsilon'_Z - iC \sin^2 \chi) \mathscr{E}_Z], \tag{107}$$

and the condition for the vanishing of the determinant of the coefficients of $\mathscr{E}_X$ and $\mathscr{E}_Z$ yields

$$\begin{vmatrix} \varepsilon_X' - iC \cos^2 \chi - \dfrac{K^2 c^2}{\nu^2} & - iC \sin \chi \cos \chi \\ - iC \sin \chi \cos \chi & \varepsilon_Z' - iC \sin^2 \chi - \dfrac{K^2 c^2}{\nu^2} \end{vmatrix} = 0. \tag{108}$$

The roots of Eqn (108) for small values of C are given by

$$\frac{c^2 K_1{}^2}{\nu^2} \approx \varepsilon_X' - iC \cos^2 \chi + \frac{C^2 \sin^2 2\chi}{4(\varepsilon_Z' - \varepsilon_X')} \tag{109}$$

and

$$\frac{c^2 K_3{}^2}{\nu^2} \approx \varepsilon_Z' - iC \sin^2 \chi - \frac{C^2 \sin^2 2\chi}{4(\varepsilon_Z' - \varepsilon_X')}, \tag{110}$$

where the numerical subscript on K identifies a particular root. Approximate expressions for the corresponding amplitudes can be found by substituting these solutions into Eqns (106) or (107), thus,

$$\frac{\mathscr{E}_{1X}}{\mathscr{E}_{1Z}} = \frac{2i(\varepsilon_X' - \varepsilon_Z')}{C \sin 2\chi} = i\eta, \tag{111}$$

and

$$\frac{\mathscr{E}_{3X}}{\mathscr{E}_{3Z}} = \frac{- C \sin 2\chi}{2i(\varepsilon_X' - \varepsilon_Z')} = \frac{i}{\eta}, \tag{112}$$

where $\eta \equiv \dfrac{2(\varepsilon_X' - \varepsilon_Z')}{C \sin 2\chi}$ depends directly on the birefringence of the crystal.

If the incident light is now polarized at an angle θ with respect to the X axis, its electric vector is represented by

$$\mathscr{E} = \begin{pmatrix} \mathscr{E}^0 \cos\theta \cdot e^{2\pi i\nu t} \\ 0 \\ \mathscr{E}^0 \sin\theta \cdot e^{2\pi i\nu t} \end{pmatrix} \tag{113}$$

at $Y = 0$, where the beam enters the crystal. Neglecting reflection losses, the waves inside the crystal are then given by

$$\mathscr{E}_1 = \begin{pmatrix} \mathscr{E}^0\, i\eta \sin\theta \cdot e^{2\pi i(\nu t - K_1 Y)} \\ 0 \\ \mathscr{E}^0 (\sin\theta - i\eta\cos\theta)\, e^{2\pi i(\nu t - K_1 Y)} \end{pmatrix} \tag{114}$$

and

$$\mathscr{E}_3 = \begin{pmatrix} \mathscr{E}^0 (\cos\theta - i\eta\sin\theta)\, e^{2\pi i(\nu t - K_3 Y)} \\ 0 \\ -\ \mathscr{E}^0\, i\eta\cos\theta \cdot e^{2\pi i(\nu t - K_3 Y)} \end{pmatrix}. \tag{115}$$

If the crystal has thickness l, the electric fields at the face where the radiation emerges can be found by putting $Y = l$ in Eqns (114) and (115). Multiplying each element in Eqns (114) and (115) by its complex conjugate and adding, one finds an approximate expression for the intensity of the emergent beam,

$$\mathscr{I} \approx \frac{1}{8\pi}\, \mathscr{E}^{02} (\cos^2\theta \cdot e^{-\alpha_X l} + \sin^2\theta \cdot e^{-\alpha_Z l}), \tag{116}$$

where terms in η^2 have been neglected. The absorption coefficients are found to be

$$\alpha_X \approx \frac{2\pi\nu}{c} \frac{C\cos^2\chi}{\sqrt{\varepsilon_X'}} \quad \text{and} \quad \alpha_Z \approx \frac{2\pi\nu}{c} \frac{C\sin^2\chi}{\sqrt{\varepsilon_Z'}}. \tag{117}$$

These results allow the dichroic ratio to be written

$$D \equiv \frac{\alpha_X}{\alpha_Z} = \left(\frac{\varepsilon_Z'}{\varepsilon_X'}\right)^{1/2} \cot^2\chi. \tag{118}$$

Equation (116) has alternate maximum and minimum values at $\theta = \frac{1}{2}\pi s$, where $s = 0, 1, 2, \ldots$. Thus, maximum or minimum absorption intensity depends on the angle of polarization with respect to the crystallographic axes and is independent of the orientation of the transition moment. This result is contrary to what has often been supposed in the interpretation of polarized spectra of crystals.

Equation (116) can be rewritten in terms of the absorbance

$$\mathfrak{a}(\theta) \equiv -\ln \frac{\mathscr{I}}{\mathscr{I}_0} = -\ln (\mathscr{T}_X \cos^2 \theta + \mathscr{T}_Z \sin^2 \theta), \qquad (119)$$

where $\mathscr{T}_X$ and $\mathscr{T}_Z$ are the principal transmittances. Thus, the orientation of the dielectric ellipsoid can be found experimentally from Eqn. (119). On the other hand, the orientation of the transition moment is determined from the experimental dichroic ratio using Eqn (118).

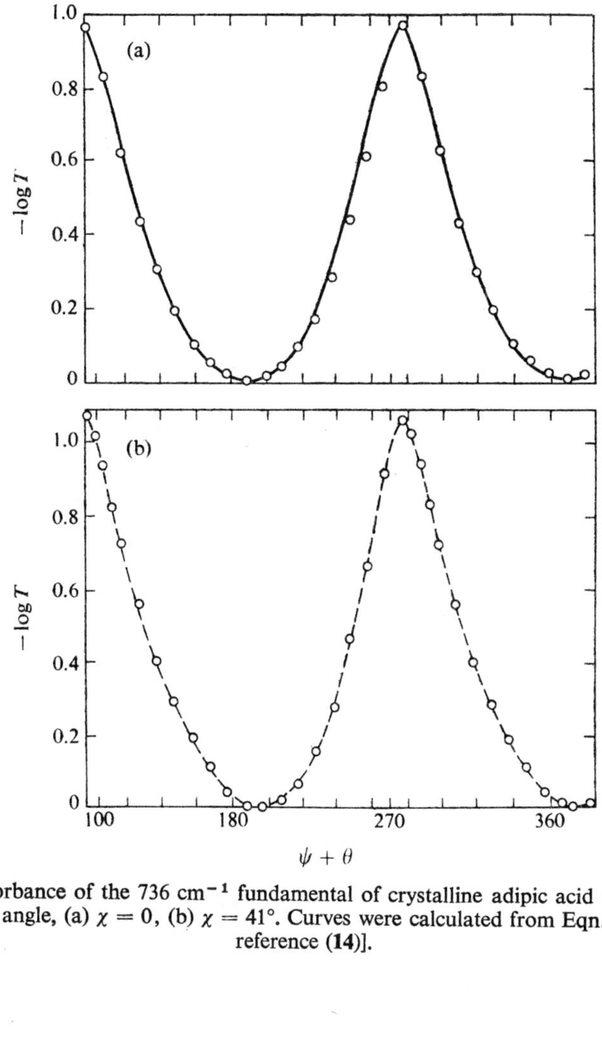

FIG. 6. Absorbance of the 736 cm^{-1} fundamental of crystalline adipic acid as a function of polarizer angle, (a) $\chi = 0$, (b) $\chi = 41°$. Curves were calculated from Eqn. (119) [From reference (**14**)].

The above analysis of infrared dichroism has been applied by Susi (**14**) in a study of monoclinic crystals of adipic acid, $HOOC(CH_2)_4COOH$. Samples were mounted with the crystallographic b axis colinear with the infrared beam. Spectra were obtained as a function of the angle $\psi + \theta$, the angle between the electric vector of the incident radiation and the crystallographic c axis (see Fig. 5). A typical result is illustrated in Fig. 6, where the maximum absorbance of the 736 cm^{-1} fundamental of adipic acid is plotted versus $\psi + \theta$ and compared with the theoretical result given by Eqn (119).

The maxima in the curve of Fig. 6 are found at angles of 98°, 278°, ... for $\theta = 0°$, 180°, ... in Eqn (119). Hence, the result $\psi = 98°$ (or 278°) gives the direction of the principal dielectric axis with respect to the crystallographic c axis at the frequency of this fundamental, 736 cm^{-1}. The ratio of the value at a maximum to that at a minimum in Fig. 6 can then be identified with the dichroic ratio defined by Eqn (118). From his experimental data, Susi estimated a value of $D \geqslant 120$ for this band of adipic acid. Hence, from Eqn (118), assuming that $\varepsilon_Z'/\varepsilon_X' \approx 1$, one obtains $|\chi| \leqslant 5°$. In this case, then, it is concluded that the transition moment associated with the CH_2 rocking of this compound is at an angle of $98 \pm 5°$ with respect to the c axis of the crystal. Similar results have been obtained for several of the vibrational fundamentals of this crystal.

Equations (118) and (119) can also be applied to the case of uniaxial crystals if the propagation direction is chosen to be perpendicular to the optical axis. Obviously, in a uniaxial system no dichroism will be observed if the propagation direction is parallel to the axis.

VII. Raman Scattering

As indicated in Chap. 2, Sec. XIII, Raman scattering arises from the radiating dipole moment induced in a system by the electric field of incident electromagnetic radiation. In a crystal the electric field of a plane, monochromatic wave of frequency ν_e propagating in a direction $\mathbf{K}_e$ is represented by the expression

$$\mathscr{E} = \mathscr{E}^0 \, e^{-2\pi i(\mathbf{K}_e \cdot \mathbf{r} - \nu_e t)}. \tag{120}$$

The frequency of this exciting radiation is usually chosen to be in the visible region of the spectrum. Thus, ν_e is very large compared to the frequencies of any of the crystal vibrations. If the crystal is essentially transparent at the exciting frequency, its (real) refractive index is equal to n_e, and the wave vector is found from Eqn (21) to have the magnitude

$$K_e = \frac{n_e}{c} \nu_e. \tag{121}$$

The dipole moment induced in the system by the exciting radiation is given by

$$\mathfrak{m} = \boldsymbol{\alpha}\mathscr{E}, \tag{122}$$

where $\boldsymbol{\alpha}$ is the polarizability tensor [See Eqn (65), Chap. 2].
Equation (122) describes polarization of the crystal, which will, in general, be a function of the instantaneous positions of the atoms in the lattice. Hence, the polarizability can be expanded in a set of normal coordinates of the crystal, *viz.*,

$$\boldsymbol{\alpha} = \boldsymbol{\alpha}_0 + \sum_{k} \left(\frac{\partial \boldsymbol{\alpha}}{\partial Q_k}\right)_0 Q_k + \tfrac{1}{2} \sum_{k,k'} \left(\frac{\partial^2 \boldsymbol{\alpha}}{\partial Q_k \partial Q_{k'}}\right) Q_k Q_{k'} + \dots . \tag{123}$$

The linear term in Q_k is responsible for first-order scattering, while the quadratic and higher terms account for the second- and higher-order effects.

Since the lattice waves are represented by

$$Q_k = A_k \,\mathrm{e}^{\pm 2\pi i(\mathbf{k}_k \cdot \mathbf{r} - \nu_k t)} \tag{124}$$

for each normal coordinate, Q_k, the induced dipole becomes, to first order,

$$\mathfrak{m}' = \boldsymbol{\alpha}_0 \mathscr{E}^0 \,\mathrm{e}^{-2\pi i(\mathbf{K}_e \cdot \mathbf{r} - \nu_e t)} + \sum_{k} \left(\frac{\partial \boldsymbol{\alpha}}{\partial Q_k}\right)_0 A_k \mathscr{E}^0 \,\mathrm{e}^{2\pi i[(\mathbf{K}_e \pm \mathbf{k}_k)\cdot \mathbf{r} - (\nu_e \mp \nu_k)t]}. \tag{125}$$

The scattered light is then of frequency $\nu_e \mp \nu_k$ and propagates in a direction given by the wave vector $\mathbf{K}_e \pm \mathbf{k}_k$. When ν_k is an optical lattice frequency, the scattering processes is referred to as the Raman effect. When acoustical frequencies are involved, the term *Brillouin scattering* is employed. The frequencies given by $\nu_e \mp \nu_k$ are called the *Stokes* or *anti-Stokes* frequencies depending on the choice of sign.

The scattering processes can be analysed by considering the geometry of Fig. 7, where the coordinates are assumed to coincide with principal axes of the crystal. If $\mathbf{K}_e$ and $\mathbf{K}_s$ are, respectively, the wave vectors of the exciting

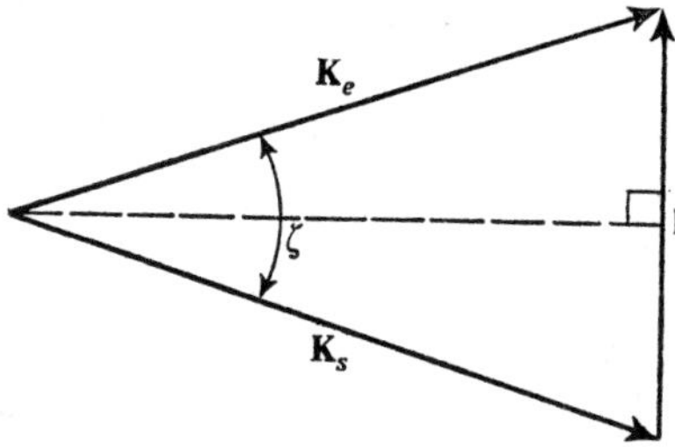

FIG. 7. Relationship between the wave vectors of the exciting and scattered radiation and the wave vector of the lattice wave.

and scattered radiation, and $\mathbf{K}$ is a wave vector of a given latitce wave (phonon), the conservation of energy and momentum are assured by the relations

$$\nu_e = \nu_s + \nu_0 \tag{126}$$

and

$$\mathbf{K}_e = \mathbf{K}_s + \mathbf{k} \tag{127}$$

Since $\nu_0 \ll \nu_e$, $K_e \approx K_s$ and the magnitude of the wave vector of the scattering lattice wave is given by

$$k = 2K_e \sin\frac{\zeta}{2}. \tag{128}$$

Since Raman scattering is usually at an angle of 90° with respect to the exciting radiation, $\zeta = 90°$, and Eqn (128) becomes

$$k = \sqrt{2}\, K_e, \tag{129}$$

and the scattering wave in a crystal makes an angle of 135° with respect to the exciting beam. Thus, as the lattice waves have directional properties, the nature of the Raman spectrum depends on the orientation of the crystal with respect to the exciting beam.

The analysis of the effect of polarization of the incident and scattered light, as well as the orientation, can most conveniently be carried out using the group-theoretical methods developed in Chapters 2 and 4. As a simple example, consider a tetragonal crystal of class $\mathscr{D}_{4h}$. Reference to the character table for this point group indicates that only vibrations of species A_{1g}, B_{1g}, B_{2g}, and E_g are Raman-active. A given polarizability component such as $\alpha_{XY} = \alpha_{YX}$, couples exciting radiation polarized in the X or Y directions with induced oscillating dipole moments parallel to Y or X, respectively, since, from Eqn (122) $\mathfrak{m}_X = \alpha_{XY}\mathscr{E}_Y$ and $\mathfrak{m}_Y = \alpha_{YX}\mathscr{E}_X$. Hence, if a crystal of this class is excited by a beam polarized parallel to the X axis and the scattered light is observed with a polarizer set parallel to the Y axis, only vibrations of species B_{2g} will contribute to the detected light intensity. Other combinations of the polarizations of the exciting and scattered radiation will involve other components of the polarizability tensor, and, hence, be sensitive to vibrations of other symmetries.

The polarized Raman spectra of MnF_2 are shown in Fig. 8 in order to illustrate the principle outlined above. The crystal class of this compound is $\mathscr{D}_{4h}$ and it is found to have one Raman-active fundamental of each of the four species, as shown in Table 1. Thus, the use of polarization techniques in this case allows unambiguous assignments of these vibrational fundamentals to be made.

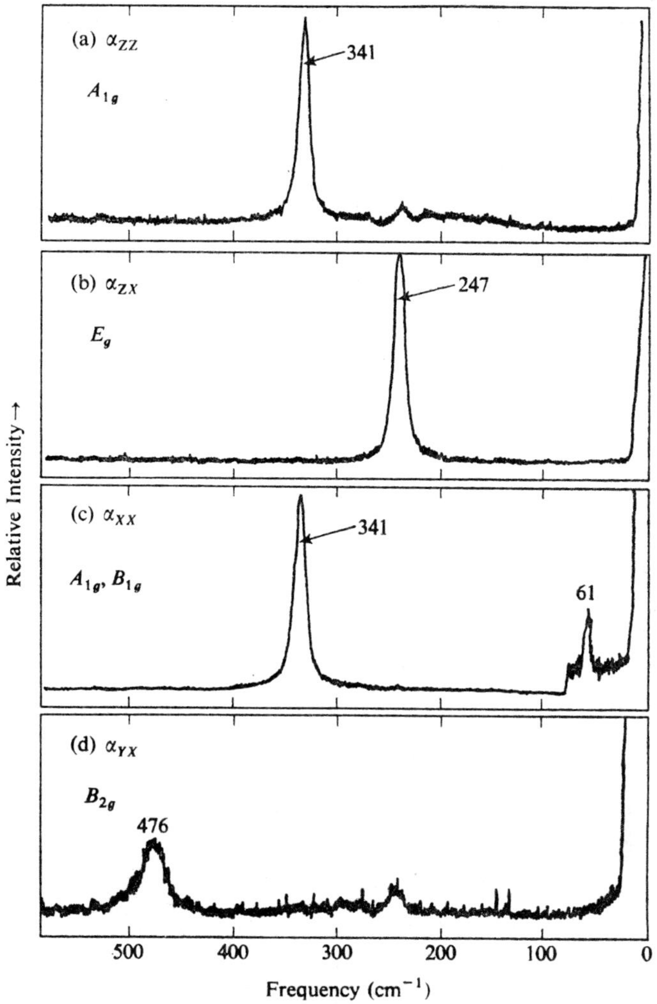

FIG. 8. Raman spectra of MnF_2. [From reference (**16**)].

TABLE 1

Species	Frequency[a] (cm^{-1})	Polarization
A_{1_g}	341	ZZ
B_{1_g}	61	XX
B_{2_g}	476	YX
E_g	247	ZX

[a] From reference (**16**).

In the general case the intensity of a given Raman line is proportional to the so-called scattering efficiency, which can be written in the form (**15**)

$$S = A\,[\mathbf{e}_e{}^{\dagger}\,\boldsymbol{\alpha}\,\mathbf{e}_s]^2, \tag{130}$$

where $\mathbf{e}_e$ and $\mathbf{e}_s$ are unit vectors parallel to $\mathscr{E}_e$ and $\mathscr{E}_s$, respectively, and A is a constant of proportionality. The experimental polarization conditions determine the nonvanishing components of $\mathbf{e}_e$ and $\mathbf{e}_s$, while the form of $\boldsymbol{\alpha}$ depends on the crystal class and the symmetry species of the vibration being considered. A table showing the form of the polarizability tensor for each Raman-active symmetry species in each class is given in Appendix G.

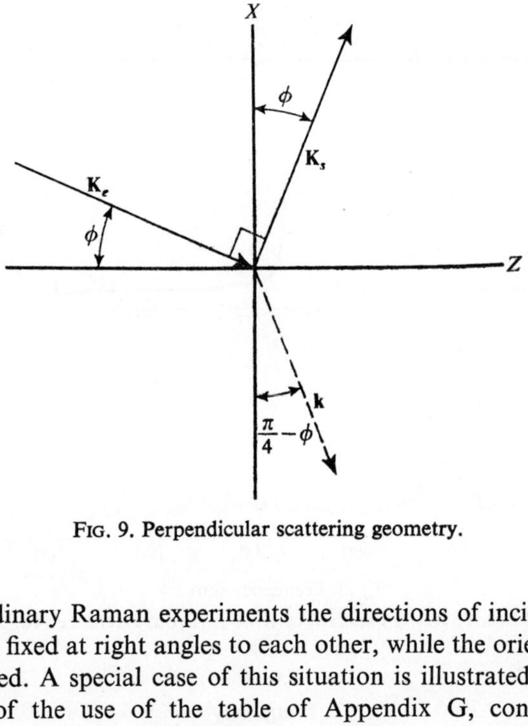

FIG. 9. Perpendicular scattering geometry.

In most ordinary Raman experiments the directions of incident and scattered light are fixed at right angles to each other, while the orientation of the crystal is varied. A special case of this situation is illustrated in Fig. 9. As an example of the use of the table of Appendix G, consider again a crystal of the $\mathscr{D}_{4h}$ class. The scattering tensor for a vibration of symmetry species B_{1g}, for example, has the form

$$\boldsymbol{\alpha}^{(B_{1g})} = \begin{pmatrix} c & 0 & 0 \\ 0 & -c & 0 \\ 0 & 0 & 0 \end{pmatrix}. \tag{131}$$

If the exciting light is nonpolarized, $\mathbf{e}_e$ has the general form

$$\mathbf{e}_e = \begin{pmatrix} e_{eX} \\ e_{eY} \\ e_{eZ} \end{pmatrix}. \tag{132}$$

On the other hand the scattered light can be represented by

$$\mathbf{e}_s^{(\parallel)} = \begin{pmatrix} \sin\phi \\ 0 \\ -\cos\phi \end{pmatrix} \quad \text{and} \quad \mathbf{e}_s^{(\perp)} = \begin{pmatrix} 0 \\ 1 \\ 0 \end{pmatrix} \tag{133}$$

for components polarized parallel or perpendicular to the scattering plane. Substitution of these matrices into Eqn (130) yields the following expressions for the scattered intensity:

$$\begin{aligned} S_{\parallel}^{(B_{1g})} &= A\, e_{eX}^{\ 2}\, c^2 \sin^2\phi \\ &= A[e_e^{(\parallel)}]^2\, c^2 \sin^2\phi \cos^2\phi = \tfrac{1}{4} A[e_e^{(\parallel)}]^2\, c^2 \sin^2 2\phi \end{aligned} \tag{134}$$

and

$$S_{\perp}^{(B_{1g})} = A\, e_{eY}^{\ 2}\, c^2 = A[e_e^{(\perp)}]^2\, c^2. \tag{135}$$

Here the symbols $\mathbf{e}_e^{(\parallel)}$ and $\mathbf{e}_e^{(\perp)}$ represent the components of the exciting electric field which are, respectively, parallel and perpendicular to the XZ plane (see Fig. 9).

For A_{1g}-species vibrations of crystals of this class the polarizability tensor has the form

$$\boldsymbol{\alpha}^{(A_{1g})} = \begin{pmatrix} a & 0 & 0 \\ 0 & a & 0 \\ 0 & 0 & b \end{pmatrix} \tag{136}$$

and the intensities are found to be

$$\begin{aligned} S^{(A_{1g})} &= A(\mathbf{e}_{eX}\, a \sin\phi - \mathbf{e}_{eZ}\, b \cos\phi)^2 \\ &= A[e_e^{(\parallel)}]^2\, (a-b)^2 \sin^2\phi \cos^2\phi = \tfrac{1}{4} A[e_e^{(\parallel)}]^2\, (a-b)^2 \sin^2 2\phi. \end{aligned} \tag{137}$$

and

$$S^{(A_{1g})} = A\, e_{eY}^{\ 2}\, a^2 = A[e_e^{(\perp)}]^2\, a^2. \tag{138}$$

Thus, for vibrations of either species the intensity of the light scattered parallel to the scattering plane will depend on the crystal orientation, while the component perpendicular to this plane will not.

VIII. Raman Spectra of Naphthalene Single Crystals (17–19)

The structure of crystalline naphthalene was discussed in Chap. 4, Sec. IX. The space group is $\mathscr{C}_{2h}{}^5$ with two molecules per unit cell. The results of the factor group analysis are summarized and correlated with the molecular vibrations in Table 8 of Chapter 4.

As was pointed out in that discussion, the predicted correlation splitting of vibrational fundamentals is not detectable in either the infrared or Raman spectra of this system. Thus, the so-called oriented-gas model (**20**, **21**) would be expected to be applicable to the interpretation of the spectra of the crystal. In this model all coupling between the internal vibrations of the two molecules in the unit cell is neglected. Hence, the polarizability of the crystal can be estimated from the components of the polarizability tensor of the free molecule and a knowledge of the orientations of the molecules in the unit cell.

As indicated in Chapter 2, the intensities of Raman fundamentals of a free molecule are determined by the values of the components of the tensors

$$\boldsymbol{\alpha}'^{(k)} = \begin{pmatrix} \alpha_{xx}{}'^{(k)} & \alpha_{xy}{}'^{(k)} & \alpha_{xz}{}'^{(k)} \\ \alpha_{yx}{}'^{(k)} & \alpha_{yy}{}'^{(k)} & \alpha_{yz}{}'^{(k)} \\ \alpha_{zx}{}'^{(k)} & \alpha_{zy}{}'^{(k)} & \alpha_{zz}{}'^{(k)} \end{pmatrix}, \tag{139}$$

where $\alpha_{xx}{}'^{(k)} \equiv \partial\alpha_{xx}'/\partial Q_k$, etc., and Q_k is the kth normal coordinate of the molecule.

The orientations of the two molecules in the unit cell, as determined by X-ray diffraction, are specified by the direction–cosine matrices

$$\boldsymbol{\Phi}_1 = \begin{pmatrix} a_x & a_y & a_z \\ b_x & b_y & b_z \\ c_x & c_y & c_z \end{pmatrix} = \begin{pmatrix} 0{\cdot}8399 & -0{\cdot}4379 & -0{\cdot}3207 \\ -0{\cdot}4425 & -0{\cdot}2103 & -0{\cdot}8718 \\ 0{\cdot}3143 & -0{\cdot}8741 & -0{\cdot}3704 \end{pmatrix} \tag{140}$$

and

$$\boldsymbol{\Phi}_2 = \begin{pmatrix} a_x & a_y & a_z \\ -b_x & -b_y & -b_z \\ c_x & c_y & c_z \end{pmatrix}, \tag{141}$$

where the numerical values were determined by Cruickshank (**22**). The tensor of Eqn (139) can then be transformed to crystal-fixed coordinates, yielding

$$\boldsymbol{\alpha}_1{}'^{(k)} = \boldsymbol{\Phi}_1 \, \boldsymbol{\alpha}'^{(k)} \, \boldsymbol{\Phi}_1{}^{\dagger} \tag{142}$$

and

$$\alpha_2'^{(k)} = \Phi_2 \, \alpha'^{(k)} \, \Phi_2^{\dagger} \tag{143}$$

for the first and second molecule, respectively. Using the oriented–gas model, in which the additivity of polarizabilities is assumed, the polarizability of the unit cell as a function of the molecular normal coordinates becomes

$$\alpha = \sum_k (\alpha_1'^{(k)} Q_{k,1} + \alpha_2^{(k)} Q_{k,2}). \tag{144}$$

From the factor-group analysis of crystalline naphthalene illustrated in Sec. IX, Chap. 4, it is apparent that the molecular normal coordinates can be combined following the unit cell symmetry. Thus, for the Raman-active species of the unit-cell group, the symmetrized normal coordinates are given by

$$\mathrm{Q}_k^{(A_g)} = \frac{1}{\sqrt{2}} (Q_{k,1} + Q_{k,2}) \tag{145}$$

and

$$\mathrm{Q}_k^{(B_g)} = \frac{1}{\sqrt{2}} (Q_{k,1} - Q_{k,2}). \tag{146}$$

The inverse of this orthogonal transformation yields expressions for $Q_{k,1}$ and $Q_{k,2}$, which, when substituted in Eqn (144), lead to a relation for the polarizability of the unit cell,

$$\alpha = \frac{1}{\sqrt{2}} \sum_k [\Phi_1 \, \alpha'^{(k)} \, \Phi_1^{\dagger} + \Phi_2 \, \alpha'^{(k)} \, \Phi_2^{\dagger}] \, \mathrm{Q}_k^{(A_g)}$$

$$+ \frac{1}{\sqrt{2}} \sum_k [\Phi_1 \, \alpha'^{(k)} \, \Phi_1^{\dagger} - \Phi_2 \, \alpha'^{(k)} \, \Phi_2^{\dagger}] \, \mathrm{Q}_k^{(B_g)}, \tag{147}$$

and, hence, to the corresponding derivatives,

$$\alpha_k'^{(A_g)} = \frac{\partial \alpha}{\partial \mathrm{Q}_k^{(A_g)}} = \frac{1}{\sqrt{2}} [\Phi_1 \, \alpha'^{(k)} \, \Phi_1^{\dagger} + \Phi_2 \, \alpha'^{(k)} \, \Phi_2^{\dagger}] \tag{148}$$

and

$$\alpha_k'^{(B_g)} = \frac{\partial \alpha}{\partial \mathrm{Q}_k^{(B_g)}} = \frac{1}{\sqrt{2}} [\Phi_1 \, \alpha'^{(k)} \, \Phi_1^{\dagger} - \Phi_2 \, \alpha'^{(k)} \, \Phi_2^{\dagger}]. \tag{149}$$

These matrices take the forms

$$\boldsymbol{\alpha}_{k}'^{(A_g)} = \begin{pmatrix} a & 0 & d \\ 0 & b & 0 \\ d & 0 & c \end{pmatrix} \tag{150}$$

and

$$\boldsymbol{\alpha}_{k}'^{(B_g)} = \begin{pmatrix} 0 & e & 0 \\ e & 0 & f \\ 0 & f & 0 \end{pmatrix}, \tag{151}$$

in agreement with the general result given in Appendix G. In this example explicit expressions for the various elements of the tensors $\boldsymbol{\alpha}_{k}'^{(A_g)}$ and $\boldsymbol{\alpha}_{k}'^{(B_g)}$ can be found by direct substitution of Eqns (140) and (141) into Eqns (148) and (149). These results are given in Table 2.

The form of the tensors of Eqns (150) and (151) indicate immediately that A_g-species fundamentals of naphthalene should be observed only in polarized spectra of types (XX), (YY), (ZZ), and (XZ), while those of B_g species should appear only in (XY)- and (YZ)-polarized spectra. Furthermore, from Table 2 it is evident that for a Raman line derived from a non-totally symmetric molecular species, the relative intensities of corresponding lines in the various polarized spectra depend only on the elements of the tensors $\boldsymbol{\Phi}_1$ and $\boldsymbol{\Phi}_2$. Since these elements can be easily calculated from the known crystal structure, assignment of the Raman fundamentals can be made on the basis of the relative line intensities calculated from the entries in Table 2. In the case of A_g-species molecular vibrations the expressions are more complicated. However, these fundamentals have been assigned from polarization studies of the liquid and of solutions.

The above method of analysis allows the internal, Raman-active fundamentals of naphthalene to be divided into four types. First, the totally symmetric fundamentals (species A_g) are identified from the spectra of solutions. Then, from the polarized Raman spectra the remaining lines are classified as A_g or B_g under the unit-cell symmetry. Finally, comparison of the observed intensities with the entries in the various columns of Table 2 allows the fundamentals to be assigned on the basis of the molecular symmetry (species B_{1g}, B_{2g}, or B_{3g}).

As a specific illustration of this method, consider the polarized Raman spectra of naphthalene single crystals in the lattice-vibration region. These spectra (**19**), which are shown in Fig. 10, exhibit six different lines, in agreement with the factor-group analysis which predicts six rotatory Raman–active fundamentals. From the spectra it is apparent that the lines at 51, 74, and 109 cm^{-1} are of species A_g, while those at 46, 71, and

TABLE 2

Crystal symmetry	Polarization	Molecular symmetry[a] A_g	B_{1g}	B_{2g}	B_{3g}
A_g	$a(XX)$	$a_x^2 \alpha_{xx}' + a_y^2 \alpha_{yy}' + a_z^2 \alpha_{zz}'$	$2a_x a_y \alpha_{xy}'$	$2a_x a_z \alpha_{zx}'$	$2a_y a_z \alpha_{yz}'$
	$b(YY)$	$b_x^2 \alpha_{xx}' + b_y^2 \alpha_{yy}' + b_z^2 \alpha_{zz}'$	$2b_x b_y \alpha_{xy}'$	$2b_x b_z \alpha_{zx}'$	$2b_y b_z \alpha_{yz}'$
	$c(ZZ)$	$c_x^2 \alpha_{xx}' + c_y^2 \alpha_{yy}' + c_z^2 \alpha_{zz}'$	$2c_x c_y \alpha_{xy}'$	$2c_x c_z \alpha_{zx}'$	$2c_y c_z \alpha_{yz}'$
	$d(ZX)$	$a_x c_x \alpha_{xx}' + a_y c_y \alpha_{yy}' + a_z c_z \alpha_{zz}'$	$(a_x c_y + a_y c_x)\alpha_{xy}'$	$(a_x c_z + a_z c_x)\alpha_{zx}'$	$(a_y c_z + a_z c_y)\alpha_{yz}'$
B_g	$e(XY)$	$a_x b_x \alpha_{xx}' + a_y b_y \alpha_{yy}' + a_z b_z \alpha_{zz}'$	$(a_x b_y + a_y b_x)\alpha_{xy}'$	$(a_x b_z + a_z b_x)\alpha_{zx}'$	$(a_y b_z + a_z b_y)\alpha_{yz}'$
	$f(YZ)$	$b_x c_x \alpha_{xx}' + b_y c_y \alpha_{yy}' + b_z c_z \alpha_{zz}'$	$(b_x c_y + b_y c_x)\alpha_{xy}'$	$(b_x c_z + b_z c_x)\alpha_{zx}'$	$(b_y c_z + b_z c_y)\alpha_{yz}'$

[a] In this table the superscript (ℓ) has been dropped on the elements $\alpha_{xx}'^{(\ell)}$, etc. [See Eqns (148) and (149)].

125 cm^{-1} can be assigned to species B_g. These lines can then be assigned to specific molecular species by comparison of the relative experimental intensities with the calculated quantities in Table 3. Although quantitative agreement is not obtained, the intensity considerations allow unambiguous assignments of these fundamentals to be made. Because of the sensitivity of the calculated intensities to the molecular orientations, the values given in Table 3 vary considerably depending on the X-ray data used.

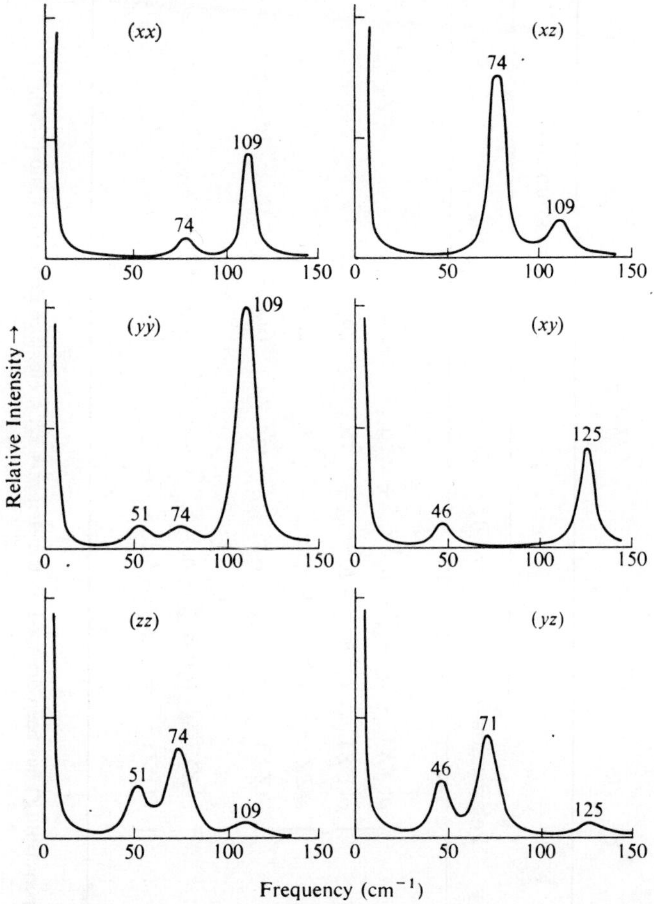

FIG. 10. Polarized Raman spectra of crystalline naphthalene at 20°C. [After Suzuki *et al.* **(19)**].

The results obtained in this example are summarized in Table 4, where approximate descriptions of the librational motions have been included. Identification of the rotation axes can be made by reference to the character table for the molecular point group, $\mathscr{D}_{2h}$, given in Appendix C.

TABLE 3

Crystal symmetry	Intensity	Molecular symmetry B_{1g}	B_{2g}	B_{3g}
A_g	A_{XX}	0·53	0·29	0·08
	A_{YY}	0·03	0·59	0·13
	A_{ZZ}	0·29	0·05	0·42
	A_{XZ}	0·36	0·17	0·01
B_g	A_{XY}	0·00	0·35	0·20
	A_{YZ}	0·20	0·01	0·45

However, it should be noted that since a molecule in a crystalline field does not necessarily rotate about its principal axes, the descriptions given here are only approximate. A specific model of the intermolecular forces must be considered in order to determine the exact form of the librational motion (**23, 24**).

TABLE 4

Unit-cell species→		A_g				B_g		Approximate description
Polarization→		XX	YY	ZZ	XZ	XY	YZ	
Molecular symmetry	B_{3g}	—	51	51	—	46	46	libration about x axis
	B_{1g}	74	74	74	74	—	71	libration about z axis
	B_{2g}	109	109	109	109	125	125	libration about y axis

IX. COMBINATIONS AND OVERTONES: MULTIPHONON PROCESSES

It was pointed out in Chapter 1 that in the harmonic approximation only vibrational fundamentals could contribute to infrared absorption or Raman scattering. However, the effect of anharmonicity resulted in the possibility of combinations or overtones arising from simultaneous changes of state by two or more quanta of vibrational energy. In crystalline solids, as well, anharmonicity of the vibrations can result in overtones or combinations, due to the so-called *multiphonon processes* (**25**).

Selection rules for infrared absorption and Raman scattering by vibrational fundamentals of a crystal were discussed in Chapter 4. There it was shown that only fundamental vibrations belonging to the totally symmetric species ($\mathbf{k} = 0$) of the translation group, $\mathfrak{T}$, could exhibit infrared or Raman activity.

In the more general case a transition from an initial vibrational state i to a final state f will be active only if $\psi_i \psi_f^*$, the product of the corresponding vibrational wave functions, belongs to the totally symmetric representation of $\mathfrak{T}$. The representation of the initial and final states are characterized, respectively, by wave vectors $\mathbf{k}_i$ and $\mathbf{k}_f$. From Eqn (23), Chap. 4 the character of the initial state is given by $\chi_i = e^{2\pi i(\mathbf{k}_i \cdot \boldsymbol{\tau}_n)}$, while for the final state the corresponding character is $\chi_f = e^{2\pi i(\mathbf{k}_f \cdot \boldsymbol{\tau}_n)}$. Thus,

$$\chi_i \chi_f^* = e^{2\pi i(\mathbf{k}_i - \mathbf{k}_f)\cdot \boldsymbol{\tau}_n}, \tag{152}$$

where χ_f^* is the complex conjugate of χ_f. Hence, the product $\chi_i \chi_f^*$ belongs to the totally symmetric representation of $\mathfrak{T}$ only when $\chi_i \chi_f^* = 1$ and, hence,

$$\mathbf{k}_i - \mathbf{k}_f = m\,\mathbf{k}_h, \tag{153}$$

where m is an integer and $\mathbf{k}_h$ is the reciprocal lattice vector defined by Eqn (65), Chap. 3.

In general, the change in wave vector accompanying the transition may involve one, two, or even several phonons. Thus, Eqn (153) can be written in the form

$$\sum_l (\pm \mathbf{k}_l) = m\,\mathbf{k}_h, \tag{154}$$

where $\mathbf{k}_l$ is the wave vector of the lth phonon and the positive and negative signs represent emission and absorption, respectively. For one- and two-phonon processes (fundamentals and binary combinations or overtones), $m = 0$. Hence, for vibrational fundamentals the selection rule

$$\mathbf{k}_i = \mathbf{k}_f = \mathbf{0} \tag{155}$$

results. This rule was derived directly from the properties of the translation group, $\mathfrak{T}$. However, as indicated in Chapter 4, not all representations of the space group for which $\mathbf{k} = \mathbf{0}$ lead to infrared or Raman activity, since the representations of the factor group must also be considered. Thus, a factor-group analysis was used to determine spectral activity for vibrational fundamentals.

In the case of combinations and overtones, further considerations enter. These multiphonon processes are determined by the more general condition of Eqn (154). For binary combinations or first overtones (two-phonon processes), Eqn (154) becomes

$$\mathbf{k}_1 \pm \mathbf{k}_2 = \mathbf{0}, \tag{156}$$

where the positive sign refers to a binary sum or harmonic and the negative sign to a difference band.

The role of binary transitions in the spectra of crystals can be most easily understood by consideration of the one-dimensional diatomic chain treated in Chapter 3. The dispersion curves for the optical and acoustical vibrations are shown in Fig. 11. The vertical arrow corresponds to the difference band $\nu_{\text{opt}} - \nu_{\text{acous}}$ for a particular value of the wave vector. Since $\mathbf{k}$ is no longer restricted, such a transition might be expected to have an essentially continuous spectrum. However, absorption maxima occur because of the form of the frequency distribution function. As shown in Chapter 3, in regions where a dispersion curve becomes horizontal, the phase velocity vanishes and the distribution function exhibits a singularity. It is primarily such singularities (van Hove singularities) which can contribute phonons to these more complex transitions. Thus, although multiphonon processes are not subject to simple selection rules, their spectral characteristics are determined by the frequency distribution (or phonon density) of the initial state. The determination of selection rules for these transitions can, in principle, be made by consideration of the symmetry of the entire space group (not the factor group), while the intensities depend on a detailed knowledge of the dispersion curves. Obviously, this problem is a very difficult one, and has been treated for only the very simplest cases **(26)**.

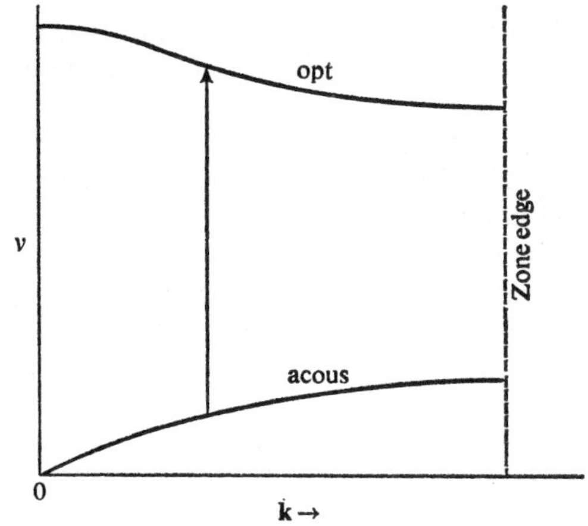

FIG. 11. Dispersion curves for the diatomic chain showing a difference transition, $\nu_{\text{opt}} - \nu_{\text{acous}}$.

X. Interaction of Electromagnetic Radiation

In the previous chapter the factor-group method was developed and applied to the problem of the determination of infrared and Raman selection rules for vibrational fundamentals. The method was based on the approximation in which the effects of the radiation field on the normal vibrations of the lattice were neglected. However, as shown earlier in the present chapter, the interaction of the electric field of the incident radiation with the instantaneous dipole moment of the crystal gives rise to dispersion effects. Furthermore, the electrical properties of the crystal lattice can produce an effective local electric field which differs from the macroscopic field of the incident radiation.

Now, consider the case of a cubic crystal having the CsCl structure (Fig. 8 of Chapter 4). The factor-group analysis, which neglects the interaction of the radiation with the crystal lattice, leads to $\mathbf{\Gamma} = 2F_{1u}$ for the reduced representation of the six degrees of freedom of this system. Since one of the F_{1u} species accounts for the acoustical motions of the lattice, the three optical vibrations are represented by the other triply degenerate, F_{1u} frequency. Thus, a single vibrational fundamental is active in the infrared spectrum and forbidden in the Raman effect. The same result is obtained for any cubic system containing a single heteronuclear atomic pair per primitive cell. In the case of homonuclear diatomic cubic systems such as diamond or germanium, the optical frequency is of species F_{2g} and, hence, is Raman-active, but forbidden in the infrared spectrum. In these examples all frequencies are triply degenerate due to the equivalence of the three crystallographic axes.

As a simple example to illustrate the effects of lattice perturbation by electromagnetic radiation, let a beam of incident light pass through a cubic crystal in a direction parallel to the C_4 symmetry axis (the 100 direction). If the lattice interacts with the incident radiation, the effective symmetry of the perturbed system is reduced from cubic to $\mathscr{C}_{4v}$. By consideration of the correlation of the species of $\mathscr{C}_{4v}$ with those of $\mathscr{O}_h$, the mapping $F_{1u} \Rightarrow A_1 + E$ is easily found. Hence, in heteronuclear diatomic, cubic crystals the optical branch is split into two components, even at the center of the Brillouin zone ($\mathbf{k} = \mathbf{0}$).

Since the perturbation considered here arises from interaction of the electric field with the oscillating dipole moment of the lattice, it can occur only for normal vibrations in which $(\partial\mathfrak{m}/\partial Q_k)_0 \neq 0$, that is, for those which give rise to infrared-active fundamentals. This result has already been obtained in analytic form using classical dispersion theory. It was shown earlier in this chapter that an electromagnetic wave can have both transverse and longitudinal components, and that, even in a cubic crystal, the corresponding frequencies can differ due to interaction with the lattice.

The frequency of a longitudinal wave can be obtained from Eqn (59), which should now be modified by replacing ν_0 by the true transverse frequency, ν_t, and e^2 by the square of the effective charge, $\mu\left(\frac{\partial \mathfrak{m}}{\partial Q_0}\right)_0^2$, as before. Then, introducing the effective field correction given by Eqn (82), one obtains the relation

$$\nu_l^2 = \nu_t^2 + \frac{1}{\pi V_c\, \varepsilon(\infty)} \left(\frac{\partial \mathfrak{m}}{\partial Q_0}\right)_0^2 \{1 + \beta\{\varepsilon(\infty) - 1]\}^2. \qquad (157)$$

Hence, at the center of the Brillouin zone, the longitudinal and transverse frequencies differ when $(\partial \mathfrak{m}/\partial Q_0)_0 \neq 0$. Therefore, in the case of the optical vibrations of homonuclear diatomic cubic crystals, $\nu_l = \nu_t = \nu_0$, and, from the Lyddane–Sachs–Teller relation [Eqn (60)], $\varepsilon(\infty) = \varepsilon(0)$. The experimental dispersion curves of germanium and NaCl, for example, are compared in Figs. 12 and 13. As expected, the degeneracies are removed in both cases when $\mathbf{k} \neq \mathbf{0}$. It should be noted that the distinction between longitudinal and transverse waves can be clearly made only for propagation directions which are parallel to a C_3 or higher symmetry axis.

A further modification of the dispersion curves for polar crystals occurs in the region where $\mathbf{k}$ is very small. This effect can be illustrated by expanding the scale of the abscissa, as shown in Fig. 14. The dashed straight

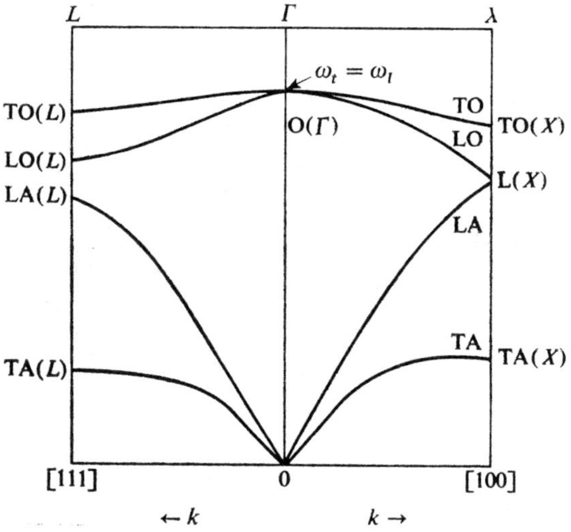

FIG. 12. Dispersion curves for germanium in the 100 and 111 directions. After Brockhouse and Iyengar (27). Note that $\omega_t \equiv \omega_l$ at the zone center. The symbols L, Γ, and X refer to points in the Brillouin zone (see Fig, 9, Chapter 8) [From reference (**28**)].

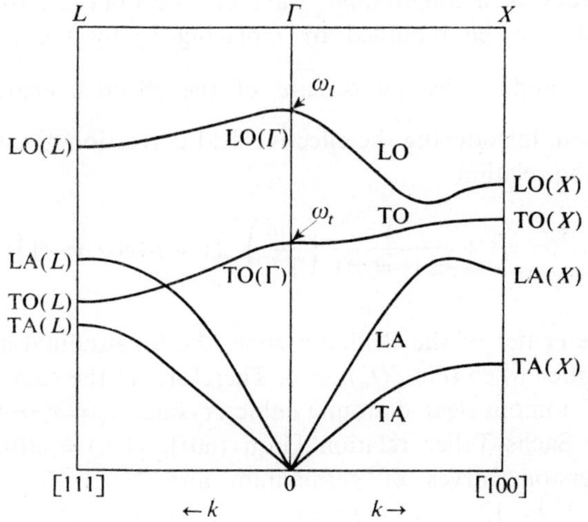

FIG. 13. Dispersion curves for NaCl in the 100 and 111 directions After Hardy and Karo (**29**). Note that $\omega_t \neq \omega_l$ at the zone center. The symbols L, Γ, and X refer to points in the Brillouin zone (see Fig. 9, Chapter 8) [From reference (**28**)].

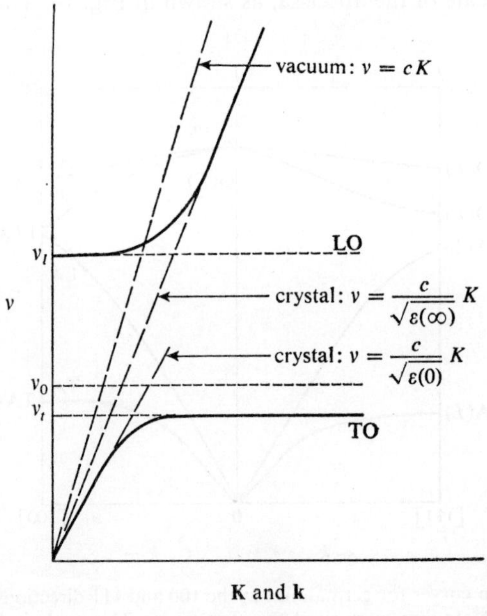

FIG. 14. Dispersion curves for polar crystals for small values of **K** and **k**.

lines represent the propagation of light of frequency ν and wave vector **K** in vacuum and in a medium of reflective index $N = \sqrt{\varepsilon(\infty)}$. The transverse and longitudinal frequencies of the lattice waves are both different from ν_0, the frequency of the free diatomic oscillator, as explained above.

If the frequency dependence of the dielectric constant is considered, the transverse optical mode of the crystal couples with the electromagnetic wave, as shown by the solid lines in Fig. 14. This interaction can be represented analytically by the expression

$$K^2 = \frac{\nu^2 N^2}{c^2} = \frac{\nu^2 \varepsilon(\nu)}{c^2} = \frac{\nu^2}{c^2}\left[\varepsilon(\infty) + \frac{\rho}{\nu_t^2 - \nu^2 - i\nu\gamma}\right], \tag{158}$$

which is easily obtained from Eqns (21) and (79). When $\nu \ll \nu_t$, the factor in brackets becomes equal to the static dielectric constant and ν is a linear function of **K**. As ν approaches ν_t, the wave vector becomes large and the ν versus **K** curve merges with the ν versus **k** curve for the transverse optical mode, and the wave vectors **K** and **k** can no longer be distinguished. In the region where $\nu \ll \nu_t$ the wave is essentially electromagnetic. However, when $\nu \approx \nu_t$, the wave is electro-mechanical in nature and at large **k** it becomes predominantly a mechanical vibration. This branch is sometimes referred to as a quasi–photon or positive *polariton* (**16**).

Between ν_t and ν_l the wave vector is imaginary. In this frequency gap waves cannot propagate in the crystal. At $\nu = \nu_l$ the condition for longitudinal waves requires that $N = 0$; therefore, $\mathbf{k} = \mathbf{0}$. As ν becomes larger than ν_t, the wave vector approaches the magnitude given by $k^2 = (\nu^2/c^2)\,\varepsilon(\infty)$, that is, the slope of the ν versus **k** curve is determined by the high-frequency dielectric constant of the crystal. This branch corresponds to a quasi-phonon (or negative polariton) of the system.

All of the effects discussed here, which result from the interaction of the lattice with electromagnetic radiation, depend directly on the nonvanishing of the quantity $(\partial \mathfrak{m}/\partial Q_0)_0$. Hence, only infrared-active modes can exhibit such effects. Furthermore, in most molecular crystals the dipole moment derivations are usually relatively small, and the resulting perturbations effects become negligible. In ionic crystals, on the other hand, infrared-active fundamentals usually correspond to very intense absorption; that is, the dipole moment derivatives are very large. The infrared-active vibrations of ionic crystals would thus be expected to be strongly perturbed by the radiation field.

There are several spectroscopic consequences of the interaction of a crystal lattice with incident electromagnetic radiation. First, although factor-group selection rules for infrared absorption remain valid for vibrational fundamentals, the observed frequencies will differ from the corresponding oscillator frequencies, as suggested by Eqn (81). The frequencies of overtones and combinations will be similarly modified, resulting in an effective-field contribution to the anharmonicity (**30, 31**).

The infrared reflection spectra of crystals were discussed earlier in this chapter. There it was shown that the Reststrahlen effect extends over the frequency region from ν_t to ν_l. Therefore, in ionic crystals infrared reflection bands are usually very broad.

From the above discussion it is evident that the Raman spectrum of an ionic crystal would also be expected to be modified by the interaction of the exciting radiation with the lattice waves. However, only those Raman-active fundamentals which are also infrared-active can be affected. Therefore, in the ten classes of crystals having centers of symmetry, the rule of mutual exclusion (see Appendix G) eliminates the possibility of perturbation effects. Crystals of the remaining 22 classes do not contain centers of symmetry and, hence, are piezoelectric.† In general, the interpretation of the first-order Raman scattering by piezoelectric crystals requires special consideration. Fundamental Raman vibrations which are also infrared-active are "anomalous" in that (1), more first-order lines are observed than are predicted by factor-group selection rules, and (2), the magnitudes and angular dependences of the scattering, as well as the depolarization ratios are not in agreement with the theory presented earlier, even when the symmetry modification produced by the excitation is considered. Both of these experimental observations can be explained by a detailed treatment of the perturbation of lattice waves by the radiation field (**31, 16**).

It has been found that in piezoelectric crystals each Raman-active fundamental predicted by factor-group analysis leads to a pair of Raman lines (at frequencies ν_l, as well as ν_t), while only the band at ν_t appears in infrared absorption. The problem of calculating angularly dependent Raman intensities and polarizations is quite involved. The reader is referred to the excellent review article by Loudon (**15**) for the details of this subject.

References

1. Born, M. and Wolf, E. "Principles of Optics", Second edition, Pergamon, Oxford (1964).
2. Longhurst, R. S. "Geometrical and Physical Optics", Second edition, Longmans, London (1967).

† Note, however, that although the point group $\mathcal{O}$ does not contain the inversion operation, components of the dipole moment vector and the polarizability tensor never fall in the same symmetry species (See Appendix C). Therefore, crystals of this class do not exhibit the spectroscopic effects characteristic of piezoelectric crystals.

3. Bell, E. E. Optical constants and their measurement. *In* "Handbuch der Physik", Vol. XXV/2a ("Licht und Materie 1a"), (L. Genzel, ed.), Springer-Verlag, Berlin (1967).
4. Lyddane, R. H., Sachs, R. G. and Teller, E. *Phys. Rev.* **59,** 673 (1941).
5. Cochran, W. and Cowley, R. A. *Phys. Chem. Solids* **23,** 447 (1962).
6. Wilkinson, G. R. *In* "Molecular Spectroscopy" (T. Hepple, ed.), p.57. Institute of Petroleum, London. Elsevier, London (1968).
7. Brown, W. F., Jr. Dielectrics. *In* "Handbuch der Physik" (S. Flügge, ed.), Vol. XVII, Springer–Verlag, Berlin (1956).
8. Born, M. and Kun Huang. "Dynamical Theory of Crystal Lattices", Oxford (1954).
9. Decius, J. C. *J. Chem. Phys.* **49,** 1387 (1968).
10. Frech, R. and Decius, J. C. *J. Chem. Phys.* **51,** 1536, 5315 (1969).
11. Person, W. B. *J. Chem. Phys.* **28,** 319 (1958).
12. Schatz, P. N. *J. Chem. Phys.* **31,** 1146 (1959).
13. Schnepp, O. *J. Chem. Phys.* **46,** 3983 (1967).
14. Susi. H. *Spectrochim.* Acta **17,** 1257 (1961).
15. Loudon, R. *Advan. Phys.* **13,** 423 (1964).
16. Porto, S. P. S. Light scattering with laser sources, Chapter A–1. *In* "Light-Scattering Spectra of Solids" (G. B. Wright, ed.), Springer–Verlag, Berlin (1969).
17. Kastler, A. and Rousset, A. *J. Phys. Radium* **2,** 49 (1941).
18. Rousset, A. "La diffusion de la lumière par les molécules rigides", C.N.R.S. and Gauthier–Villars, Paris (1947).
19. Suzuki, M., Yokoyama, T. and Ito, M. *Spectrochim. Acta* **24A,** 1091 (1968).
20. Ambrose, E. J., Elliot, A. and Temple, R. B. *Proc. Roy. Soc.* (*London*) **A206,** 192 (1951).
21. Pimentel, G. C. and McClellan, A. L. *J. Chem. Phys.* **20,** 270 (1952).
22. Cruickshank, D. W. J. *Acta Cryst.* **10,** 504 (1957).
23. Pawley, G. S. *Phys. Stat. Sol.* **20,** 347 (1967).
24. Weulersse, *Compt. Rend.* **264,** 327 (1967).
25. Mitra, S. S. *J. Chem. Phys.* **39,** 3031 (1963).
26. Turner, W. J. and Reese, W. E. *Phys. Rev.* **127,** 126 (1962).
27. Brockhouse, B. N. and Iyengar, P. K. *Phys. Rev.* **108,** 894 (1957).
28. Bak, Thor. A. "Phonons and Phonon Interactions", Benjamin, New York (1964).
29. Hardy, J. R. and Karo, A. M. *Phil Mag,* **5,** 859 (1960).
30. Haas, C. and Ketelaar, A. A. *Physica* **22,** 1284 (1956).
31. Poulet, H. *Ann. Phys.* (*Paris*) **10,** 908 (1955).

Chapter 6

Determination of Force Fields and Structures

The problem of calculating potential constants for a crystal is closely analogous to that discussed in Chapter 1 for an isolated molecule. From the observed vibrational frequencies and a knowledge of the geometrical structure of the crystal, a suitable set of harmonic force constants can, in principle, be found. However, as indicated in Chapter 3, in the case of crystals the **F** matrix, and hence, the vibrational frequencies are functions of the wave vector, **k**. Since spectroscopic selection rules have shown that the vibrational fundamentals are active only for $\mathbf{k} = \mathbf{0}$, the needed set of frequencies can, at best, be observed at this particular point. Nevertheless, a force–constant calculation at $\mathbf{k} = \mathbf{0}$ is important as it provides a starting point for the determination of the dispersion curves. Furthermore, the results obtained are often of considerable value in clarifying the nature of the bonding in crystalline solids.

I. Interatomic Forces in Solids

Before considering the problem of calculating force fields from the vibrational spectra of crystals, it is of interest to review briefly the origins of interatomic forces. The various interactions between atoms in a crystalline solid can be classified in the following way:

(a) Coulombic
(b) Van der Waals'
(c) Overlap
(d) Covalent
(e) Metallic bonding
(f) Hydrogen bonding.

Although it is sometimes difficult to distinguish between certain of these forces in a given case, the classification is of value in that the various types of interatomic forces indicated above have rather different properties. It will be useful to characterize briefly these various types of interaction before discussing the quantitative determination of force fields in crystals.

Coulombic forces. In ionic crystals the attractive forces arise primarily because of the electrostatic interaction between the ions, which behave essentially as point charges. Since these forces can be calculated relatively easily, they have been extensively studied. They are two-body forces in the sense that the interaction between any two ions is independent of the positions of the others. However, since the potential varies inversely with the distance of separation, Coulombic forces extend over comparatively large distances.

Van der Waals' forces. The various long-range forces which contribute to the constant a in van der Waals' equation are usually grouped in this category. They include the London dispersion forces, which arise even between two neutral, spherical atoms as a result of the mutual induction effects of their instantaneous electric moments. A correct description of these forces, which requires a quantum-mechanical treatment, leads to an attractive interaction potential which varies as $-r^{-6}$, where r is the distance between the two atoms [(**1**), p. 955].

Also contributing to van der Waals' forces are the classical interactions between multipoles due to nonspherical charge distributions. Thus, forces of the types dipole–dipole, dipole–quadrupole, quadrupole–quadrupole, etc. exist, in general, in molecular crystals. In addition, in ionic crystals the higher moments of a complex ion interact with the charge on another ion. All of these forces which arise from multipole moments can be considered simply as higher terms in the development of the electrostatic interactions, in which the first term accounts for the Coulombic contributions discussed above. Since atoms are in general polarizable, induction forces can also occur.

Overlap forces. These forces are short-range in nature and arise from the overlap of the electron clouds of two nearby atoms. They are repulsive forces which vary rapidly as a function of interatomic distance. The overlap potential is usually represented by an exponential function of r, which can be obtained by a simple quantum-mechanical calculation [(**2**), Chap. 3]. In some cases a function of the form r^{-s}, where $s = 9$ to 13 used for mathematical convenience.

Covalent forces. Intramolecular attractive forces are due primarily to covalent bonding, which arises essentially from electron exchange between bonded atoms. These forces are very strong compared to van der Waals' forces. Thus, in molecular crystals, intramolecular interactions, i.e. chemical bonds, are considerably stronger than intermolecular forces. In ionic crystals containing complex ions the distinction between ionic (Coulombic) and covalent bonding often cannot be made. Thus, following Pauling [(**3**), p. 64ff], a given bond may be described as having a certain percentage of ionic character.

Metallic bonding. Interatomic forces in metals arise from the ability of the conduction electrons to move more-or-less freely through the crystal lattice. These forces are completely nonlocalized in character. Since, the theory of metals is beyond the scope of this book, the reader is referred to other sources for discussions of metallic bonding [(**4**), Chap. 12].

Hydrogen bonding. Both intramolecular and intermolecular bonds occur having the general structure X—H···Y. Here the proton appears to be bonded to two different atoms, a result which is somewhat incompatible with old-fashioned valence theory. The theory of the hydrogen bond is complicated, although it is generally recognized that the forces are largely ionic. Hydrogen bonds are normally quite weak compared to ordinary chemical (covalent) bonds. Thus, bond energies of 50 to 100 kcal/mole are typical of covalent bonds, while hydrogen bonds usually have energies of the order of 6 kcal/mole.

II. F – G Method at k = 0

The **F** – **G** matrix method was first applied to the problem of crystal vibrations by Shimanouchi, Tsuboi, and Miyazawa (**5**). However, these authors presented the method in a rather empirical manner without indicating its relation to the more general dynamical problem. More recently Piseri and Zerbi (**6**) showed that their general treatment leads to results which, when $\mathbf{k} = \mathbf{0}$, are identical to the method of Shimanouchi, Tsuboi and Miyazawa.

When $\mathbf{k} = \mathbf{0}$, the crystal has all of the symmetry of the translation group, and Eqns (77) and (78), Chap. 3, become expressions for external and internal symmetry coordinates, respectively, for species $\boldsymbol{\Gamma}^{(1)}$. That is,

$$\boldsymbol{\Xi} \equiv \boldsymbol{\Xi}(\mathbf{0}) = \frac{1}{\sqrt{\mathcal{N}}} \sum_n \boldsymbol{\xi}_n \tag{1}$$

and

$$\boldsymbol{\Sigma} \equiv \boldsymbol{\Sigma}(\mathbf{0}) = \frac{1}{\sqrt{\mathcal{N}}} \sum_n \mathbf{S}_n. \tag{2}$$

The kinetic and potential energies in terms of these *optical coordinates* are then given by

$$2T = \dot{\boldsymbol{\Xi}}^\dagger \mathbf{G}_\Xi^{-1} \dot{\boldsymbol{\Xi}} = \dot{\boldsymbol{\Xi}}^\dagger \mathbf{M} \dot{\boldsymbol{\Xi}} \tag{3}$$

and

$$2V = \mathbf{\Sigma}^{\dagger}\,\mathbf{F}_{\Sigma}\,\mathbf{\Sigma}, \tag{4}$$

where from Eqn (81), Chap. 3,

$$\mathbf{F}_{\Sigma} = \sum_{n} \mathbf{F}_{n}. \tag{5}$$

Putting $\mathbf{k} = \mathbf{0}$ in Eqn (84), Chap. 3, the transformation from external to internal optical coordinates becomes

$$\mathbf{\Sigma} = \mathbf{B}\,\mathbf{\Xi}, \tag{6}$$

with

$$\mathbf{B} = \sum_{m} \mathbf{B}_{m}. \tag{7}$$

Thus,

$$\mathbf{F}_{\Xi} = \mathbf{B}^{\dagger}\,\mathbf{F}_{\Sigma}\,\mathbf{B}. \tag{8}$$

It is often convenient to introduce coordinates based on the symmetry of the factor group. For example, the matrix

$$\boldsymbol{\mathsf{\Xi}} = \mathbf{U\Xi} \tag{9}$$

represents a set of external symmetry coordinates which are normalized linear combinations of the elements of $\mathbf{\Xi}$. Using these external symmetry coordinates the $\boldsymbol{\mathsf{F}}_{\Xi}$ matrix can be defined by [See Eqn (49), Chap. 2]

$$\boldsymbol{\mathsf{F}}_{\Xi} \equiv \mathbf{UF}_{\Xi}\,\mathbf{U}^{\dagger} = \mathbf{UB}^{\dagger}\,\mathbf{F}_{\Sigma}\,\mathbf{BU}^{\dagger} = \boldsymbol{\mathsf{B}}^{\dagger}\mathbf{F}_{\Sigma}\boldsymbol{\mathsf{B}}, \tag{10}$$

where in the last step a "symmetrized" (but not symmetric) matrix $\boldsymbol{\mathsf{B}}$ has been defined by

$$\boldsymbol{\mathsf{B}} = \mathbf{BU}. \tag{11}$$

Similarly, the $\boldsymbol{\mathsf{G}}_{\Xi}$ matrix becomes

$$\boldsymbol{\mathsf{G}}_{\Xi} \equiv \mathbf{UG}_{\Xi}\,\mathbf{U}^{\dagger} = \mathbf{UM}^{-1}\,\mathbf{U}^{\dagger}, \tag{12}$$

and the secular determinant for $\mathbf{k} = \mathbf{0}$ becomes

$$|\boldsymbol{\mathsf{G}}_{\Xi}\boldsymbol{\mathsf{F}}_{\Xi} - \mathbf{E}\lambda_{k}| = 0. \tag{13}$$

To illustrate this method consider once again the linear diatomic chain with $f = f'$. When $k = 0$, the two atoms move either in phase or out of phase, as shown in Fig. 6, Chap. 3. The former motion is the acoustical vibration, which now has a frequency of zero. Thus, two symmetry coordinates.

$$\Xi_{\text{acous}} = \frac{1}{\sqrt{2}}(\Delta x_1 + \Delta x_2) \tag{14}$$

and

$$\Xi_{\text{opt}} = \frac{1}{\sqrt{2}}(\Delta x_1 - \Delta x_2) \tag{15}$$

can be written, where Ξ_{acous} is now, in effect, a redundant coordinate resulting from the condition $k = 0$. Thus, the transformation to symmetry coordinates is accomplished using the matrix

$$\mathbf{U} = \begin{pmatrix} 1/\sqrt{2} & 1/\sqrt{2} \\ 1/\sqrt{2} & -1/\sqrt{2} \end{pmatrix} \tag{16}$$

and, from Eqn (93), Chap. 3, with $k = 0$,

$$\mathbf{B} = \begin{pmatrix} -1 & 1 \\ 1 & -1 \end{pmatrix}. \tag{17}$$

Hence,

$$\mathbf{B} = \begin{pmatrix} -1 & 1 \\ 1 & -1 \end{pmatrix}\begin{pmatrix} 1/\sqrt{2} & 1/\sqrt{2} \\ 1/\sqrt{2} & -1/\sqrt{2} \end{pmatrix} = \begin{pmatrix} 0 & -\sqrt{2} \\ 0 & \sqrt{2} \end{pmatrix}. \tag{18}$$

Using the matrix $\mathbf{F}_\Sigma$ given by Eqn (5) and substituting into Eqn (10), one finds

$$\mathbf{F}_\Xi = \mathbf{B}^\dagger \mathbf{F}_\Sigma \mathbf{B} = \begin{pmatrix} 0 & 0 \\ -\sqrt{2} & \sqrt{2} \end{pmatrix}\begin{pmatrix} f & 0 \\ 0 & f \end{pmatrix}\begin{pmatrix} 0 & -\sqrt{2} \\ 0 & \sqrt{2} \end{pmatrix} = \begin{pmatrix} 0 & 0 \\ 0 & 4f \end{pmatrix}. \tag{19}$$

From Eqn (95), Chap. 3, the inverse kinetic-energy matrix is given by

$$\mathbf{G}_\Xi = \begin{pmatrix} \mu_1 & 0 \\ 0 & \mu_2 \end{pmatrix}. \tag{20}$$

with $\mu_1 = 1/\text{M}_1$ and $\mu_2 = 1/\text{M}_2$. In symmetry coordinates

$$\mathbf{G}_\Xi = \mathbf{U}\mathbf{G}_\Xi\mathbf{U}^\dagger = \begin{pmatrix} 1/\sqrt{2} & 1/\sqrt{2} \\ 1/\sqrt{2} & -1/\sqrt{2} \end{pmatrix}\begin{pmatrix} \mu_1 & 0 \\ 0 & \mu_2 \end{pmatrix}\begin{pmatrix} 1/\sqrt{2} & 1/\sqrt{2} \\ 1/\sqrt{2} & -1/\sqrt{2} \end{pmatrix}$$

$$= \begin{pmatrix} (\mu_1 + \mu_2)/2 & (\mu_1 - \mu_2)/2 \\ (\mu_1 - \mu_2)/2 & (\mu_1 + \mu_2)/2 \end{pmatrix}. \tag{21}$$

The eigenvalues of the product $\mathbf{G}_{\Xi}\,\mathbf{F}_{\Xi}$ are then the roots of the secular equation

$$\begin{vmatrix} 0-\lambda & 2f(\mu_1-\mu_2) \\ 0 & 2f(\mu_1+\mu^2)-\lambda \end{vmatrix} = 0, \tag{22}$$

namely,

$$\lambda_{\text{acous}} = 0 \tag{23}$$

and

$$\lambda_{\text{opt}} = 2f(\mu_1+\mu_2), \tag{24}$$

in agreement with Eqn (44), Chap. 3, for the case $k = 0$.

As a second example, consider the sodium chloride crystal. The dodecahedral Wigner–Seitz cell for this structure, which was shown in Fig. 10, Chap. 4, is reproduced in Fig. 1, where the atoms of sodium and chlorine have been numbered 1 and 2, respectively. In Section VII, Chap. 4, it was shown that the six vibrational degrees of freedom of this primitive cell are described by the representation $\mathbf{\Gamma}_{\text{vib}} = 2F_{1u}$, where one of the triply degenerate modes is the acoustical vibration. In this case the external optical coordinates are given by

$$\Xi^{\dagger} = (\Delta X_1\ \Delta X_2\ \Delta Y_1\ \Delta Y_2\ \Delta Z_1\ \Delta Z_2), \tag{25}$$

where each element ΔX_1, ΔX_2, etc., represents the normalized sum over corresponding Cartesian displacement coordinates in the crystal.

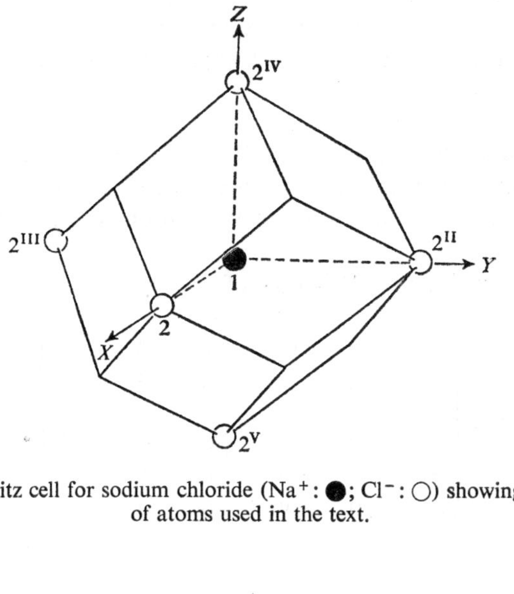

FIG. 1. Wigner–Seitz cell for sodium chloride (Na^+: ●; Cl^-: ○) showing the numbering of atoms used in the text.

Using the methods of Section XII, Chapt. 2, external symmetry coordinates can be constructed *via* the transformation $\boldsymbol{\Xi} = \mathbf{U}\boldsymbol{\Xi}$, where

$$\mathbf{U} = \begin{pmatrix} 1/\sqrt{2} & 1/\sqrt{2} & 0 & 0 & 0 & 0 \\ 1/\sqrt{2} & -1/\sqrt{2} & 0 & 0 & 0 & 0 \\ 0 & 0 & 1/\sqrt{2} & 1/\sqrt{2} & 0 & 0 \\ 0 & 0 & 1/\sqrt{2} & -1/\sqrt{2} & 0 & 0 \\ 0 & 0 & 0 & 0 & 1/\sqrt{2} & 1/\sqrt{2} \\ 0 & 0 & 0 & 0 & 1/\sqrt{2} & -1/\sqrt{2} \end{pmatrix}. \tag{26}$$

The internal and external optical coordinates are related by the matrix **B**, which is given by

$$\mathbf{B} = \begin{pmatrix} -1 & 1 & 0 & 0 & 0 & 0 \\ 1 & -1 & 0 & 0 & 0 & 0 \\ 0 & 0 & -1 & 1 & 0 & 0 \\ 0 & 0 & 1 & -1 & 0 & 0 \\ 0 & 0 & 0 & 0 & -1 & 1 \\ 0 & 0 & 0 & 0 & 1 & -1 \end{pmatrix}, \tag{27}$$

where the internal optical coordinates are the elements of the matrix

$$\boldsymbol{\Sigma}^{\dagger} = (\Delta r_{12}\ \Delta r_{12^{\mathrm{I}}}\ \Delta r_{12^{\mathrm{II}}}\ \Delta r_{12^{\mathrm{III}}}\ \Delta r_{12^{\mathrm{IV}}}\ \Delta r_{12^{\mathrm{V}}}). \tag{28}$$

Here again, each optical coordinate Δr_{12}, $\Delta r_{12^{\mathrm{I}}}$, etc, represents the normalized sum over all corresponding internal coordinates in the crystal. The elements of the **B** matrix are essentially the direction cosines of the vectors $\mathbf{s}_t$ considered in Section III, Chap. 1. Furthermore, the summation over unit cells represented by Eqn (7) has been automatically made by simply dropping the superscripts which distinguish translationally equivalent atoms. Thus, 2^{I} becomes identical to 2, etc. It should be noted, however, that this procedure cannot be extended to the internal coordinates, as the coordinates Δr_{12} and $\Delta r_{12^{\mathrm{I}}}$, for example, are not translationally equivalent.

The symmetrized matrix **B** is now of the form

$$\mathbf{B} = \mathbf{BU} = \begin{pmatrix} 0 & -\sqrt{2} & 0 & 0 & 0 & 0 \\ 0 & \sqrt{2} & 0 & 0 & 0 & 0 \\ 0 & 0 & 0 & -\sqrt{2} & 0 & 0 \\ 0 & 0 & 0 & \sqrt{2} & 0 & 0 \\ 0 & 0 & 0 & 0 & 0 & -\sqrt{2} \\ 0 & 0 & 0 & 0 & 0 & \sqrt{2} \end{pmatrix}. \tag{29}$$

Assuming only nearest-neighbor interactions, the potential-energy matrix $\mathbf{F}_\Sigma$ is diagonal, viz.,

$$\mathbf{F}_\Sigma = \begin{pmatrix} f & & & & & 0 \\ & f & & & & \\ & & f & & & \\ & & & f & & \\ & & & & f & \\ 0 & & & & & f \end{pmatrix}, \tag{30}$$

where f is the force constant characteristic of Na^+—Cl^- stretching. Using the external symmetry coordinates as the basis, the potential-energy matrix becomes

$$\mathbf{F}_\Xi = \mathbf{B}^\dagger \mathbf{F}_\Sigma \mathbf{B} = \begin{pmatrix} 0 & 0 & 0 & 0 & 0 & 0 \\ 0 & 4f & 0 & 0 & 0 & 0 \\ 0 & 0 & 0 & 0 & 0 & 0 \\ 0 & 0 & 0 & 4f & 0 & 0 \\ 0 & 0 & 0 & 0 & 0 & 0 \\ 0 & 0 & 0 & 0 & 0 & 4f \end{pmatrix}. \tag{31}$$

The inverse kinetic-energy matrix in Cartesian coordinates is just the diagonal matrix

$$\mathbf{G} = \begin{pmatrix} \mu_1 & & & & & 0 \\ & \mu_2 & & & & \\ & & \mu_1 & & & \\ & & & \mu_2 & & \\ & & & & \mu_1 & \\ 0 & & & & & \mu_2 \end{pmatrix}, \tag{32}$$

as before [Eqn (20)]. Thus,

$$\mathbf{G} = \mathbf{U}\mathbf{G}\mathbf{U}^\dagger$$

$$= \begin{pmatrix} (\mu_1+\mu_2)/2 & (\mu_1-\mu_2)/2 & 0 & 0 & 0 & 0 \\ (\mu_1-\mu_2)/2 & (\mu_1+\mu_2)/2 & 0 & 0 & 0 & 0 \\ 0 & 0 & (\mu_1+\mu_2)/2 & (\mu_1-\mu_2)/2 & 0 & 0 \\ 0 & 0 & (\mu_1-\mu_2)/2 & (\mu_1+\mu_2)/2 & 0 & 0 \\ 0 & 0 & 0 & 0 & (\mu_1+\mu_2)/2 & (\mu_1-\mu_2)/2 \\ 0 & 0 & 0 & 0 & (\mu_1-\mu_2)/2 & (\mu_1+\mu_2)/2 \end{pmatrix}, \tag{33}$$

and the secular determinant becomes

$$\begin{vmatrix} 0-\lambda & 2f(\mu_1-\mu_2) & 0 & 0 & 0 & 0 \\ 0 & 2f(\mu_1+\mu_2)-\lambda & 0 & 0 & 0 & 0 \\ 0 & 0 & 0-\lambda & 2f(\mu_1-\mu_2) & 0 & 0 \\ 0 & 0 & 0 & 2f(\mu_1+\mu_2)-\lambda & 0 & 0 \\ 0 & 0 & 0 & 0 & 0-\lambda & 2f(\mu_1-\mu_2) \\ 0 & 0 & 0 & 0 & 0 & 2f(\mu_1+\mu_2)-\lambda \end{vmatrix} = 0, \quad (34)$$

whose triple roots are

$$\lambda_{\text{acous}} = 0 \quad (35)$$

and

$$\lambda_{\text{opt}} = 2f(\mu_1 + \mu_2). \quad (36)$$

It should be noted that each of the three factors of the determinant of Eqn (34) is associated with a different Cartesian direction. Furthermore, the optical vibration, which is now triply degenerate, has a frequency identical to that of the linear lattice at $k = 0$ [See Eqn (24)]. Thus, the $\mathbf{k} = \mathbf{0}$, optical modes of an NaCl-type lattice can be described as relative motions between the anion and cation sublattices of the crystal.

The absorption maximum in NaCl is observed at $164\ \text{cm}^{-1}$ (**7**). Hence, using Eqn (36) the force constant for Na^+—Cl^- stretching is found to be $f = 0{\cdot}11 \times 10^5$ dyn/cm.

III. Force Constants of Perovskite Fluorides

The infrared spectra of the series of compounds KMF_3, where M = Ni, Mg, and Zn, have been investigated by Nakagawa, Tsuchida, and Shimanouchi (**8**). These compounds exhibit the regular perovskite structure shown in Fig. 2. The space group is $Pm3m = \mathcal{O}_h^{\ 1}$. The factor-group analysis of the unit cell containing five atoms yields

$$\Gamma = 4F_{1u} + F_{2u}, \quad (37)$$

where one of the F_{1u} species corresponds to the three degenerate acoustic vibrations.

The potential energy is most easily written in terms of a set of internal optical coordinates

$$\mathbf{S}^{\dagger} = (\Delta r_{12}\ldots\Delta r_{14}\ \Delta\alpha_{214}\ldots\Delta\alpha_{4^{\text{I}}12^{\text{I}}}\ \Delta q_{23}\ldots\Delta q_{3^{\text{I}}4^{\text{I}}}$$

$$\Delta q'_{25}\ldots\Delta q'_{45^{\text{V}}}\ \Delta q''_{15}\ldots\Delta q''_{15^{\text{VII}}}) \quad (38)$$

defined in Fig. 2. These 50 internal coordinates are related to the 15 optical Cartesian displacement coordinates by the matrix **B** given in Table 1.

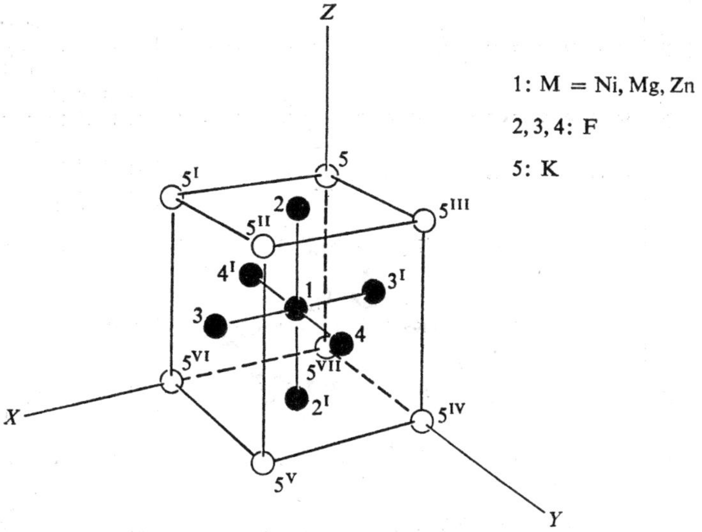

FIG. 2. Structure of a perovskite crystal showing the numbering of atoms used in the text.

If a valence potential function is assumed for the covalent structure $(MF_3^-)_n$ and central-force terms are added to take into account nonbonded F···F interactions and interionic contributions of the type $F\cdots K^+$ and $M\cdots K^+$, the potential, which is then of the Urey–Bradley type, becomes

$$2V = \sum_{t=1}^{6} K(\Delta r_{\mathrm{MF}}^{(t)})^2 + \sum_{t=1}^{12} H(\Delta\alpha_{\mathrm{FMF}}^{(t)})^2 + \sum_{t=1}^{12} F(\Delta q_{\mathrm{F...F}}^{(t)})^2$$

$$+ \sum_{t=1}^{12} f_1(\Delta q_{\mathrm{K...F}}^{\prime(t)})^2 + \sum_{t=1}^{8} f_2(\Delta q_{\mathrm{K...M}}^{\prime\prime(t)})^2. \quad (39)$$

The $\mathbf{F}_\Sigma$ matrix is then diagonal with K appearing six times, H twelve times, etc. Using the **B** matrix of Table 1 the $\mathbf{F}_\Xi$ matrix becomes

$$\mathbf{F}_\Xi = \mathbf{B}^\dagger \mathbf{F}_\Sigma \mathbf{B}. \quad (40)$$

In the $\mathbf{F}_\Xi$ matrix shown in Table 2 each of the three blocks corresponds to a Cartesian direction. Furthermore, the blocks differ only in the order of the basis coordinates. Since the **G** matrix in the Cartesian coordinate system is diagonal, the secular determinant will be factored as in the previous example,

TABLE 1

	B	ΔX_1	ΔX_2	ΔX_3	ΔX_4	ΔX_5	ΔY_1	ΔY_2	ΔY_3	ΔY_4	ΔY_5	ΔZ_1	ΔZ_2	ΔZ_3	ΔZ_4	ΔZ_5
	Δr_{12}	0	0	0	0	0	0	0	0	0	0	-1	1	0	0	0
	$\Delta r_{12'}$	0	0	0	0	0	0	0	0	0	0	1	-1	0	0	0
Δr_{MF}	Δr_{13}	-1	0	1	0	0	0	0	0	0	0	0	0	0	0	0
	$\Delta r_{13'}$	1	0	-1	0	0	0	0	0	0	0	0	0	0	0	0
	Δr_{14}	0	0	0	0	0	-1	0	0	1	0	0	0	0	0	0
	$\Delta r_{14'}$	0	0	0	0	0	1	0	0	-1	0	0	0	0	0	0
	$\Delta\alpha_{214}$	0	0	0	0	0	$1/r$	$-1/r$	0	0	0	$1/r$	0	0	$-1/r$	0
	$\Delta\alpha_{214'}$	0	0	0	0	0	$-1/r$	$1/r$	0	0	0	$1/r$	0	0	$-1/r$	0
	$\Delta\alpha_{314}$	$1/r$	0	0	$-1/r$	0	$1/r$	0	$-1/r$	0	0	0	0	0	0	0
	$\Delta\alpha_{314'}$	$1/r$	0	0	$-1/r$	0	$-1/r$	0	$1/r$	0	0	0	0	0	0	0
	$\Delta\alpha_{3'14}$	$-1/r$	0	0	$1/r$	0	$1/r$	0	$-1/r$	0	0	0	0	0	0	0
	$\Delta\alpha_{3'14'}$	$-1/r$	0	0	$1/r$	0	$-1/r$	0	$1/r$	0	0	0	0	0	0	0
$\Delta\alpha_{FMF}$	$\Delta\alpha_{213}$	$1/r$	$-1/r$	0	0	0	0	0	0	0	0	$1/r$	0	$-1/r$	0	0
	$\Delta\alpha_{213'}$	$-1/r$	$1/r$	0	0	0	0	0	0	0	0	$1/r$	0	$-1/r$	0	0
	$\Delta\alpha_{312'}$	$1/r$	$-1/r$	0	0	0	0	0	0	0	0	$-1/r$	0	$1/r$	0	0
	$\Delta\alpha_{3'12'}$	$-1/r$	$1/r$	0	0	0	0	0	0	0	0	$-1/r$	0	$1/r$	0	0
	$\Delta\alpha_{412'}$	0	0	0	0	0	$1/r$	$-1/r$	0	0	0	$-1/r$	0	0	$1/r$	0
	$\Delta\alpha_{4'12'}$	0	0	0	0	0	$-1/r$	$1/r$	0	0	0	$-1/r$	0	0	$1/r$	0
	Δq_{23}	0	$-1/\sqrt{2}$	$1/\sqrt{2}$	0	0	0	0	0	0	0	0	$1/\sqrt{2}$	$-1/\sqrt{2}$	0	0
	$\Delta q_{23'}$	0	$1/\sqrt{2}$	$-1/\sqrt{2}$	0	0	0	0	0	0	0	0	$1/\sqrt{2}$	$-1/\sqrt{2}$	0	0
	$\Delta q_{2'3}$	0	$-1/\sqrt{2}$	$1/\sqrt{2}$	0	0	0	0	0	0	0	0	$-1/\sqrt{2}$	$1/\sqrt{2}$	0	0
	$\Delta q_{2'3'}$	0	$1/\sqrt{2}$	$-1/\sqrt{2}$	0	0	0	0	0	0	0	0	$-1/\sqrt{2}$	$1/\sqrt{2}$	0	0
$\Delta q_{F\cdots F}$	Δq_{24}	0	0	0	0	0	0	$-1/\sqrt{2}$	0	$1/\sqrt{2}$	0	0	$1/\sqrt{2}$	0	$-1/\sqrt{2}$	0
	$\Delta q_{24'}$	0	0	0	0	0	0	$1/\sqrt{2}$	0	$-1/\sqrt{2}$	0	0	$1/\sqrt{2}$	0	$-1/\sqrt{2}$	0
	$\Delta q_{2'4}$	0	0	0	0	0	0	$-1/\sqrt{2}$	0	$1/\sqrt{2}$	0	0	$-1/\sqrt{2}$	0	$1/\sqrt{2}$	0

	$\Delta q_{2^{I}4^{I}}$	0	0	0	0	0	0	$1/\sqrt{2}$	0	$-1/\sqrt{2}$	0	0	$-1/\sqrt{2}$	0	$1/\sqrt{2}$	0
	Δq_{34}	0	0	$1/\sqrt{2}$	$-1/\sqrt{2}$	0	0	0	$-1/\sqrt{2}$	$1/\sqrt{2}$	0	0	0	0	0	0
	$\Delta q_{34^{I}}$	0	0	$1/\sqrt{2}$	$-1/\sqrt{2}$	0	0	0	$1/\sqrt{2}$	$-1/\sqrt{2}$	0	0	0	0	0	0
	$\Delta q_{3^{I}4}$	0	0	$-1/\sqrt{2}$	$1/\sqrt{2}$	0	0	0	$-1/\sqrt{2}$	$1/\sqrt{2}$	0	0	0	0	0	0
	$\Delta q_{3^{I}4^{I}}$	0	0	$-1/\sqrt{2}$	$1/\sqrt{2}$	0	0	0	$1/\sqrt{2}$	$-1/\sqrt{2}$	0	0	0	0	0	0
	$\Delta q'_{25}$	0	$1/\sqrt{2}$	0	0	$-1/\sqrt{2}$	0	$1/\sqrt{2}$	0	0	$-1/\sqrt{2}$	0	0	0	0	0
	$\Delta q'_{25^{I}}$	0	$-1/\sqrt{2}$	0	0	$1/\sqrt{2}$	0	$1/\sqrt{2}$	0	0	$-1/\sqrt{2}$	0	0	0	0	0
	$\Delta q'_{25^{II}}$	0	$-1/\sqrt{2}$	0	0	$1/\sqrt{2}$	0	$-1/\sqrt{2}$	0	0	$1/\sqrt{2}$	0	0	0	0	0
	$\Delta q'_{25^{III}}$	0	$1/\sqrt{2}$	0	0	$-1/\sqrt{2}$	0	$-1/\sqrt{2}$	0	0	$1/\sqrt{2}$	0	0	0	0	0
	$\Delta q'_{35^{I}}$	0	0	0	0	0	0	0	$1/\sqrt{2}$	0	$-1/\sqrt{2}$	0	0	$-1/\sqrt{2}$	0	$1/\sqrt{2}$
$\Delta q'_{K\cdots F}$	$\Delta q'_{35^{II}}$	0	0	0	0	0	0	0	$-1/\sqrt{2}$	0	$1/\sqrt{2}$	0	0	$-1/\sqrt{2}$	0	$1/\sqrt{2}$
	$\Delta q'_{35^{V}}$	0	0	0	0	0	0	0	$-1/\sqrt{2}$	0	$1/\sqrt{2}$	0	0	$1/\sqrt{2}$	0	$-1/\sqrt{2}$
	$\Delta q'_{35^{VI}}$	0	0	0	0	0	0	0	$1/\sqrt{2}$	0	$-1/\sqrt{2}$	0	0	$1/\sqrt{2}$	0	$-1/\sqrt{2}$
	$\Delta q'_{45^{II}}$	0	0	0	$-1/\sqrt{2}$	$1/\sqrt{2}$	0	0	0	0	0	0	0	0	$-1/\sqrt{2}$	$1/\sqrt{2}$
	$\Delta q'_{45^{III}}$	0	0	0	$1/\sqrt{2}$	$-1/\sqrt{2}$	0	0	0	0	0	0	0	0	$-1/\sqrt{2}$	$1/\sqrt{2}$
	$\Delta q'_{45^{IV}}$	0	0	0	$1/\sqrt{2}$	$-1/\sqrt{2}$	0	0	0	0	0	0	0	0	$1/\sqrt{2}$	$-1/\sqrt{2}$
	$\Delta q'_{45^{V}}$	0	0	0	$-1/\sqrt{2}$	$1/\sqrt{2}$	0	0	0	0	0	0	0	0	$1/\sqrt{2}$	$-1/\sqrt{2}$
	$\Delta q''_{15}$	$1/\sqrt{3}$	0	0	0	$-1/\sqrt{3}$	$1/\sqrt{3}$	0	0	0	$-1/\sqrt{3}$	$-1/\sqrt{3}$	0	0	0	$1/\sqrt{3}$
	$\Delta q''_{15^{I}}$	$-1/\sqrt{3}$	0	0	0	$1/\sqrt{3}$	$1/\sqrt{3}$	0	0	0	$-1/\sqrt{3}$	$-1/\sqrt{3}$	0	0	0	$1/\sqrt{3}$
	$\Delta q''_{15^{II}}$	$-1/\sqrt{3}$	0	0	0	$1/\sqrt{3}$	$-1/\sqrt{3}$	0	0	0	$1/\sqrt{3}$	$-1/\sqrt{3}$	0	0	0	$1/\sqrt{3}$
$\Delta q''_{K\cdots M}$	$\Delta q''_{15^{III}}$	$1/\sqrt{3}$	0	0	0	$-1/\sqrt{3}$	$-1/\sqrt{3}$	0	0	0	$1/\sqrt{3}$	$-1/\sqrt{3}$	0	0	0	$1/\sqrt{3}$
	$\Delta q''_{15^{IV}}$	$1/\sqrt{3}$	0	0	0	$-1/\sqrt{3}$	$-1/\sqrt{3}$	0	0	0	$1/\sqrt{3}$	$1/\sqrt{3}$	0	0	0	$-1/\sqrt{3}$
	$\Delta q''_{15^{V}}$	$-1/\sqrt{3}$	0	0	0	$1/\sqrt{3}$	$-1/\sqrt{3}$	0	0	0	$1/\sqrt{3}$	$1/\sqrt{3}$	0	0	0	$-1/\sqrt{3}$
	$\Delta q''_{15^{VI}}$	$-1/\sqrt{3}$	0	0	0	$1/\sqrt{3}$	$1/\sqrt{3}$	0	0	0	$-1/\sqrt{3}$	$1/\sqrt{3}$	0	0	0	$-1/\sqrt{3}$
	$\Delta q''_{15^{VII}}$	$1/\sqrt{3}$	0	0	0	$-1/\sqrt{3}$	$1/\sqrt{3}$	0	0	0	$-1/\sqrt{3}$	$1/\sqrt{3}$	0	0	0	$-1/\sqrt{3}$

TABLE 2

F	ΔX_1	ΔX_2	ΔX_3	ΔX_4	ΔX_5
ΔX_1	$2K + \frac{8H}{r^2} + \frac{8f_2}{3}$	$-\frac{4H}{r^2}$	$-2K$	$-\frac{4H}{r^2}$	$-\frac{8f_2}{3}$
ΔX_2		$\frac{4H}{r^2} + 2F + 2f_1$	$-2F$	0	$-2f_1$
ΔX_3			$2K + 4F$	$-2F$	0
ΔX_4		Symmetric		$\frac{4H}{r^2} + 2F + 2f_1$	$-2f_1$
ΔX_5					$4f_1 + \frac{8}{3} f_2$

	ΔY_1	ΔY_2	ΔY_3	ΔY_4	ΔY_5
ΔY_1	$2K + \frac{8H}{r^2} + \frac{8f_2}{3}$	$-\frac{4H}{r^2}$	$-\frac{4H}{r^2}$	$-2K$	$-\frac{8f_2}{3}$
ΔY_2		$\frac{4H}{r^2} + 2F + 2f_1$	0	$-2F$	$-2f_1$
ΔY_3			$\frac{4H}{r^2} + 2F + 2f_1$	$-2F$	$-2f_1$
ΔY_4		Symmetric		$2K + 4F$	0
ΔY_5					$2f_1 + \frac{8}{3} f_2$

	ΔZ_1	ΔZ_2	ΔZ_3	ΔZ_4	ΔZ_5
ΔZ_1	$2K + \frac{8H}{r^2} + \frac{8f_2}{3}$	$-2K$	$-\frac{4H}{r^2}$	$-\frac{4H}{r^2}$	$-\frac{8f_2}{3}$
ΔZ_2		$2K + 4F$	$-2F$	$-2F$	0
ΔZ_3		Symmetric	$\frac{4H}{r^2} + 2F + 2f_1$	0	$-2f_1$
ΔZ_4				$\frac{4H}{r^2} + 2F + 2f_1$	$-2f_1$
ΔZ_5					$4f_1 + \frac{8}{3} f_2$

and only one factor need be considered. This result illustrates a general property of cubic crystal systems, namely, each component of a given triply degenerate species, is associated with a Cartesian direction. Thus, if only triply degenerate species are being treated, the vibrational problem can always be formulated in one dimension.†

A given block of the above $\mathbf{F}_\Xi$ matrix, and hence, the resulting factor in the secular determinant, contains four F_{1u}-species and one F_{2u}-species vibrations. One of the former will yield a zero root for the acoustical mode. Further factoring of the secular determinant can be accomplished using the external symmetry coordinates $\boldsymbol{\Xi} = \mathbf{U}\boldsymbol{\Xi}$ defined by the matrix

$$\mathbf{U} = \begin{pmatrix} 1 & 0 & 0 & 0 & 0 \\ 0 & 1 & 0 & 0 & 0 \\ 0 & 0 & 0 & 0 & 1 \\ 0 & 0 & 1/\sqrt{2} & 1/\sqrt{2} & 0 \\ 0 & 0 & 1/\sqrt{2} & -1/\sqrt{2} & 0 \end{pmatrix}, \tag{41}$$

where the basis coordinates $\boldsymbol{\Xi}_Z^\dagger = (\Delta Z_1\ \Delta Z_2 \ldots \Delta Z_5)$ have been chosen. The potential energy matrix is then given by

$$\mathbf{F}_\Xi = \mathbf{U}\mathbf{F}_\Xi\,\mathbf{U}^\dagger$$

$$= \begin{pmatrix} 2K + \frac{8H}{r^2} + \frac{8f_2}{3} & -2K & -\frac{8f_2}{3} & -\frac{8H}{\sqrt{2}r^2} & 0 \\ -2K & 2K+4F & 0 & -\frac{4F}{\sqrt{2}} & 0 \\ -\frac{8f_2}{3} & 0 & 4f_1 + \frac{8f_2}{3} & -\frac{4f_1}{\sqrt{2}} & 0 \\ -\frac{8H}{\sqrt{2}r^2} & -\frac{4F}{\sqrt{2}} & -\frac{4f_1}{\sqrt{2}} & \frac{4H}{r^2} + 2F + 2f_1 & 0 \\ 0 & 0 & 0 & 0 & \frac{4H}{r^2} + 2F + 2f_1 \end{pmatrix} \tag{42}$$

for each Cartesian direction, and it is seen that in this case little is gained by the use of symmetry coordinates.

† The reader is warned, however, that when nondegenerate or doubly degenerate vibrations are involved, all three Cartesian directions must be included.

In the work of Nakagawa, Tsuchida, and Shimanouchi the secular determinant based on the $\mathbf{F}_{\Xi}$ matrix of Table 2 was solved numerically to fit the observed frequencies shown in Table 3. The experimental data were obtained from the infrared absorption spectra by the same authors and from the reflection spectra of Perry (**9**).

The five force constants appearing in Eqn (42) were determined in the following manner. The constant F representing nonbonded fluorine–fluorine interactions was estimated using a Lennard–Jones (6–12) potential and the data on a number of fluorine compounds (**10**).

TABLE 3

Species		$KNiF_3$		$KMgF_3$		$KZnF_3$	
		calc	obs	calc	obs	calc	obs
F_{1u}	ν_1	455	446	477	478	435	430
	ν_2	250	255	285	300	235	206
	ν_3	144	153	160	156	134	142
F_{2u}	ν_4	214	inactive in IR	234	inactive in IR	196	inactive in IR

The values of f_1 and f_2 were held common to the three compounds in the series, as the equilibrium distances q' and q'' show little variation. These force constants for the series, and K and H for each compound were then adjusted to fit the observed frequencies. Only a single set of force constants was found for which K was larger than both H and the interionic constants f_1 and f_2. These constants are given in Table 4.

TABLE 4

	$KNiF_3$	$KMgF_3$	$KZnF_3$
$K(\mathrm{M}-\mathrm{F})$	0·80	0·60	0·75
$r^2H(\mathrm{FMF})$	0·063	0·083	0·038
$F(\mathrm{F}\cdots\mathrm{F})$	0·05	0·05	0·05
$f_1(\mathrm{K}^+\cdots\mathrm{F})$	0·09	0·09	0·09
$f_2(\mathrm{K}^+\cdots\mathrm{M})$	0·03	0·03	0·03

The value $f_1 = 0{\cdot}09$ md/Å for the $K^+ \cdots F$ bond is typical of purely ionic compounds ($f = 0{\cdot}11$ md/Å for NaCl). On the other hand, the values of K for the M—F bonds are an order of magnitude higher, indicating a significant amount of covalent character. These values are nevertheless, about one sixth as large as values obtained for essentially covalent metal–fluorine bonds (**11, 12**).

IV. Separation of Internal and External Vibrations

In the method illustrated above the vibrations of a primitive cell were considered in much the same way as the vibrations of a polyatomic molecule. The condition $\mathbf{k} = \mathbf{0}$ resulted in three redundancy conditions, where the redundant coordinates corresponded to the three acoustical vibrations of zero frequency. In the case of a molecular crystal containing σ identical molecules per primitive cell, with N atoms per molecule, this approach to the vibrational problem requires the solution of a secular determinant of order $3N\sigma - 3$ for the optical frequencies. The separation of these modes of motion into vibrations which are internal to a given molecule and those which constitute the "lattice vibrations" often affords a considerable simplification of the problem.

The kinetic energy resulting from the motion of the atoms of a given primitive cell, n, can be written in the form

$$2T_n = \sum_{i=1}^{\sigma} \dot{\boldsymbol{\xi}}_{ni}^{\dagger} \mathbf{M}_i \dot{\boldsymbol{\xi}}_{ni}, \tag{43}$$

where $\boldsymbol{\xi}_{ni}$ is the Cartesian displacement-coordinate matrix for the ith molecule, and $\mathbf{M}_i = \mathbf{G}_i^{-1}$ is its diagonal mass matrix, as defined previously [Eqn (10), Chap. 1]. In the absence of a crystal field, the potential energy of the nth unit cell is given in the harmonic approximation by

$$2V_n^0 = \sum_{i=1}^{\sigma} \boldsymbol{\xi}_{ni}^{\dagger} \mathbf{F}_{\xi}^0 \boldsymbol{\xi}_{ni}. \tag{44}$$

At $\mathbf{k} = \mathbf{0}$ all translationally equivalent atoms in the crystal move in phase and with the same amplitude, that is, $\boldsymbol{\xi}_{1i} = \boldsymbol{\xi}_{2i} \ldots \boldsymbol{\xi}_{\mathcal{N}i}$. Therefore, Eqn (77), Chap. 3, becomes

$$\boldsymbol{\Xi}_i = \frac{1}{\sqrt{\mathcal{N}}} \sum_n \boldsymbol{\xi}_{ni} = \sqrt{\mathcal{N}}\, \boldsymbol{\xi}_i,$$

where the unit-cell index, n, has been dropped. The total kinetic energy of the crystal is then given by

$$2T = \mathcal{N} \,.\, 2T_n = \mathcal{N} \sum_i \dot{\boldsymbol{\xi}}_i^{\dagger} \mathbf{M}_i \dot{\boldsymbol{\xi}}_i = \sum_i \dot{\boldsymbol{\Xi}}_i^{\dagger} \mathbf{M}_i \dot{\boldsymbol{\Xi}}_i. \tag{45}$$

Similarly, the total potential energy becomes

$$2V = \sum_{i=1}^{\sigma} \Xi_i^{\dagger} \mathbf{F}_{\Xi}^{0} \Xi_i + 2V', \tag{46}$$

where V' is the perturbation due to intermolecular interactions in the crystal.

If the normal coordinates of the ith isolated molecule have been determined, it becomes possible to define optical normal coordinates $\mathbf{Q}_i^0$ such that Eqns (45) and (46) become

$$2T = \sum_i \dot{\mathbf{Q}}_i^{0\dagger} \dot{\mathbf{Q}}_i^0 \tag{47}$$

and

$$2V = \sum_i \mathbf{Q}_i^{0\dagger} \mathbf{\Lambda}_i^0 \mathbf{Q}_i^0 + 2V', \tag{48}$$

respectively, where

$$\Xi_i = \mathscr{L}_i^0 \mathbf{Q}_i^0. \tag{49}$$

Here $\mathbf{Q}_i^0$ includes the six external degrees of freedom of the ith molecule. Hence, the diagonal matrix $\mathbf{\Lambda}_i^0$ which is of rank $3N$, contains six zero frequencies.

The external forces which govern the translatory and rotatory vibrations of a molecule in a crystal are derivable from the potential

$$2V' = \sum_i \mathbf{\Upsilon}_i^{\dagger} \mathbf{F}_{ii}' \mathbf{\Upsilon}_i + \sum_{i,j}' \mathbf{\Upsilon}_i^{\dagger} \mathbf{F}_{ij}' \mathbf{\Upsilon}_j, \tag{50}$$

where the elements of the matrix $\mathbf{\Upsilon}_i$ are optical Cartesian displacement coordinates of the N atoms with respect to space-fixed axes. Since $\mathbf{\Upsilon}_i$ is a linear function of the optical displacements, Ξ_i, relative to molecular-fixed axes, a transformation matrix $\mathbf{\Phi}_i$ can be defined whose elements are sets of direction cosines which specify the orientation of the ith molecule with respect to a lattice–coordinate system X, Y, Z. Thus,

$$\mathbf{\Upsilon}_i = \mathbf{\Phi}_i \Xi_i \tag{51}$$

for the ith molecule and $\mathbf{\Phi}_i$ is composed of N identical submatrices of the form

$$\mathbf{\Phi}_{i,\alpha} = \begin{pmatrix} a_x^{\,i} & a_y^{\,i} & a_z^{\,i} \\ b_x^{\,i} & b_y^{\,i} & b_z^{\,i} \\ c_x^{\,i} & c_y^{\,i} & c_z^{\,i} \end{pmatrix}, \tag{52}$$

one for each atom, α.

From Eqns (48) and (50) the total potential energy becomes

$$2V = \sum_i \mathbf{\Xi}_i^\dagger \mathbf{F}_\Xi^0 \mathbf{\Xi}_i + \sum_i \mathbf{\Upsilon}_i^\dagger \mathbf{F}_{ii}' \mathbf{\Upsilon}_i + \sum_{i,j}' \mathbf{\Upsilon}_i^\dagger \mathbf{F}_{ij}' \mathbf{\Upsilon}_j, \tag{53}$$

and substituting Eqn (51),

$$\begin{aligned} 2V &= \sum_i \mathbf{\Xi}_i^\dagger \mathbf{F}_\Xi^0 \mathbf{\Xi}_i + \sum_i \mathbf{\Xi}_i^\dagger \mathbf{\Phi}_i^\dagger \mathbf{F}_{ii}' \mathbf{\Phi}_i \mathbf{\Xi}_i + \sum_{i,j}' \mathbf{\Xi}_i^\dagger \mathbf{\Phi}_i^\dagger \mathbf{F}_{ij}' \mathbf{\Phi}_j \mathbf{\Xi}_j \\ &= \sum_i \mathbf{Q}_i^\dagger \mathbf{\Lambda}_i^0 \mathbf{Q}_i^0 + \sum_i \mathbf{Q}_i^{0\dagger} \mathscr{L}_i^{0\dagger} \mathbf{\Phi}_i^\dagger \mathbf{F}_{ii}' \mathbf{\Phi}_i \mathscr{L}_i^0 \mathbf{Q}_i^0 \\ &\quad + \sum_{i,j}' \mathbf{Q}_i^{0\dagger} \mathscr{L}_i^{0\dagger} \mathbf{\Phi}_i^\dagger \mathbf{F}_{i,j}' \mathbf{\Phi}_j \mathscr{L}_j^0 \mathbf{Q}_j^0. \end{aligned} \tag{54}$$

As indicated above, $\mathbf{Q}_i^0$ contains the normal coordinates for external motion of the ith molecule. It can, therefore, be partitioned as follows,

$$\mathbf{Q}_i^0 = \begin{pmatrix} \mathbf{Q}_i^{0(\text{in})} \\ \cdots\cdots \\ \mathbf{Q}_i^{0(\text{ex})} \end{pmatrix}. \tag{55}$$

Similarly,

$$\mathscr{L}_i^0 = (\mathscr{L}_i^{0(\text{in})} \vdots \mathscr{L}_i^{0(\text{ex})}), \tag{56}$$

and

$$\mathbf{\Phi}_1 \mathscr{L}_i^0 \mathbf{Q}_i^0 = \mathbf{\Phi}_i \mathscr{L}_i^{0(\text{in})} \mathbf{Q}_i^{0(\text{in})} + \mathbf{\Phi}_i \mathscr{L}_i^{0(\text{ex})} \mathbf{Q}_i^{0(\text{ex})}. \tag{57}$$

Equation (54) is then approximately equivalent to

$$\begin{aligned} 2V \approx{}& \sum_i \mathbf{Q}_i^{0(\text{in})\dagger} \mathbf{\Lambda}_i^{0(\text{in})} \mathbf{Q}_i^{0(\text{in})} + \sum_i \mathbf{Q}_i^{0(\text{in})\dagger} \mathscr{L}_i^{0(\text{in})\dagger} \mathbf{\Phi}_i^\dagger \mathbf{F}_{ii}' \mathbf{\Phi}_i \mathscr{L}_i^{0(\text{in})} \mathbf{Q}_i^{0(\text{in})} \\ &+ \sum_{i,j}' \mathbf{Q}_i^{0(\text{in})\dagger} \mathscr{L}_i^{0(\text{in})\dagger} \mathbf{\Phi}_i^\dagger \mathbf{F}_{ij}' \mathbf{\Phi}_j \mathscr{L}_j^{0(\text{in})} \mathbf{Q}_j^{0(\text{in})} \\ &+ \sum_i \mathbf{Q}_i^{0(\text{ex})\dagger} \mathscr{L}_i^{0(\text{ex})\dagger} \mathbf{\Phi}_i^\dagger \mathbf{F}_{ii}' \mathbf{\Phi}_i \mathscr{L}_i^{0(\text{ex})} \mathbf{Q}_i^{0(\text{ex})} \\ &+ \sum_{i,j}' \mathbf{Q}_i^{0(\text{ex})\dagger} \mathscr{L}_i^{0(\text{ex})\dagger} \mathbf{\Phi}_i^\dagger \mathbf{F}_{ij}' \mathbf{\Phi}_j \mathscr{L}_j^{0(\text{ex})} \mathbf{Q}_j^{0(\text{ex})}, \end{aligned} \tag{58}$$

where

$$\sum_i \mathbf{Q}_i^{0(\text{ex})\dagger} \mathbf{\Lambda}_i^{0(\text{ex})} \mathbf{Q}_i^{0(\text{ex})} = 0,$$

and cross terms between internal and external coordinates have been neglected.

In order to interpret the above expression, consider first a primitive cell containing only one molecule. In this case the two double summations of

Eqn (58) vanish and the potential energy becomes

$$2V = \mathbf{Q}^{0(\text{in})\dagger} \mathbf{\Lambda}^{0(\text{in})} \mathbf{Q}^{0(\text{in})} + \mathbf{Q}^{0(\text{in})\dagger} \mathscr{L}^{0(\text{in})\dagger} \mathbf{F}' \mathscr{L}^{0(\text{in})} \mathbf{Q}^{0(\text{in})}$$
$$+ \mathbf{Q}^{0(\text{ex})\dagger} \mathscr{L}^{0(\text{ex})\dagger} \mathbf{\Phi}^{\dagger} \mathbf{F}' \mathbf{\Phi} \mathscr{L}^{0(\text{ex})} \mathbf{Q}^{0(\text{ex})}. \quad (59)$$

The first term in this equation represents the potential energy of a free molecule whose $3N - 6$ vibrational frequencies are the elements of $\mathbf{\Lambda}^{0(\text{in})}$. The last two terms are the direct result of the external forces which bind the molecule to its site in a definite equilibrium orientation. The symmetry of the external force-constant matrix $\mathbf{F}'$ is determined by the site group. The possible effects of the last two terms of Eqn (59) include modification of the vibrational frequencies of the free molecule, changes in selection rules resulting in the appearance of previously forbidden bands, and splitting of degenerate vibrations.

In many cases intramolecular forces arise from covalent bonding, while intermolecular interaction results from the much weaker van der Waals' forces. Thus, the frequencies of lattice vibrations are ordinarily much lower than those of internal vibrations and the method of approximate separation of high and low frequencies can be used [(**13**), p. 311]. The lower frequencies of translation and hindered rotation (libration) are given approximately from the roots of the secular determinant

$$|\mathscr{L}^{0(\text{ex})\dagger} \mathbf{\Phi}^{\dagger} \mathbf{F}' \mathbf{\Phi} \mathscr{L}^{0(\text{ex})} - \mathbf{E}\lambda_{k}^{(\text{ex})}| = 0. \quad (60)$$

However, with but a single molecule in the unit cell, the three translational motions are the acoustic vibrations of the cell, corresponding to the three zero roots of Eqn (60).

The frequencies of the internal motions are determined by the energies

$$2T_{\text{in}} = \dot{\mathbf{Q}}^{0(\text{in})\dagger} \dot{\mathbf{Q}}^{0(\text{in})} \quad (61)$$

and

$$2V_{\text{in}} = \mathbf{Q}^{0(\text{in})\dagger} \mathbf{\Lambda}_0^{(\text{in})} \mathbf{Q}^{0(\text{in})} + \mathbf{Q}^{0(\text{in})\dagger} \mathscr{L}^{0(\text{in})\dagger} \mathbf{\Phi}^{\dagger} \mathbf{F}' \mathbf{\Phi} \mathscr{L}^{0(\text{in})} \mathbf{Q}^{0(\text{in})}. \quad (62)$$

Furthermore, since the second term in Eqn (62) is small compared to the first, it can be considered as a perturbation. To first order, the internal frequencies of the molecule are then given by

$$\mathbf{\Lambda} = \mathbf{\Lambda}^0 + \mathscr{L}^{0(\text{in})\dagger} \mathbf{\Phi}^{\dagger} \mathbf{F}' \mathbf{\Phi} \mathscr{L}^{0(\text{in})}, \quad (63)$$

and the second term in Eqn (63) accounts for the frequency shifts which are observed when a gaseous molecule is condensed into the crystalline state.

If there are two or more molecules in a unit cell, coupling of internal motions is accounted for by the third term of Eqn (58). This term is the origin of the correlation or Davydov splitting which was discussed qualitatively in Chapter 4 using the factor group method. There, it was pointed out that the splitting of nondegenerate internal vibrations can only be explained by correlation coupling. Hence, in such cases the magnitudes of experimentally observed splittings can be directly related to the intermolecular potential function by treating the third term of Eqn (58) as a perturbation. Some examples of this method will be illustrated later in the present chapter.

The frequencies of the lattice modes of crystals containing more than one molecule per primitive cell can be approximately calculated using the last two terms of Eqn (58). Thus, the lattice frequencies are essentially the eigenvalues of the dynamical matrix

$$\mathbf{D}^{(\text{ex})} = \begin{pmatrix} \mathscr{L}^{0(\text{ex})\dagger}\,\mathbf{\Phi}_1^{\dagger}\,\mathbf{F}_{11}'\,\mathbf{\Phi}_1\,\mathscr{L}^{0(\text{ex})} & \mathscr{L}^{0(\text{ex})\dagger}\,\mathbf{\Phi}_1^{\dagger}\,\mathbf{F}_{12}'\,\mathbf{\Phi}_2\,\mathscr{L}^{0(\text{ex})} & \cdots\cdots \\ \mathscr{L}^{0(\text{ex})\dagger}\,\mathbf{\Phi}_2^{\dagger}\,\mathbf{F}_{21}'\,\mathbf{\Phi}_1\,\mathscr{L}^{0(\text{ex})} & \mathscr{L}^{0(\text{ex})\dagger}\,\mathbf{\Phi}_2^{\dagger}\,\mathbf{F}_{22}'\,\mathbf{\Phi}_2\,\mathscr{L}^{0(\text{ex})} & \\ \vdots & & \mathscr{L}^{0(\text{ex})\dagger}\,\mathbf{\Phi}_\sigma^{\dagger}\,\mathbf{F}_{\sigma\sigma}'\,\mathbf{\Phi}_\sigma\,\mathscr{L}^{0(\text{ex})} \end{pmatrix}. \tag{64}$$

This method has been used in particular, by Shimanouchi and his coworkers to evaluate intermolecular forces in organic compounds. In many cases, the resulting force constants can be compared with values estimated from models of repulsive forces between atoms or dipole–dipole interaction between bonds in adjacent molecules.

In practice, the application of Eqn (64) requires the construction of the eigenvectors $\mathscr{L}^{0(\text{ex})}$ for the translational and rotational motions of a free molecule. The kinetic energy of these external degrees of freedom is obtained From Eqn (7), Chap. 1, in the form

$$2\,T_{\text{ex}} = \dot{\mathbf{R}}^2 M + \sum_\alpha m_\alpha(\boldsymbol{\omega} \times \mathbf{r}_\alpha)\cdot(\boldsymbol{\omega} \times \mathbf{r}_\alpha), \tag{65}$$

where $M = \sum_\alpha m_\alpha$ is the total mass of the molecule. If

$$\mathbf{P}_R \equiv \frac{\partial T_{\text{ex}}}{\partial \dot{\mathbf{R}}} = M\,\dot{\mathbf{R}}$$

is the momentum arising from translation of the center of gravity of the molecule, the first term in Eqn (65) becomes

$$\mathbf{P}_R^{\dagger}\,M^{-1}\,\mathbf{P}_R = \mathbf{P}_R^{\dagger}\,\mathbf{G}_R\,\mathbf{P}_R = \mathbf{P}_R^{\dagger}\,\mathscr{L}_R^{0}\,\mathscr{L}_R^{0\dagger}\,\mathbf{P}_R, \tag{66}$$

and, since $\mathbf{M}$ is diagonal, the desired transformation is

$$\mathscr{L}_R{}^0 = \mathscr{L}_R{}^{0\dagger} = \begin{pmatrix} 1/\sqrt{M} & 0 & 0 \\ 0 & 1/\sqrt{M} & 0 \\ 0 & 0 & 1/\sqrt{M} \end{pmatrix} \tag{67}$$

for the translational part of the kinetic energy.

The rotational kinetic energy of the free molecule is given from Eqn. (8), Chap. 1, by

$$2\,T_{\text{rot}} = \sum_\alpha m_\alpha(\boldsymbol{\omega} \times \mathbf{r}_\alpha)\cdot(\boldsymbol{\omega} \times \mathbf{r}_\alpha) = \boldsymbol{\omega}^\dagger\,\mathbf{I}\,\boldsymbol{\omega} = \mathbf{P}_\omega{}^\dagger\,\mathbf{I}^{-1}\,\mathbf{P}_\omega, \tag{68}$$

where $\mathbf{I}$ is the moment-of-inertia tensor and $\mathbf{P}_\omega$ is the angular-momentum conjugate to $\boldsymbol{\omega}$. The vector product $\boldsymbol{\omega} \times \mathbf{r}_\alpha$ appearing in Eqn (68) can be written in matrix form as

$$\begin{pmatrix} 0 & z^0 & -y^0 \\ -z^0 & 0 & x^0 \\ y^0 & -x^0 & 0 \end{pmatrix} \begin{pmatrix} \omega_x \\ \omega_y \\ \omega_z \end{pmatrix} = \mathbf{C}\,\boldsymbol{\omega}, \tag{69}$$

where x^0, y^0, z^0 give the equilibrium position of the αth atom in the molecule-fixed coordinate system. The rotational angular momentum is then

$$\mathbf{P}_\omega = \sum_\alpha m_\alpha\,\mathbf{r}_\alpha \times \dot{\mathbf{r}}_\alpha = \sum_\alpha \mathbf{r}_\alpha \times \mathbf{p}_\alpha, \tag{70}$$

where $\mathbf{p}_\alpha$ is the momentum conjugate to $\mathbf{r}_\alpha$. In matrix form,

$$\mathbf{P}_{\omega,\alpha} = \mathbf{C}_\alpha{}^\dagger\,\mathbf{p}_\alpha \tag{71}$$

for the contribution of the αth atom to the angular momentum. The rotational kinetic energy then becomes

$$2\,T_{\text{rot}} = \sum_\alpha \mathbf{p}_\alpha{}^\dagger\,\mathbf{C}_\alpha\,\mathbf{I}^{-1}\,\mathbf{C}_\alpha{}^\dagger\,\mathbf{p}_\alpha = \sum_\alpha \mathbf{p}_\alpha{}^\dagger\,\mathbf{G}_\alpha\,\mathbf{p}_\alpha, \tag{72}$$

where $\mathbf{G}_\alpha = \mathbf{C}_\alpha\,\mathbf{I}^{-1}\,\mathbf{C}_\alpha{}^\dagger = \mathscr{L}_{\omega\alpha}{}^0\,\mathscr{L}_{\omega\alpha}{}^{0\dagger}$, and, if the molecular-fixed coordinates x, y, z are coincident with the principal axes of the molecule,

$$\mathscr{L}_{\omega\alpha}{}^0 = \mathbf{C}_\alpha\,\mathbf{I}^{-1/2}, \tag{73}$$

where

$$\mathbf{I}^{-1/2} = \begin{pmatrix} 1/\sqrt{I_x} & 0 & 0 \\ 0 & 1/\sqrt{I_y} & 0 \\ 0 & 0 & 1/\sqrt{I_z} \end{pmatrix}. \tag{74}$$

Combining this result with the translational part of the eigenvector given by Eqn (67) and the definition of **C** given in Eqn (69), one finds

$$\mathscr{L}^{0(\mathrm{ex})} = (\mathscr{L}_R{}^0 \vdots \mathscr{L}_{\omega 1}{}^0 \ldots \mathscr{L}_{\omega N}{}^0)$$

$$= \left(\begin{array}{ccc:ccc:c} 1/\sqrt{M} & 0 & 0 & 0 & z_1{}^0/\sqrt{I_y} & -y_1{}^0/\sqrt{I_z} & \\ 0 & 1/\sqrt{M} & 0 & -z_1{}^0/\sqrt{I_x} & 0 & x_1{}^0/\sqrt{I_z} & \cdots \\ 0 & 0 & 1/\sqrt{M} & y_1{}^0/\sqrt{I_x} & -x_1{}^0/\sqrt{I_y} & 0 & \end{array}\right.$$

$$\left.\begin{array}{c:ccc} & 0 & z_N{}^0/\sqrt{I_y} & -y_N{}^0/\sqrt{I_z} \\ \cdots & -z_N{}^0/\sqrt{I_x} & 0 & x_N{}^0/\sqrt{I_z} \\ & y_N{}^0/\sqrt{I_x} & -x_N{}^0/\sqrt{I_y} & 0 \end{array}\right), \qquad (75)$$

where N is the number of atoms in the molecule. Equation (75) may often be simplified. For example, for planar molecules $z_\alpha{}^0 = 0$ for all values of α if the z axis is perpendicular to the molecular plane. Two of the moments of inertia are equal if the molecule is a symmetric-top rotor, etc.

V. Lattice Vibrations of Benzene and Naphthalene

Crystalline benzene belongs to the orthorhombic system. Its crystallographic unit cell contains four molecules and is of space-group symmetry $Pbca \equiv \mathscr{D}_{2h}{}^{15}$. As the Bravais lattice is of the primitive type, the crystallographic unit cell is identical to the primitive cell.

Figure 3 shows the positions and orientations of the molecules in the unit cell and the resulting symmetry elements. The four differently orientated molecules are located at sites 0, 0, 1/2; 1/2, 1/2, 1/2; 0, 1/2, 0; and 1/2, 0, 0 having symmetry of point group $\mathscr{C}_i$. Their orientations with respect to the crystallographic axes are specified by the matrices

$$\mathbf{\Phi}_1 = \begin{pmatrix} a_x & a_y & a_z \\ b_x & b_y & b_z \\ c_x & c_y & c_z \end{pmatrix} = \begin{pmatrix} 0{\cdot}64758 & -0{\cdot}27828 & 0{\cdot}70937 \\ 0{\cdot}16742 & 0{\cdot}96014 & 0{\cdot}22382 \\ -0{\cdot}74338 & -0{\cdot}02618 & 0{\cdot}66836 \end{pmatrix},$$

$$\mathbf{\Phi}_2 = \begin{pmatrix} a_x & a_y & a_z \\ -b_x & -b_y & -b_z \\ -c_x & -c_y & -c_z \end{pmatrix},$$

$$\mathbf{\Phi}_3 = \begin{pmatrix} -a_x & -a_y & -a_z \\ b_x & b_y & b_z \\ c_x & c_y & c_z \end{pmatrix},$$

and

$$\mathbf{\Phi}_4 = \begin{pmatrix} -a_x & -a_y & -a_z \\ -b_x & -b_y & -b_z \\ c_x & c_y & c_z \end{pmatrix}, \tag{76}$$

which are defined by Eqn (51). The form of these matrices can be verified by consideration of the factor group operations shown in Fig. 3, while the numerical values of $\mathbf{\Phi}_i$ are determined from the X-ray data of Cox (**14**).

The results of the factor-group analysis of crystalline benzene are given in Table 5. The correlation diagram relating the factor-group to the site group and the point group ($\mathscr{D}_{6h}$) of an isolated molecule is shown in Table 6. Infrared and Raman activity is indicated by underlines or underdashes, following the convention adopted earlier. The symmetry species of the various components of the dipole moment and the polarizability are indicated in Table 5. This information can be used to predict infrared dichroism and Raman polarizations, as was shown in Chapter 5.

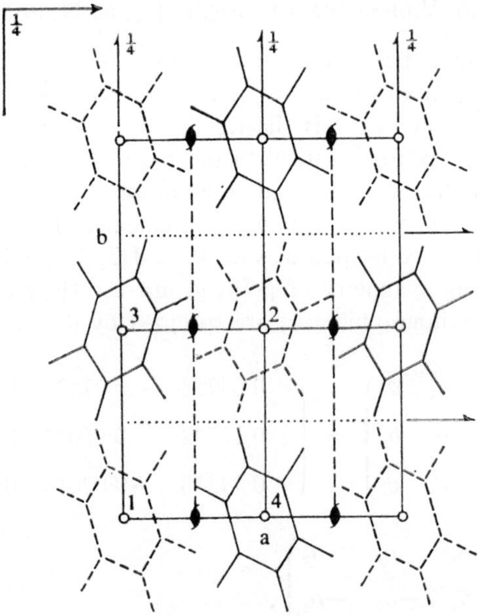

FIG. 3. Structure of crystalline benzene (Space group $Pbca \equiv \mathscr{D}_{2h}^{15}$).

TABLE 5

$\mathcal{D}_{2h}^{15}$	$n_{tot}^{(\gamma)}$	$n_{acous}^{(\gamma)}$	$n_{trans}^{(\gamma)}$	$n_{rot}^{(\gamma)}$	$n_{vib}^{(\gamma)}$	I R	Raman
A_g	18	0	0	3	15	—	$\alpha_{XX}, \alpha_{YY}, \alpha_{ZZ}$
B_{1g}	18	0	0	3	15	—	α_{YZ}
B_{2g}	18	0	0	3	15	—	α_{ZX}
B_{3g}	18	0	0	3	15	—	α_{XY}
A_u	18	0	3	0	15	—	—
B_{1u}	18	1	2	0	15	T_X	—
B_{2u}	18	1	2	0	15	T_Y	—
B_{3u}	18	1	2	0	15	T_Z	—
degrees of freedom	144	3	9	12	120	—	—

TABLE 6

$\mathcal{D}_{6h}$ (molecule)	$\mathcal{C}_i$ (site)	$\mathcal{D}_{2h}^{15}$ (unit cell)
2 A_{1g}		$A_g(15 + 3r)$
1 A_{2g}		$B_{1g}(15 + 3r)$
0 B_{1g}	A_g	$B_{2g}(15 + 3r)$
2 B_{2g}		$B_{3g}(15 + 3r)$
1 E_{1g}		
4 E_{2g}		
0 A_{1u}		
1 A_{2u}		$A_u(15 + 3t)$
2 B_{1u}	A_u	$B_{1u}(15 + 2t)$
2 B_{2u}		$B_{2u}(15 + 2t)$
3 E_{1u}		$B_{3u}(15 + 2t)$
2 E_{2u}		

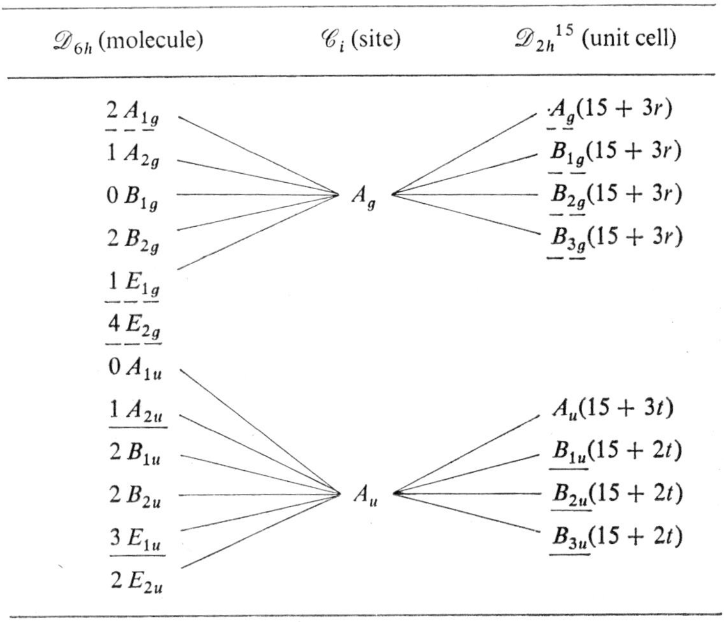

In order to describe the lattice vibrations of benzene it is necessary to introduce a set of intermolecular coordinates which form the basis of the particular interactions to be considered. In their treatment of the lattice vibrations of benzene, Harada and Shimanouchi (**15**) included those hydrogen repulsion terms which are functions of the coordinates q_1 through q_5 defined in Fig. 4. Thus, for the nth unit cell

$$2\,V_n' = \mathbf{R}^\dagger\,\mathbf{F}_R'\,\mathbf{R} \qquad (77)$$

where $\mathbf{R}^\dagger = (\Delta q_1 \ldots \Delta q_5)$ and $\mathbf{F}'$ is assumed to be a diagonal matrix of force constants f_1 through f_5.

The potential energy of Eqn (77) can be rewritten in terms of the positions of each pair of interacting atoms. For example, for the interaction of the αth atom of molecule i with the βth atom of molecule j, the

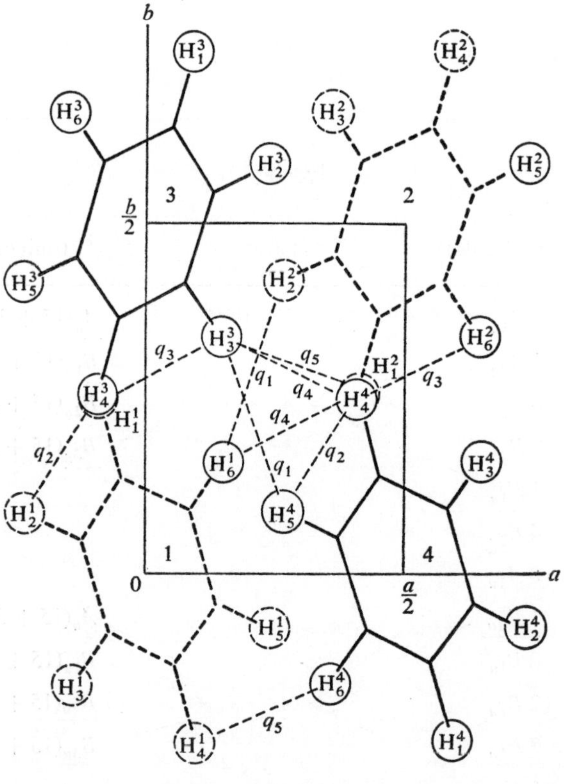

FIG. 4. Bravais cell of crystalline benzene showing intermolecular coordinates [After Harada and Shimanouchi (**15**)].

relative displacement coordinate which forms the corresponding element of $\mathbf{R}$ is given by

$$\mathbf{R}_{ij}^{\alpha\beta} = \mathbf{\Omega}_{ij}^{\alpha\beta} \begin{pmatrix} \mathbf{\Upsilon}_i^{\alpha} \\ \hdashline \mathbf{\Upsilon}_j^{\beta} \end{pmatrix} = \mathbf{\Omega}_{ij}^{\alpha\beta} \begin{pmatrix} \Phi_i \Xi_i^{\alpha} \\ \hdashline \Phi_j \Xi_j^{\beta} \end{pmatrix}, \tag{78}$$

where only the αth and βth submatrices of $\mathbf{\Upsilon}_i$ and $\mathbf{\Upsilon}_j$ need be used. The matrix $\mathbf{\Omega}_{ij}^{\alpha\beta}$ is a simple function of the crystal structure. For example, from Fig. 4

$$q_1^2 = [a(H_6^1) - a(H_2^2)]^2 + [b(H_6^1) - b(H_2^2)]^2 + [c(H_6^1) - c(H_2^2)]^2, \tag{79}$$

and, for small displacements,

$$\Delta q_1 = \frac{1}{q_1}\left(a(H_6^1) - a(H_2^2) \quad b(H_6^1) - b(H_2^2) \ldots c(H_6^1) - c(H_2^2)\right) \begin{pmatrix} \Delta a(H_6^1) \\ \Delta b(H_6^1) \\ \Delta c(H_6^1) \\ \hdashline \vdots \\ \Delta c(H_2^2) \end{pmatrix}. \tag{80}$$

The numerical application of this method involves the determination of the five force constants $f_1 \ldots f_5$ from the eight observed lattice frequencies given in Table 7 using a least-squares fitting procedure. The resulting force constants are given in Table 8, while the complete set of calculated frequencies is compared with the experimental data in Table 7. The agreement is quite good in this example. Furthermore, the force constants vary monotonically as functions of hydrogen–hydrogen distance.

The same set of intermolecular force constants was used by Harada and Shimanouchi to estimate the splittings of internal vibrations in crystalline benzene by evaluating the terms in the second and third summation of Eqn (58). This calculation depends directly on a knowledge of the eigenvectors $\mathscr{L}^{0(\text{in})}$ for the free molecule, which of course requires a complete normal coordinate treatment. The calculated splittings are found to be the right order of magnitude, although, in general, good quantitative agreement is not obtained.

One of the more favorable results of this calculation is for the out-of-plane bending mode of species A_{2u} which is observed at 675 cm^{-1} in the infrared spectrum of the vapor. From Table 6 it is seen that this fundamental is split into four components in the spectrum of the crystal, three of which are infrared-active. The calculated and observed frequencies of this mode are compared in Fig. 5.

In principle, both experimentally observed lattice frequencies and internal-mode splittings can be used to determine the intermolecular force field in a crystal. However, internally consistent experimental data are not often available, as few far-infrared or Raman studies have been made at the low sample temperatures at which correlation splittings can be readily observed. Furthermore, the computational difficulties become significant for molecules as large as benzene.

TABLE 7

Symmetry species	ν_{obs}[a] (cm^{-1})	ν_{cal}[a] (cm^{-1})
A_g	35	28
	63	57
	—	79
B_{1g}	35	33
	—	70
	105	102
B_{2g}	63	56
	69	69
	—	76
B_{3g}	—	60
	60	72
	105	100
A_u	—	26
	—	49
	—	58
B_{1u}	—	34
	—	58
B_{2u}	—	31
	—	57
B_{3u}	—	30
	—	44

[a] From reference (**15**).

TABLE 8

Coordinate	Interatomic distance (Å)	Force constant[a] (md/Å)
q_1	2·628	0·016
q_2	2·698	0·012
q_3	2·764	0·010
q_4	2·767	0·0096
q_5	2·826	0·0080

[a] From reference (**15**).

The structure of crystalline naphthalene was discussed in Chapter 4. Assignment of the six Raman-active lattice vibrations was considered in Chapter 5. In addition, the three infrared-active lattice vibrations predicted by factor-group analysis have been observed in the far-infrared region by Harada and Shimanouchi (**15**). The nine lattice frequencies are listed in Table 9.

Using a model of hydrogen–hydrogen repulsion similar to that described above for benzene, Harada and Shimanouchi have determined a set of intermolecular force constants which provides reasonably good agreement between the calculated and experimental frequencies. These authors found that the addition of force constants for intermolecular hydrogen–carbon interactions resulted in a significant improvement in the calculated results.

The hydrogen–hydrogen force constants determined in these studies of crystalline benzene and naphthalene, as well as similar constants obtained

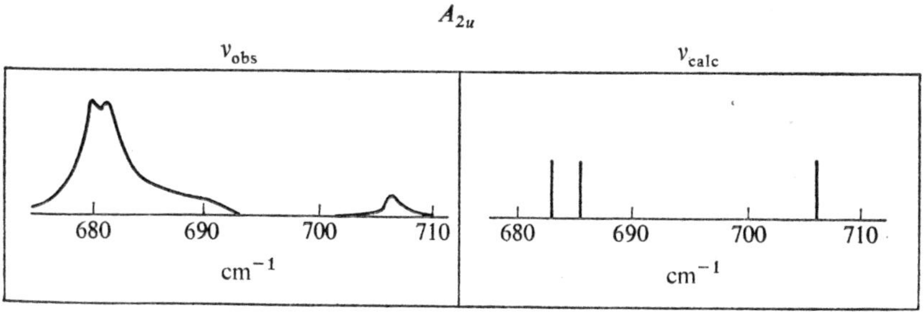

FIG. 5. Observed and calculated frequencies of the A_{2u} modes ω_1 of crystalline benzene [After Harada and Shimanouchi (**15**)].

from the spectrum of crystalline polyethylene (see Table 14 of Chapter 7), are plotted as a function of the H–H distance in Fig. 6. These results can be compared with the constants derived from an intermolecular potential function of the type

$$V'(q_l) = -\frac{A}{q_l^{\,6}} + B\,\mathrm{e}^{-Cq_l}, \tag{81}$$

where various values of the parameters A, B, and C have been suggested (**16**).

The second derivative of Eqn (81) can be identified with the hydrogen–hydrogen interaction associated with the intermolecular coordinate q_l. Thus,

$$F_{ij}^{\alpha\beta} \equiv \left(\frac{\partial^2 V'}{\partial q_l^{\,2}}\right)_0 = -\frac{42A}{q_l^{\,0 8}} + BC^2\,\mathrm{e}^{-Cq_l^0}, \tag{82}$$

where q_l^0 is the equilibrium value of the lth element of the matrix **R** of Eqn (78). With $A = 0$, $B = 12{\cdot}2$ mdyn/Å and $C = 3{\cdot}53$ Å^{-1}, Eqn (82) gives the repulsive part of the de Boer potential. This function, which is shown for comparison in Fig. 6, is often used to obtain approximate hydrogen–hydrogen force constants. The addition of the attractive contribution to the force constant $A > 0$ in Eqn (82) does not appear to improve significantly the agreement with the points shown in Fig. 6. Thus, it is usually concluded

TABLE 9

Symmetry species	ν_{obs}[a] (cm^{-1})	ν_{cal}[a] (cm^{-1})
A_u	98	104
	53	58
B_u	66	76
A_g	127	119
	76	86
	54	62
B_g	109	93
	74	77
	46	47

[a] From reference (**15**).

that intermolecular interactions involving hydrogen atoms depend primarily on the repulsive forces between them.

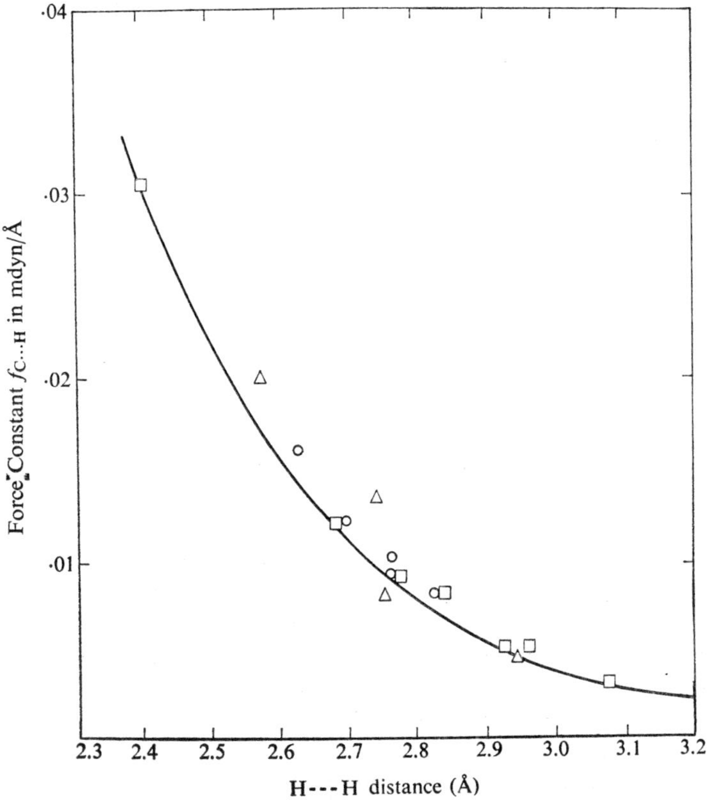

FIG. 6. Intermolecular H—H potential function. Experimental points: ○ benzene, □ naphthalene, △ polyethylene. Curve calculated from the de Boer potential (See text).

VI. Spectroscopic Properties of Hydrogen Bonds

The vibrational spectrum of a simple hydrogen bond can be characterized by considering the degrees of freedom of the structure X—H·· Y. In most cases the atoms X and Y are at least an order of magnitude heavier than hydrogen and the vibrational modes of this system are described approximately by:

1. An X—H stretching mode of frequency ν_s:

$$\mathrm{X{-}\overset{\cdot\longrightarrow}{H}\cdots Y},$$

where $3500 \geqslant \nu_s \geqslant 1700\,\mathrm{cm}^{-1}$.

2. An X—H···Y bending mode:

$$\overset{\uparrow}{\text{X—H}\cdots\text{Y}},$$

whose frequency $1700 \geqslant \nu_b \geqslant 800\ \text{cm}^{-1}$ is doubly degenerate if the symmetry of rotation about the XHY bonds is three-fold or higher. If the XHY bond is bent, or, if the groups X or Y lower the symmetry of the system, ν_b will, in principle, be split into a doublet.

3. A stretching mode of frequency ν_{XY}:

$$\overleftarrow{\text{X}}\text{—H}\cdots\overrightarrow{\text{Y}},$$

where $600 \geqslant \nu_{XY} \geqslant 50\ \text{cm}^{-1}$.

4. A torsional mode with $\nu_\tau \geqslant 500\ \text{cm}^{-1}$ which occurs when X and Y are nonlinear polyatomic groups.

Vibrations of types (1) and (2) occur in the absence of hydrogen bonding, while those of types (3) and (4) are associated with the formation of a hydrogen bond. Although the appearance of (3) and (4) could be used as a criterion for hydrogen-bond formation, it is usually much more convenient and reliable to use the observed shifts of ν_s and ν_b which accompany it. Studies of a large number of hydrogen-bonded systems have shown that ν_s decreases in frequency, ν_b shifts to higher frequencies, and the width and intensity of the ν_s absorption band increases when a molecule X—H forms a hydrogen bond with Y. Furthermore, these effects vary monotonically with the strength of the hydrogen bond.

A simple example of hydrogen bonding in a linear system is provided by gyanoacetylene, HCCCN. The infrared spectra **(17)** of this molecule in the caseous, liquid, and solid states are compared in schematic form in Fig. 7.

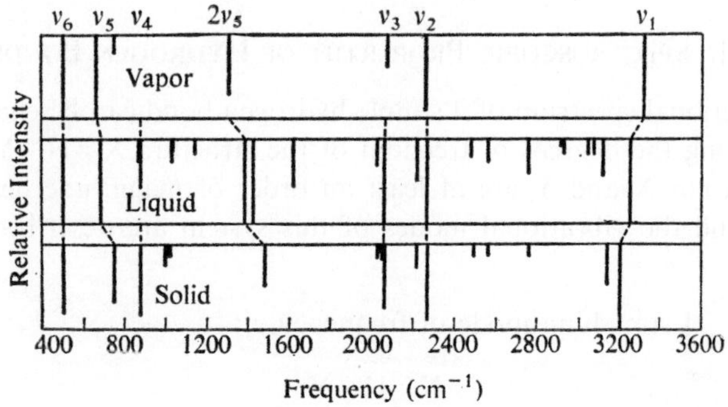

FIG. 7. Comparison of the spectra of cyanoacetylene [From reference **(17)**].

It is seen that $\nu_s = \nu_1$ shifts from 3328·5 to 3204 cm^{-1} in passing from the gas to the solid, while $\nu_b = \nu_5$ increases from 663·0 cm^{-1} to 748·5. The first overtone of ν_b also increases, as expected. Other absorption bands are essentially independent of the physical state of the molecule, although ν_3, which is approximately a C $\equiv$ N stretching vibration, decreases by 12 cm^{-1} in passing from the vapor to the solid.

Figure 8 shows the region of the C—H stretching fundamental, ν_s, of cyanoacetylene. The broad absorption band with a superimposed peak is characteristic of hydrogen-bonded systems. Although several theories have been suggested to account for the broadness of ν_s, none is completely convincing. It seems, however, that vibrational anharmonicity of the hydrogen bond is largely responsible for this effect. It should be noted that the bending fundamental of frequency ν_b does not broaden significantly with the formation of a hydrogen bond.

An X-ray study (**18**) of crystalline HCCCN has shown that the molecules form linear chains of the type ···H—CCCN···H—CCCN···H—CCCN··· . The N···H—C distance is 3·27 Å.

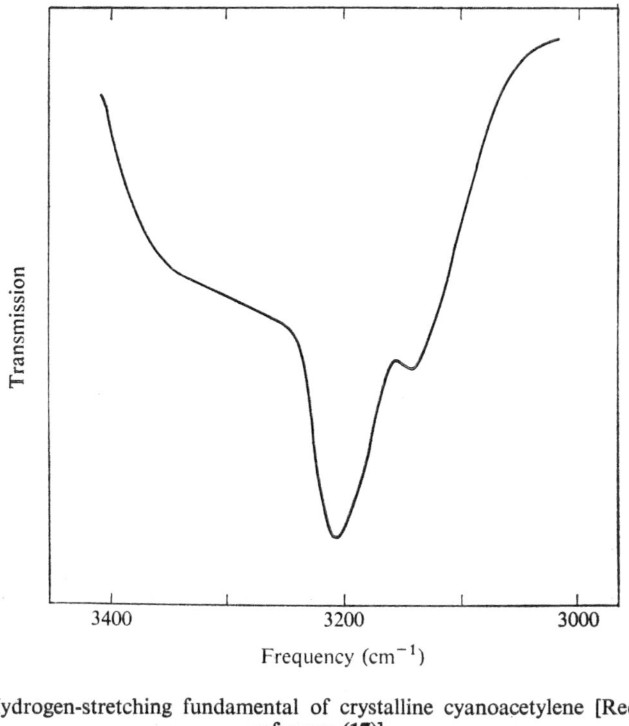

FIG. 8. Hydrogen-stretching fundamental of crystalline cyanoacetylene [Redrawn from reference (**17**)].

A number of empirical relations has been established among various properties of the hydrogen bond (**19**). For example, studies of the spectra of hydrogen-bonded solids indicate that the shift of ν_s correlates reasonably well with the equilibrium X—H···Y distance, R_0, as shown in Fig. 9. Although the form of the curve cannot be determined from these data, it is clear that a monotonic relationship exists and that ν_s approaches zero asymptotically as R_0 becomes infinite.

Although not all aspects of hydrogen-bond formation are understood in detail, several models have been proposed in attempts to account for the behavior of hydrogen-bonded systems. One of the more successful models is that of Lippincott and Schroeder (**20, 21**), which is based on the potential function derived by Lippincott (**22**). For an ordinary chemical bond such as X—H this potential is usually written in the form

$$V = D[1 - \exp(-\tfrac{1}{2}n\overline{\Delta r}^2/r)], \tag{83}$$

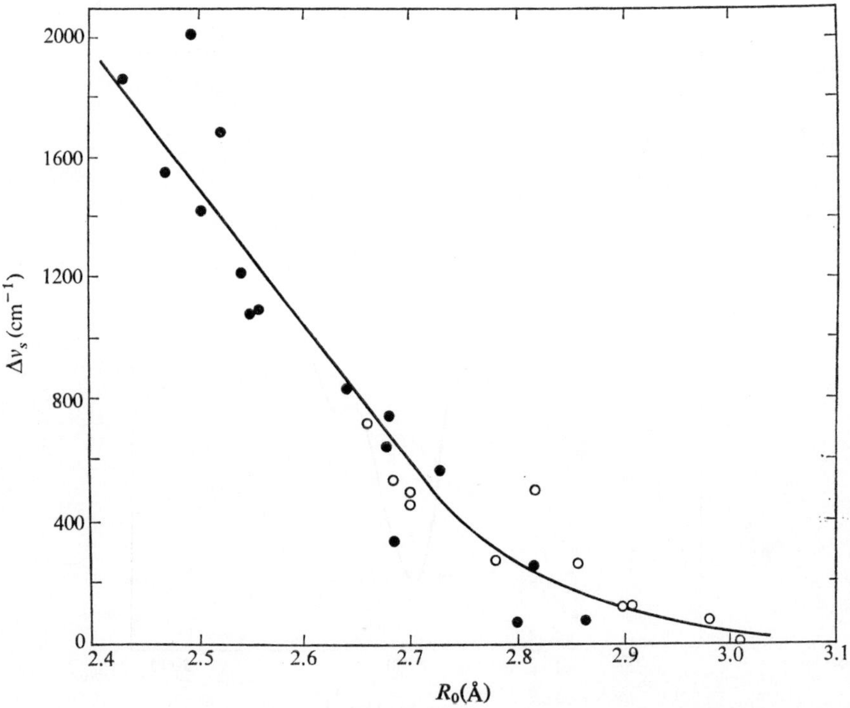

FIG. 9. R_0 *vs.* $\Delta\nu_s$ for O···H—O bonds in solids. ○ from Glemser and Hartert (**24**); ● from Pimentel and Sederholm (**25**) [After Pimentel and McClellan (**19**)].

where D is the bond dissociation energy, $\Delta r = r - r_0$, and $n = k_0 r_0/D$. Here k_0 is the bond-stretching force constant and r_0 is the equilibrium value of the bond length.

Equation (83) was modified by Moulton and Kromhout (**23**) to include angular deformation of the X—H bond. In their model the potential becomes

$$V = D[1 - (\cos^2 \theta) \exp(-\tfrac{1}{2} n \overline{\Delta r}^2 / r)], \qquad (84)$$

where θ is the angle defined in Fig. 10. This function was used to describe the vibrations of a hydrogen bond X—H···Y by employing a potential of the form of Eqn (84) for the interaction of the hydrogen atom with each of the atoms X and Y and adding repulsion and electrostatic (and/or dispersion) terms to represent the forces between X and Y. The potential function for the hydrogen-bonded system X—H···Y then takes the form

$$\begin{aligned} V = {} & D\{1 - (\cos^2 \theta) \exp[-\tfrac{1}{2} n (r - r_0)^2 / r]\} \\ & + D'\{1 - (\cos^2 \theta') \exp[-\tfrac{1}{2} n' (R - r - r_0')^2 / (R - r)]\} \\ & + A[\exp(-bR) - (R_0/2R) \exp(-bR_0)], \end{aligned} \qquad (85)$$

where

$$\begin{aligned} A = {} & n' D' R_0 [1 - r_0'^2 / (R_0 - r)^2 \\ & \times \exp[-\tfrac{1}{2} n' (R_0 - r - r_0')^2 / (R_0 - r)] / [(2bR_0 - 1) \exp(-bR_0)]. \end{aligned}$$

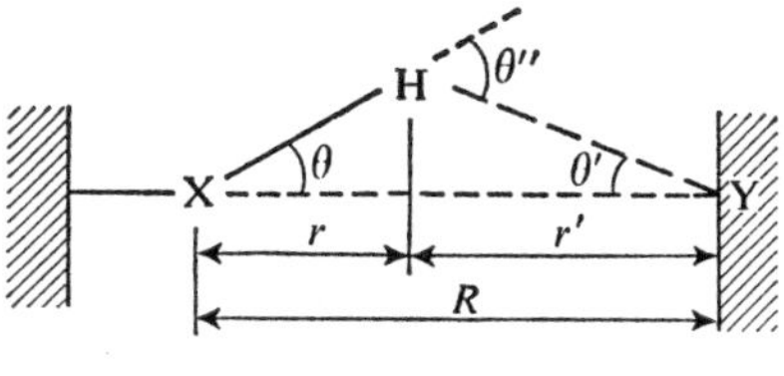

FIG. 10. Coordinate system used to describe the hydrogen bond.

The potential function of Eqn (85) has been used to interpret the vibrational spectra of acetylenic compounds hydrogen-bonded to oxygen and nitrogen bases (**26**). The model shown in Fig. 10 is assumed to be linear in its equilibrium configuration, and X—H is replaced by ≡C—H. The C—H stretching and bending frequencies are then given approximately by

$$4\pi^2 \nu_s^2 \equiv \lambda_s \approx \mu_H (k_H' + k_H) + \mu_C \mu_H k_H^2 / [\mu_H (k_H' + k_H) - \mu_C (k_C + k_H)] \qquad (86)$$

and

$$4\pi^2 \nu_b^2 \equiv \lambda_b \approx [\mu_H \rho_H^2 + (\rho_C + \rho_H)^2 \mu_C] H + [\mu_C \rho_H^2 + (\rho_H + \rho_H')^2 \mu_H] H', \qquad (87)$$

respectively, where the approximate separation of high and low frequencies has been used. In Eqns (86) and (87) μ_C and μ_H are the appropriate reciprocal masses, and the quantities ρ_H, ρ_H', and ρ_C are the reciprocals of the equilibrium values of the lengths of the corresponding bonds, C—H, H···Y, and C ≡ C. The stretching force constants k_H and k_H' are associated respectively with the C—H and H···Y bonds, while the bending constants H and H′ refer to changes in the angles C ≡ C—H and C—H···Y, respectively. Values of these various force constants were determined as functions of R_0, the equilibrium value of C—Y distance, using the Lippincott–Moulton model.

As the hydrogen-bending and -stretching frequencies of gaseous 1-heptyne, a typical acetylenic compound, have been measured, it is convenient to determine suitable values of the constants n and D in the Lippincott–Moulton model using these observed quantities. As $\mu_C \ll \mu_H$, Eqn (86) reduces to

$$4\pi^2(\nu_s^0)^2 \equiv \lambda_s^0 \approx k_H^0(\mu_H + \mu_C) \tag{88}$$

for the case of an unperturbed ≡ C—H bond ($k_H' = 0$). The use of the observed hydrogen stretching frequency 3330 cm^{-1} yields the constant $k_H = 6{\cdot}03 \times 10^5$ dyn/cm. For the bending mode of the free molecule, $H' = 0$, and Eqn (87) reduces to

$$4\pi^2(\nu_b^0)^2 \equiv \lambda_b^0 \approx [\mu_H\rho_H^2 + (\rho_C + \rho_H)^2\,\mu_C]H^0. \tag{89}$$

Taking the observed value of the bending frequency, $\nu_b^0 = 630{\cdot}5$ cm^{-1}, along with a C ≡ C bond length of 1·204 Å, one obtains the force constant

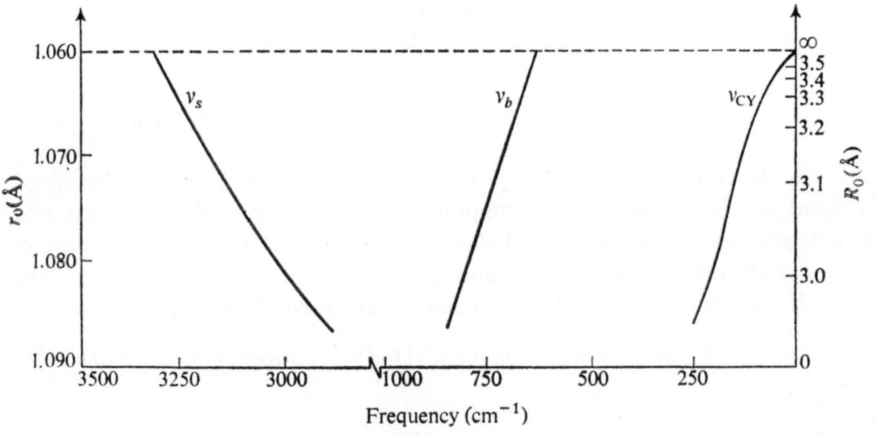

FIG. 11. Vibrational frequencies of ≡ C—H···Y calculated from the Lippincott–Moulton model [From reference (**26**)].

$H^0 = 2{\cdot}03 \times 10^{-12}$ dyn-cm. But for the free molecule ($r = r_0$ and $R_0 = \infty$),

$$k_H{}^0 = nD/r_0, \tag{90}$$

while from Eqn (85) one finds

$$H^0 \equiv \partial^2 V/\partial\theta^2|_{r_0} = 2D. \tag{91}$$

Thus the quantity $n = 2r_0 k_H{}^0/H^0 = 62{\cdot}9 \times 10^8\ \text{cm}^{-1}$ is calculated from the observed vibrational frequencies of the free molecule.

By making the substitution $\theta' = \theta'' - \theta$, one finds

$$H' \equiv \partial^2 V/\partial\theta''^2|_{r_0} = 2\,D' \exp\left[-\tfrac{1}{2}n'(R_0 - r - r_0')^2/(R_0 - r)\right], \tag{92}$$

and

$$H \equiv \partial^2 V/\partial\theta^2|_{r_0} = 2\,D \exp\left[-\tfrac{1}{2}n(r - r_0)^2/r\right] + H'. \tag{93}$$

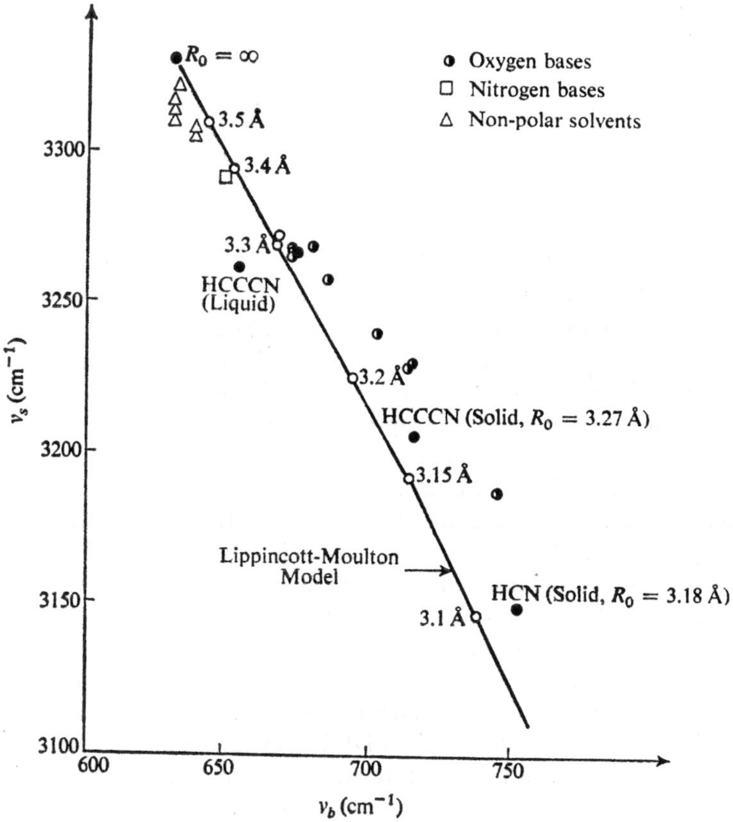

FIG. 12. Hydrogen-stretching frequency ν_s *vs.* bending frequency ν_b of 1-heptyne complexes [From reference (**26**)].

The bending force constants can be computed for the perturbed molecule and the corresponding bending frequency, ν_b, calculated approximately using Eqn (87).

The stretching and bending frequencies of the hydrogen-bonded system, as calculated by the method outlined above, are plotted in Fig. 11 as functions of r_0 and R_0. The decrease in stretching frequency, accompanied by an increase in the bending frequency with increasing hydrogen-bonding is shown by these results.

The shift in ν_s is approximately a linear function of r_0, in agreement with the empirical results, while the asymptotic variation of ν_s with R_0 is also clearly evident (see Fig. 11).

To provide a direct test of the Lippincott–Moulton model the frequency of the hydrogen-stretching mode is plotted against that of the bending mode in Fig. 12. The experimental points are obtained from studies of 1-heptyne in liquid solutions (**25**), and from the spectra of pure crystalline HCCCN and HCN. The values of R_0, which are shown on the calculated curve, are known from X-ray results in the case of the pure solids. It is apparent from Fig. 12 that the calculated values of R_0 are somewhat lower than the experimental ones. In liquid solutions R_0 can be considered to be an "effective distance of approach" of a solvent molecule.

VII. Structures of Calcium Amides

X-ray analyses of the structures of many inorganic amides have been reported, particularly by Juza and co-workers (**27**). Although X-ray methods do not, in general, locate the positions of the hydrogen atoms, these positions can often be inferred from the coordinates of the nitrogen atoms, or simply by consideration of the space-group symmetry of the crystal.

It has been recently shown that the calcuim amide exists in two different crystalline modifications (**28, 29**). The structure reported by Juza and Schumacher will be referred to as the α phase. Initial infrared studies of $Ca(NH_2)_2$ were made on the phase which is now designated β. A more recent infrared investigation indicates that the transition $\beta \rightarrow \alpha$, which occurs spontaneously at temperatures above 80°C, is accompanied by marked changes in the infrared spectrum (**30**).

The α phase is of the anatase type, a tetragonal unit cell containing four $Ca(NH_2)_2$ formulas per unit cell. The space group is $I4_1/amd \equiv \mathscr{D}_{4h}{}^{19}$. From Appendix E it is found that the eight NH_2^- ions must occupy sites of either $\mathscr{D}_{2h}$ or $\mathscr{C}_{2v}$ symmetry. In the structure reported by Juza and Schumacher the latter sites have been chosen. Thus, from the International Tables (p. 245) it is seen that the nitrogen atoms lie along the vertical axes defined by 0, 0, Z; 0, $\frac{1}{2}$, $\frac{1}{4} + Z$; etc. Hence, the C_2 axis of an NH_2^- ion must be coincident with vertical lines in the unit cell. Only two structures satisfy this condition. One of them is shown in Fig. 13, while the other, which results in opposing, colinear N—H bonds in adjacent ions, is rejected on electrostatic grounds.

Because the space group of α-$Ca(NH_2)_2$ is of type *I*, the primitive cell must have one half the volume of the crystallographic unit cell of Fig. 13. Consideration of the primitive cell shown in Appendix F allows the atoms to be classified as equivalent or nonequivalent under primitive translations. The results are indicated in Fig. 13. A factor-group analysis carried out on the primitive cell yields the results of Table 10. The site-group and factor-group analyses are compared in Table 11, where it is apparent that none of the three internal modes of NH_2^- should be split in the infrared spectrum of the solid. The infrared spectra of both α-$Ca(NH_2)_2$ and α-$Ca(ND_2)_2$ are in agreement with this prediction. The hydrogen-stretching region of the spectrum of α-$Ca(NH_2)_2$ is shown in Fig. 14 as an example. Unfortunately, neither the far-infrared spectrum nor Raman spectrum of this compound has been observed.

Turning now to the β phase of $Ca(NH_2)_2$, the X-ray diffraction results indicate a cubic structure of the NaCl type. However, the true symmetry

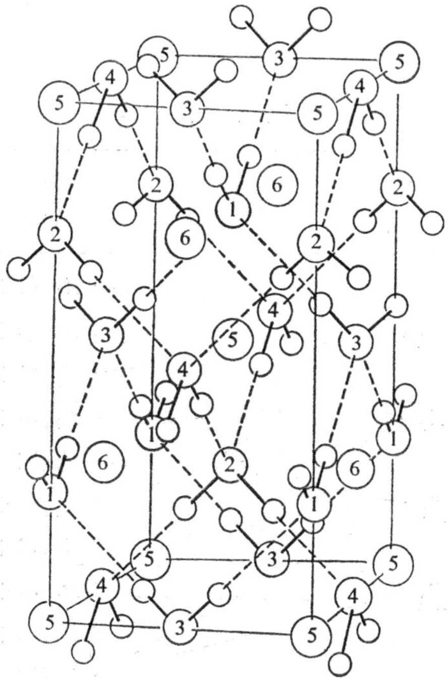

FIG 13. Crystallographic unit cell of α-$Ca(NH_2)_2$ [After (30)].

TABLE 10

$\mathscr{D}_{4h}^{19}$	$n_{\text{tot}}^{(\gamma)}$	$n_{\text{acous}}^{(\gamma)}$	$n_{\text{trans}}^{(\gamma)}$	$n_{\text{rot}}^{(\gamma)}$	$n_{\text{vib}}^{(\gamma)}$
A_{1g}	3	0	1	0	2
A_{2g}	1	0	0	1	0
B_{1g}	4	0	2	0	2
B_{2g}	1	0	0	1	0
E_g	6	0	3	2	1
A_{1u}	1	0	0	1	0
A_{2u}	4	1	1	0	2
B_{1u}	1	0	0	1	0
B_{2u}	3	0	1	0	2
E_u	6	1	2	2	1
degrees of freedom	42	3	15	12	12

TABLE 11

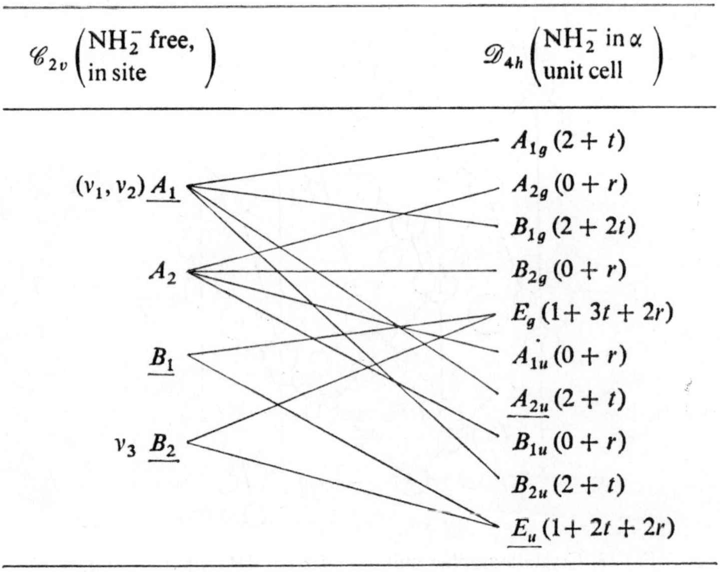

of this phase must be lower in order to accomodate the NH_2^- ions. Since, these ions have been approximately located by the X-ray study, and, the sites at which they are located cannot have symmetries higher than $\mathscr{C}_{2v}$, a unique structure can be determined from symmetry considerations, as in the case of the α phase. The resulting structure is shown in Fig. 15. If all

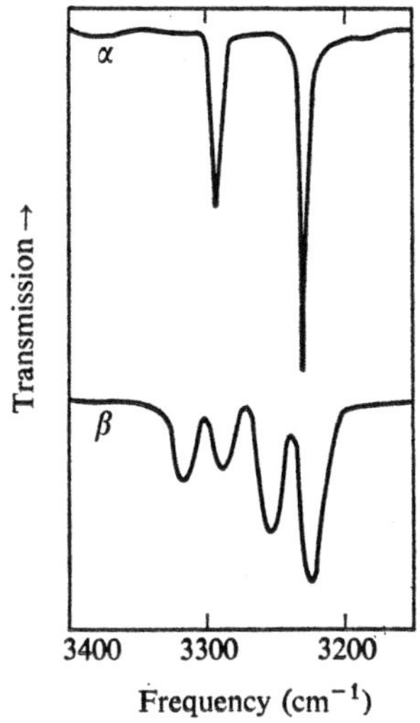

FIG. 14. Infrared spectra of α- and β-$Ca(NH_2)_2$ in the hydrogen-stretching region [From reference (30)].

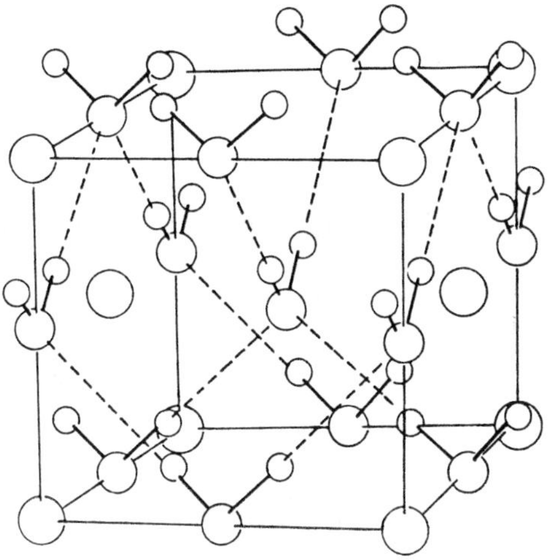

FIG. 15. Crystallographic unit cell of β-$Ca(NH_2)_2$. The cation sites are filled statistically (See text) [From reference (30).]

cation sites are occupied (or vacant), the space group is $\mathscr{C}_{4v}{}^4 \equiv P4_2\,nm$, and the cell of Fig. 15 has twice the volume of the unit cell. However, if electrical neutrality of the cell dictates the most probable cation configuration, the primitive unit cell will contain only one cation and its symmetry will be reduced to $\mathscr{C}_{2v}{}^{11}(Cmm2)$.

The above results are summarized by the correlation diagram of Table 12. The presence of two anions per primitive cell results in correlation splitting of the A_1-species fundamentals of the NH_2^- group. However, in the absence of perturbation by the cation only one component of each doublet is infrared-active. On the other hand, if one of the cation sites is vacant, both components become active. Furthermore, the two NH_2^- ions are then no longer on equivalent sites and further frequency shifts can occur. In the case of the B_2-species vibrations of NH_2^-, the two molecules in the unit cell contribute to a single doubly degenerate, infrared-active vibration, while further perturbation by the anion distribution again results in an infrared-active doublet. These predictions are in agreement with the spectrum shown in Fig. 14. It should be pointed out that the original analysis (**30**) of this problem was in error because of the use of the crystallographic, rather, than the primitive cell. Thus, although the proposed structures of α- and β-$Ca(NH_2)_2$ appear to be correct, the analysis of the splitting in the spectrum of the β phase should be revised.

TABLE 12

Site	Unit cell	
$\mathscr{C}_{2v}[NH_2^-]$	$\mathscr{C}_{4v}{}^4\,[Ca_2\,(NH_2)_2{}^{2+}]$	$\mathscr{C}_{2v}{}^{11}\,[Ca\,\square\,(NH_2)_2]$
$(\nu_1, \nu_2)\,A_1$ (*IR*)	$2\,A_1$ (*IR*)	$4\,A_1$ (*IR*)
	$0\,A_2$	
	$2\,B_1$	
A_2	$0\,B_2$	$0\,A_2$
B_1 (*IR*)		$1\,B_1$ (*IR*)
	$1\,E$ (*IR*)	
$(\nu_3)B_2$ (*IR*)		$1\,B_2$ (*IR*)

It might be suggested that the NH_2^- groups in β-$Ca(NH_2)_2$ are randomly oriented. Although a general theory of the vibrational spectra of crystals containing randomly oriented groups has not been developed, the analysis by Whalley and Bertie (**31**) indicates that in such cases all internal vibrations become active. Furthermore, as the effective unit-cell size becomes ambiguous in such disordered crystals, the resultant varied and multiple splitting of each fundamental would yield a single very broad absorption feature corresponding to each internal vibration. The spectrum shown in Fig. 14 is not, therefore, consistent with a model of randomly oriented amide groups in β-$Ca(NH_2)_2$.

VIII. Structure of Crystalline Trifluoroacetonitrile

As illustrated in the earlier sections of this chapter, the interpretation of the infrared and Raman spectra of crystals depends on a knowledge of their structures, as determined, ordinarily, by X-ray diffraction. In certain favorable cases, such as the inorganic amides, symmetry considerations, are sufficient to locate hydrogen atoms, whose positions are ordinarily difficult to determine by X-ray analysis. Vibrational spectroscopy can then provide a check on the proposed structure.

In general, although infrared and Raman spectroscopies provide valuable information regarding the nature of the bonding in crystalline solids, they cannot be considered as methods of structure determination. Nevertheless, in certain cases crystal structures have been determined, or, at least, reduced to a few possibilities using only spectroscopic methods.

In a recent example Shurvell and Faniran (**32**) have reported a very complete spectroscopic study of crystalline trifluoroacetonitrile which leads to a determination of the crystal class, and the factor group, as well as the molecular site group, even in the absence of diffraction data. In the gas phase CF_3CN belongs to the point group $\mathscr{C}_{3v}$. Hence, using the methods of Chapter 2 it is found that the $3N - 6 = 12$ vibrational fundamentals are described by the structure $\mathbf{\Gamma}_{vib} = 4A_1 + 4E$, where all eight fundamentals are both infrared- and Raman-active. Although the symmetry of the site can lead to splitting of the doubly degenerate vibrations, only correlation splitting arising from the presence of two or more molecules per unit cell can result in splitting of the A_1-species fundamentals. In this case, then, infrared and Raman spectroscopic studies can provide direct information regarding the crystal structure.

The vibrational spectra of CF_3CN are summarized in Table 13, where it is apparent that fundamentals of species A_1 of the isolated molecule appear as doublets in both the infrared and Raman spectra of the crystal. On the other hand, the E-species vibrations become triplets in the infrared spectrum

TABLE 13

Species in Molecular group		IR-active fundamentals gas[a]	IR-active fundamentals solid[a]	Raman-active fundamentals gas[b]	Raman-active fundamentals liquid[c]	Raman-active fundamentals solid[a]	Approximate description
A_1	ν_1	2274·1	2283·4 2280·3	2277	2274	2277·4 2276·7	C ≡ N stretch
	ν_2	1227·2	1212·8 1209·6	1226	1222	1212·9 1209·4	Sym. CF_3 stretch
	ν_3	801·7	807·4 804·6	818	803	809·4 804·6	C—C stretch
	ν_4	521·0	522·0 520·2	520	521	523·2 520·2	Sym. CF_3 bend
E	ν_5	1214·3	1194·0 1190·2 1181·3 —	1212	1192	1196·8 1193·2 1189·9 1186·3	Asym. CF_3 stretch
	ν_6	618·3	622·9 621·1 620·0 —	620	620	621·2 620·3 619·9 619·0	Asym. CF_3 bend
	ν_7	462·3	465·5 — 463·1 —	460	463	462·5 461·4 459·9 458·2	CH_3 rock
	ν_8	196·0	—	187	192	204·8 198·4 187·9 183·8	C—C ≡ N bend

[a] reference (32) [b] reference (33) [c] reference (34)

and quadruplets in the Raman spectrum of the crystalline solid. In general, these splittings are accompanied by frequency shifts from the corresponding vibrations of the gaseous molecule.

In order to consider the possible crystal structures of CF_3CN it first should be recalled that the site group must be a subgroup of both the molecular point group and the unit-cell group. The possible subgroups of $\mathscr{C}_{3v}$ are found from Appendix D to be $\mathscr{C}_1$, $\mathscr{C}_s$, and $\mathscr{C}_3$. The last possibility is ruled out, as it implies a structure belonging to the trigonal or higher crystal systems, while in general, molecules of $\mathscr{C}_{3v}$ symmetry belong to either the monoclinic or orthorhombic systems. Furthermore, predicted infrared and Raman spectra determined from factor groups in the trigonal system are much simpler than the spectra observed in the present example.

The possible factor groups of the monoclinic and orthorhombic systems are found from Appendix E to be $\mathscr{C}_2$, $\mathscr{C}_s$, $\mathscr{C}_{2h}$, $\mathscr{D}_2$, $\mathscr{C}_{2v}$, and $\mathscr{D}_{2h}$. Since all of the observed frequencies coincide to within experimental error in the infrared and Raman spectra, the rule of mutual exclusion can be used to eliminate factor groups having centers of symmetry. Thus, the point groups $\mathscr{C}_{2h}$ and $\mathscr{D}_{2h}$ can be dropped from the above list. Of the four remaining groups only $\mathscr{C}_{2v}$ can account for a different number of components on the E-species fundamentals in the infrared and Raman spectra. The site group is, therefore, $\mathscr{C}_s$, and there must be two molecules per primitive cell. The correlation of these groups is shown in Table 14.

The space group is necessarily one of those whose factor group is $\mathscr{C}_{2v}$ and which contains sites of the type $\mathscr{C}_s(2)$. From Appendix E the possibilities are found to be limited to $\mathscr{C}_{2v}^1$, $\mathscr{C}_{2v}^2$, $\mathscr{C}_{2v}^4$, or $\mathscr{C}_{2v}^7$. A possible structure corresponding to each space group is shown in Fig. 16. Thus,

TABLE 14

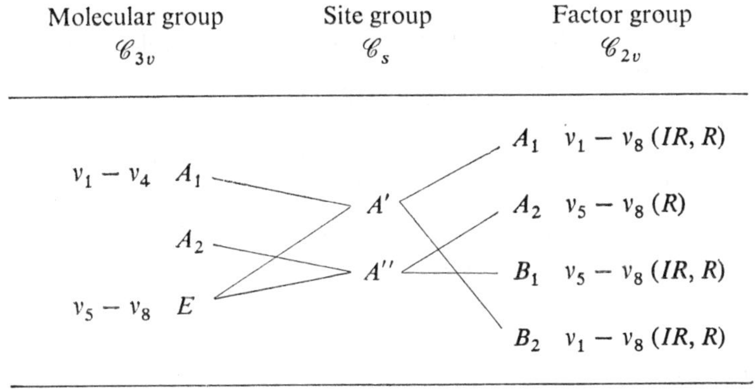
Molecular group
$\mathscr{C}_{3v}$
Site group
$\mathscr{C}_s$
Factor group
$\mathscr{C}_{2v}$
$\nu_1 - \nu_4$ A_1
A_2
$\nu_5 - \nu_8$ E
A'
A''
A_1 $\nu_1 - \nu_8$ (IR, R)
A_2 $\nu_5 - \nu_8$ (R)
B_1 $\nu_5 - \nu_8$ (IR, R)
B_2 $\nu_1 - \nu_8$ (IR, R)

although the possible space groups cannot be reduced further using spectroscopic data alone, this study is an excellent example of the use of vibrational spectroscopy in the determination of crystal structure, even in the absence of X-ray data. Unfortunately, few cases are as favorable as this one.

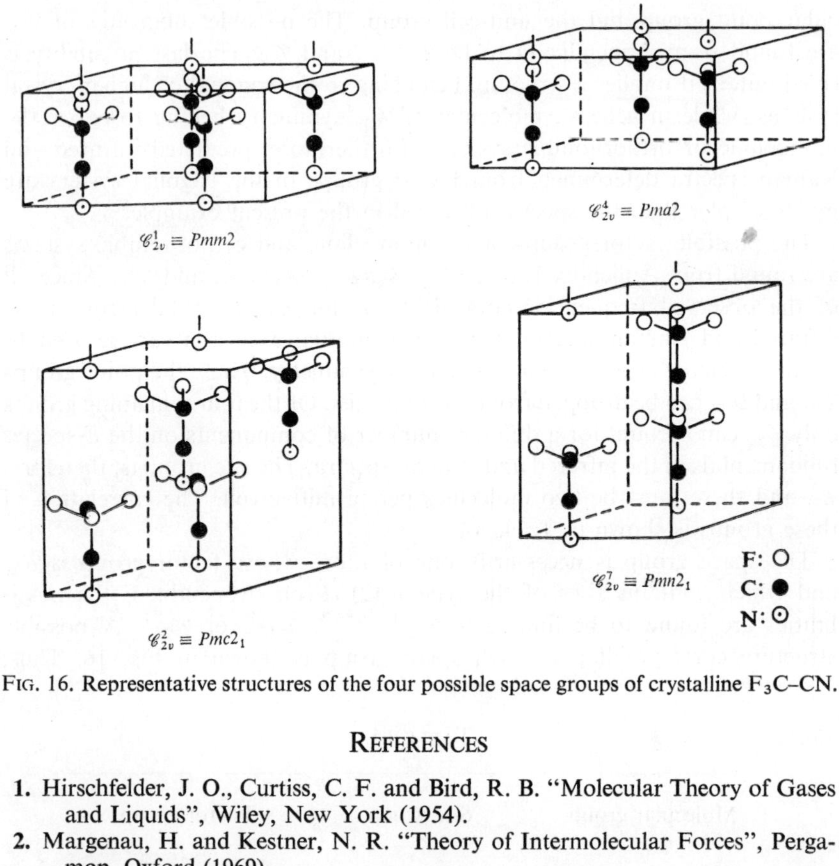

FIG. 16. Representative structures of the four possible space groups of crystalline F_3C–CN.

REFERENCES

1. Hirschfelder, J. O., Curtiss, C. F. and Bird, R. B. "Molecular Theory of Gases and Liquids", Wiley, New York (1954).
2. Margenau, H. and Kestner, N. R. "Theory of Intermolecular Forces", Pergamon, Oxford (1969).
3. Pauling, L. "Nature of the Chemical Bond", Third edition, Cornell, Ithaca (1960).
4. Coulson, C. A. "Valence", Second Edition, Oxford (1961).
5. Shimanouchi, T., Tsuboi, M. and Miyazawa, T. (1961). *J. Chem. Phys.* **35,** 1597 (1961).
6. Piseri, L. and Zerbi, G. *J. Mol. Spectry.* **26,** 254 (1968).
7. Barnes, R. B. and Czerny, M. *Z. Physik.* **72,** 447 (1931).
8. Nakagawa, I., Tsuchida, A. and Shimanoushi, T. *J. Chem. Phys.* **47,** 982 (1967).
9. Perry, C. H. *Japan. J. App. Phys. Suppl. 1,* **4,** 564 (1965).
10. Shimanouchi, T., Nakagawa, I., Hiraishi, J. and Ishii. M. *J. Mol. Spectry.* **19,** 78 (1966).

11. Dubeau. M. Thesis, Paris (1969).
12. Shearer-Turrell, S. Thesis, Bordeaux (1970).
13. Wilson, E. B., Jr., Decius, J. C. and Cross, P. C. "Molecular Vibrations", McGraw-Hill, New York (1955).
14. Cox, E. G. *Rev. Mod. Phys.* **30,** 159 (1958).
15. Harada, I. and Shimanouchi, T. *J. Chem. Phys.* **44,** 2016 (1966).
16. Hiraishi, J. and Shimanouchi, T. *Spectrochim. Acta* **22,** 1483 (1966).
17. Turrell, G. C., Jones, W. D. and Maki, A. G. *J. Chem. Phys.* **26,** 1544 (1957).
18. Shallcross, F. V. and Carpenter, G. B. *Acta Cryst.* **11,** 490 (1958).
19. Pimental, G. C. and McClellan, A. L. "The Hydrogen Bond", Freeman, San Francisco (1960).
20. Lippincott, E. R. and Schroeder, R. *J. Chem. Phys.* **23,** 1099 (1955).
21. Schroeder, R. and Lippincott, E. R. *J. Phys. Chem.* **61,** 921 (1957).
22. Lippincott, E. R. *J. Chem. Phys.* **23,** 603 (1955).
23. Moulton, W. G. and Cromhout, R. A. *J. Chem. Phys.* **25,** 34 (1956).
24. Glemser, O. and Hartert, E. *Zeit. Anorg. Allgem. Chem.* **283,** 111 (1965).
25. Pimentel G. C. and Sederholm, C. H. *J. Chem. Phys.* **24,** 639 (1956).
26. Huong, P. V. and Turrell, G. C. *J. Mol. Spectry.* **25,** 185 (1968).
27. Juza, R. and Schumacher, H. *Zeit. Anorg. Allgem. Chem.* **324,** 278 (1962).
28. Bouclier, P., Novak, A., Portier, J. and Hagenmuller, P. *Compt. Rend.* **263,** 875 (1966).
29. Novak, A., Portier, J. and Bouclier, P. *Compt. Rend.* **261,** 445 (1965).
30. Bouclier, P., Portier, J. and Turrell, G. *J. Mol. Structure* **4,** 1 (1969).
31. Whalley, E. and Bertie, J. E. *J. Chem. Phys.* **46,** 1264 (1967).
32. Shurvell, H. F. and Faniran, J. A. *J. Mol. Spectry.* **23,** 436 (1970).
33. Gullikson, C. W. and Nielsen, J. R. *J. Mol. Spectry.* **1,** 155 (1954).
34. Edgell, W. F. and Potter, R. M. *J. Chem. Phys.* **24,** 80 (1956).

Chapter 7

Infrared and Raman Spectra of Polymers

The general problem of the vibration of three-dimensional crystalline solids was introduced in Chapter 3 by first considering the vibrations of linear chains. It was shown that the vibrational frequencies of a chain are related to the wave number by a set of dispersion curves. The wave number specifies the phase shift between corresponding vibrations of successive repeat units, or "unit cells," in these one-dimensional structures.

In the case of infinite chains the vibrational frequencies are continuous functions of the wave number. On the other hand, for chains of finite length certain boundary conditions must be introduced which limit the vibrational frequencies to discrete values. The number of allowed frequencies is found to depend on the length of the chain. These results were obtained in Chapter 3 using the model of a monatomic chain with fixed ends, as well as by imposing the cyclic boundary condition. However, they are quite general, as will be illustrated below for some more complex examples.

There are two types of chemical system which can be well represented by simple chain models. First, in crystalline solids the molecular or complex ions are often arranged in chain-like structures in which the interaction between successive molecules or ions along the chain is much stronger than the forces between chains. A second example is the case of organic polymers such as polyethylene or nylon, in the amorphous state or in solution. Even in crystalline high polymers the chain approximation is applicable, as the forces between chains are usually weak. The vibrations of these two types of system will be treated in the present chapter using a number of examples from recent publications.

I. Coupled-Oscillator Model

Consider a linear chain of parallel dipoles separated by a fixed distance d, as shown in Fig. 1. Each dipole will be assumed to have a natural oscillator frequency ν_0. If, for example, each dipole represents a diatomic molecule, its frequency is given by $\lambda_0 = 4\pi^2 \nu_0{}^2 = gf = f/\mu$, where f is its force constant and μ is its reduced mass. If the dipoles are allowed to interact, their coupling can be represented by a force constant f_1. In the event dipole–dipole forces are responsible for the coupling, f_1 can be related to dipole–moment derivatives, as shown in the following section.

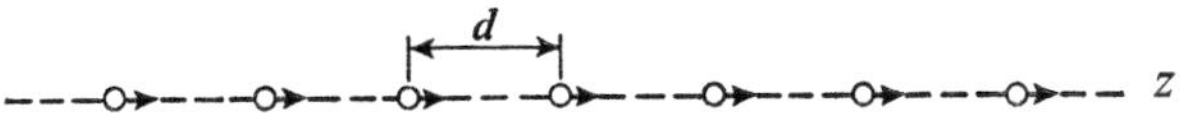

FIG. 1. Linear chain of parallel dipoles.

The **F** and **G** matrices for the infinite dipole chain take the form

$$\mathbf{F} = \begin{pmatrix} \ddots & \ddots & & & \\ \ddots & f & f_1 & & 0 \\ & f_1 & f & f_1 & \\ & & f_1 & f & f_1 \ddots \\ & 0 & & f_1 \ddots & f \ddots \end{pmatrix} \tag{1}$$

and

$$\mathbf{G} = \begin{pmatrix} \ddots & & & & \\ & g & & & 0 \\ & & g & & \\ & & & g & \\ & 0 & & & g \ddots \end{pmatrix}, \tag{2}$$

respectively. For the infinite or cyclic chain the translational symmetry allows the **F** matrix to be written

$$F(k) = f + f_1 \, \mathrm{e}^{-2\pi ikz} + f_1 \, \mathrm{e}^{+2\pi ikz}, \tag{3}$$

as in Eqn (81), Chap. 3. The corresponding **G** matrix consists of the single element g. Thus, the dispersion relation is found from the expression

$$\begin{aligned} \lambda = 4\pi^2\nu^2 &= GF(k) = g(f + f_1 \, \mathrm{e}^{-2\pi ikz} + f_1 \, \mathrm{e}^{+2\pi ikz}) \\ &= g(f + 2f_1 \cos \eta), \end{aligned} \tag{4}$$

where $\eta \equiv 2\pi kz$ is the phase shift between vibrations of adjacent dipoles.

For the infinite chain, ν is a continuous function of η. In the case of a finite chain, Eqn (4) is also valid, although η is then restricted to discrete values.

Consideration of the boundary conditions for a finite linear chain of $\mathcal{N}$ oscillators leads to the relation [See Eqn (30), Chap. 3]

$$\eta = \frac{\pi s}{n + 1} \tag{5}$$

where $s = 1, 2, \ldots, \mathcal{N}$. Thus, for example, the three vibrational frequencies, ν_{3s}, of a linear chain of three coupled oscillators are given by

$$\begin{aligned} \lambda_{31} &= 4\pi^2 \nu_{31}{}^2 = g(f + \sqrt{2} f_1), \\ \lambda_{32} &= 4\pi^2 \nu_{32}{}^2 = g f, \\ \lambda_{33} &= 4\pi^3 \nu_{33}{}^2 = g(f - \sqrt{2} f_1). \end{aligned} \tag{6}$$

With the matrices of Eqns (1) and (2) limited to rank three, the eigenvalues given in Eqn (6) can be substituted into the original secular equations $\mathbf{GFL} = \mathbf{L\Lambda}$ [See Eqn (34), Chap. 1] to obtain the eigenvectors

$$\mathbf{L} = \begin{pmatrix} 1/2 & 1/\sqrt{2} & 1/2 \\ 1/\sqrt{2} & 0 & -1/\sqrt{2} \\ 1/2 & -1/\sqrt{2} & 1/2 \end{pmatrix}. \tag{7}$$

These results will be used in the following section to account for the observed spectra of isotopically substituted carbonates and nitrates having aragonite structures (**1**).

II. Coupling of Nitrate and Carbonate Ions in Aragonites

In the aragonite-type lattice the CO_3^{-2} or NO_3^- nearest neighbors form linear chains with their C_3 symmetry axes colinear with the chain axis. Hence, the instantaneous dipole moments arising from the out-of-plane bending vibrations of these ions are parallel to the chain axes, as shown in Fig. 2, and the linear-dipole model considered above might be expected to serve as a basis for the interpretation of the spectra of these compounds.

The G-matrix element for the out-of-plane vibration of an XY_3 species is given by $g = 3\mu_X + \mu_Y$ (See Appendix A), and the frequency of the single infrared-active fundamental of an infinite chain of such oscillators becomes

$$\lambda = 4\pi^2 \nu^2 = (3\mu_X + \mu_Y)(f + 2f_1), \tag{8}$$

where the condition for infrared activity, $\eta = 0$, has been imposed on Eqn

(4). It is apparent that the frequency observable in the infrared spectrum differs from the vibrational frequency of the unperturbed oscillator because of the coupling represented by the constant f_1.

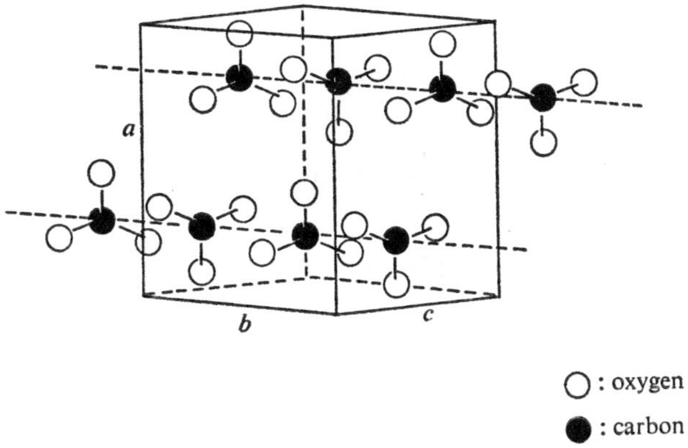

FIG. 2. Disposition of the carbonate ions in the aragonite crystal.

If f_1 is attributable to dipole–dipole coupling, it can be evaluated by considering the interaction potential of two dipoles of moments $\mathfrak{m}_i$ and $\mathfrak{m}_j$ separated by a distance d_{ij},

$$V_{ij} = -\frac{\mathfrak{m}_i\,\mathfrak{m}_j}{d_{ij}^{\,3}}(2\cos\theta_i\cos\theta_j - \sin\theta_i\sin\theta_j\cos\chi_{ij}), \tag{9}$$

where the angles θ_i, θ_j, and χ_{ij} describe the relative orientation of the two dipoles [(**1**), p. 27]. If each moment is expanded in internal coordinates, the harmonic contribution to the interaction potential becomes

$$V_{ij} = -\frac{1}{d_{ij}^{\,3}}\left(\frac{\partial\mathfrak{m}_i}{\partial S_i}\right)_0\left(\frac{\partial\mathfrak{m}_j}{\partial S_j}\right)_0(2\cos\theta_i\cos\theta_j - \sin\theta_i\sin\theta_j\cos\chi_{ij})\,S_iS_j. \tag{10}$$

Thus, for example, the force constant for the interaction of two identical, parallel dipoles separated by a distance d becomes

$$f_1 = -\frac{2}{d^3}\left(\frac{\partial\mathfrak{m}}{\partial S}\right)_0^2. \tag{11}$$

On the basis of this model, then, the coupling constant would be expected to be negative.

This model of dipole–dipole coupling in nitrates and carbonates was used by Decius to interpret the spectra of isotropic mixtures involving $^{15}NO_3^-$

and $^{13}CO_3^{-2}$. In a mixture of, say, $^{15}NO_3^-$ and $^{14}NO_3^-$ a given nitrate chain would be expected to consist of groups of one, two, three, etc., successive $^{15}NO_3^-$ ions separated by ions of the lighter species, and similar groups of $^{14}NO_3^-$ ions with heavier ions at each end. The vibrational frequencies of a group of three identical ions, for example, would be given in the first approximation by Eqns (6).

In order to take into account the interaction of the group of three ions with the terminal ions, one must consider the **G** matrix

$$\mathbf{G} = \begin{pmatrix} g' & 0 & 0 & 0 & 0 \\ 0 & g & 0 & 0 & 0 \\ 0 & 0 & g & 0 & 0 \\ 0 & 0 & 0 & g & 0 \\ 0 & 0 & 0 & 0 & g' \end{pmatrix}, \tag{12}$$

where g' is the matrix element for a terminal ion. The **F** matrix in this case becomes

$$\mathbf{F} = \begin{pmatrix} f & f_1' & 0 & 0 & 0 \\ f_1' & f & f_1 & 0 & 0 \\ 0 & f_1 & f & f_1 & 0 \\ 0 & 0 & f_1 & f & f_1' \\ 0 & 0 & 0 & f_1' & f \end{pmatrix}, \tag{13}$$

where f_1' accounts for the coupling between the group and the terminal ions. Clearly, f_1' is numerically equal to f_1. The effect of f_1' on the vibrational frequencies can be estimated using second-order perturbation theory [(3), p. 230]. The corrections to the eigenvalues are found to be

$$\lambda_{3k}'' = (L_{1k}^2 + L_{3k}^2)\frac{g\,g' f_1^2}{\lambda_{3k} - g'f}, \tag{14}$$

where the L_{ik}'s are the element of **L** given by Eqn (7). The above result can be generalized for the coupling of a chain of $\mathcal{N}$ ions of one species with terminal ions of the other. Thus,

$$\lambda_{\mathcal{N}k}'' = (L_{1k}^2 + L_{\mathcal{N}k}^2)\frac{g\,g' f_1^2}{\lambda_{\mathcal{N}k} - g'f} = (L_{1k}^2 + L_{\mathcal{N}k}^2)\,\rho_{\mathcal{N}k}. \tag{15}$$

Expressions for the unperturbed eigenvalues, their second-order corrections, and the eigenvectors are summarized in Table 1 for $\mathcal{N} = 1$ through $\mathcal{N} = 4$.

It is important to note that the sign of the second-order correction given by Eqn (15) depends on the relative magnitudes of the quantities $\lambda_{\mathcal{N}k}$ and $g'f$. Thus, for a chain of lighter ions terminated by heavier ones, $\lambda_{\mathcal{N}k} > g'f$, and $\lambda''_{\mathcal{N}k} > 0$, while if the terminal ions are lighter than the chain ions, $\lambda_{\mathcal{N}k} < 0$.

TABLE 1 [From reference (1)]

$\mathcal{N}$	k	λ	λ''[a]	L
1		gf	2ρ	1
2	1	$g(f+f_1)$	ρ	$\begin{pmatrix} 2^{-\frac{1}{2}} & 2^{-\frac{1}{2}} \\ 2^{-\frac{1}{2}} & -2^{-\frac{1}{2}} \end{pmatrix}$
	2	$g(f-f_1)$	ρ	
3	1	$g(f+2^{\frac{1}{2}}f_1)$	$\frac{1}{2}\rho$	$\begin{pmatrix} \frac{1}{2} & 2^{-\frac{1}{2}} & \frac{1}{2} \\ 2^{-\frac{1}{2}} & 0 & -2^{-\frac{1}{2}} \\ \frac{1}{2} & -2^{-\frac{1}{2}} & \frac{1}{2} \end{pmatrix}$
	2	gf	ρ	
	3	$g(f-2^{\frac{1}{2}}f_1)$	$\frac{1}{2}\rho$	
4	1	$g[f+(3+5^{\frac{1}{2}}/2)f_1]$	$0{\cdot}2764\rho$	$\begin{pmatrix} 0{\cdot}3718 & 0{\cdot}6015 & 0{\cdot}6015 & 0{\cdot}3718 \\ \cdots & \cdots & \cdots & \cdots \\ \cdots & \cdots & \cdots & \cdots \\ \cdots & \cdots & \cdots & \cdots \end{pmatrix}$
4	2[b]	$g[f+(3-5^{\frac{1}{2}}/2)f_1]$	...	
	3	$g[f-(3-5^{\frac{1}{2}}/2)f_1]$	...	
	4	$g[f-(3+5^{\frac{1}{2}}/2)f_1]$	...	

[a] $\rho = gg'(f')^2/(\lambda_{\mathcal{N}k} - g'f)$.

[b] Modes ω_{42}, ω_{43}, and ω_{44} were not calculated in detail, since their contributions to the over-all intensity are small.

The vibrational amplitudes of the terminal oscillators can also be estimated from the perturbation treatment. It is found that

$$L''_{0,k} = \frac{L_{1k}\, g' f_1}{\lambda_{\mathcal{N}k} - g'f} \tag{16}$$

and

$$L''_{\mathcal{N}+1,k} = \frac{L_{\mathcal{N}k}\, g' f_1}{\lambda_{\mathcal{N}k} - g'f}\,. \tag{17}$$

By combining these results with the approximate amplitudes of the chain oscillators (the elements of **L** in Table 1), one can obtain the forms of the various modes, as shown in Fig. 3.

As indicated in Chapter 1, the absorption intensity of each band in the infrared spectrum depends on the total change in dipole moment produced by the corresponding mode of motion. In Fig. 3 it is seen that for several of the modes (e.g., ω_{22} and ω_{32}) the phase relations are such that the dipole–moment derivates cancel. These modes will not, therefore, result in absorption. It is apparent that for each chain length it is the lowest frequency ($k = 1$) which is the strongest absorber.

The terminal ions also contribute to the total intensity of a given absorption band. Thus, consideration of the vibration ω_{31}, for example, shows that because of coupling with the end ions, the chain of ^{15}N nitrates should exhibit stronger absorption than the corresponding chain of ^{14}N nitrates. The different phase relations in the two cases are the direct result of the negative sign of f_1, since $L''_{0k} = L_{1k} g' f_1/(\lambda_{3k} - g'f)$ will have the same sign as L_{1k} when $f_1 < 0$ and $\lambda_{3k} < g'f$.

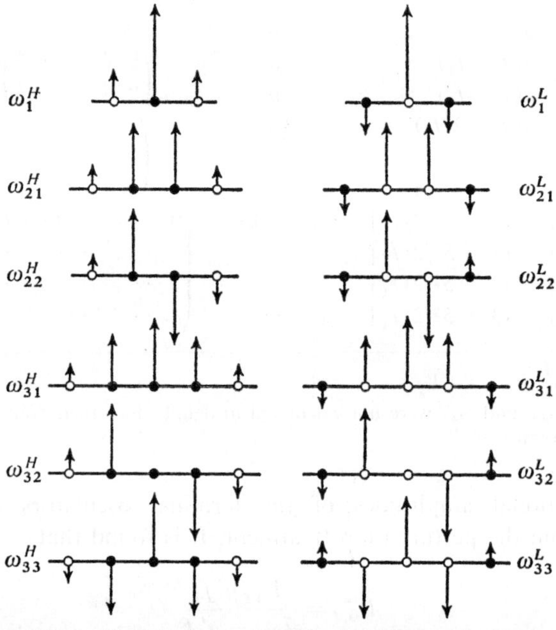

FIG. 3. Normal modes $\omega_{\mathcal{N}k}$ for one, two, and three molecules of one isotopic species surrounded by other species. Arrows indicate relative amplitudes schematically; L is the lighter isotope, and H is the heavier isotope [From reference (**1**)].

In order to make a quantitative estimate of the vibrational absorption intensities of the various chains it is necessary to evaluate the change in dipole moment arising from each mode shown in Fig. 3. Using Eqn (76), Chap. 1, column matrix **A** of elements $\left(\frac{\partial \mathfrak{m}}{\partial Q_k}\right)_0$ can be found from the column matrix $\mathfrak{P}$ of the quantities $\left(\frac{\partial \mathfrak{m}}{\partial S_i}\right)_0$ using

$$\mathbf{A} = \mathbf{L}^{\dagger} \mathfrak{P}. \tag{18}$$

Then, the intensity of the kth fundamental is proportional to

$$A_k^2 = \sum_i \mathfrak{P}_i L_{ik} \sum_j \mathfrak{P}_j L_{jk}$$

$$\approx \mathfrak{P}^2 \sum_i \sum_j L_{ik} L_{jk}$$

$$= \mathfrak{P}^2 \left(\sum_{i=0}^{\mathcal{N}+1} L_{ik} \right)^2, \tag{19}$$

where the slight effect of isotopic substitution on the vibrational amplitudes has been neglected.

Finally, in order to calculate the intensities it is necessary to determine the distribution of groups of one, two, three, etc., ions as a function of isotopic concentration. As shown by Flory (**4**), if the fraction of isotopic ions of one kind is γ, the total number of groups having $\mathcal{N}$ members is given by

$$g_{\mathcal{N}}(\gamma) = \gamma^{\mathcal{N}}(1-\gamma)^2. \tag{20}$$

Hence, the relative intensity of the kth fundamental of a chain of $\mathcal{N}$ ions is given by

$$\mathcal{I}_{\mathcal{N},k}(\gamma) = g_{\mathcal{N}}(\gamma) \left(\frac{\partial \mathfrak{m}}{\partial Q_k} \right)^2 = g_{\mathcal{N}}(\gamma)\, A_k^2$$

$$= g_{\mathcal{N}}(\gamma)\, \mathfrak{P}^2 \left(\sum_{i=0}^{\mathcal{N}+1} L_{ik} \right)^2. \tag{21}$$

From an infrared spectroscopic study of pure $K^{14}NO_3$ and $K^{15}NO_3$, Decius calculated the force constants $f = 1{\cdot}48$ md/Å and $f_1 = -0{\cdot}015$ md/Å. He then found the vibrational frequencies of the various modes of groups of one, two, three, etc., ions of each isotopic species using the results given in Table 1. Finally, he applied Eqn (21) to determine relative intensities using Eqn (20) to evaluate $g_{\mathcal{N}}(\gamma)$ for various isotopic mixtures.

The calculated results are compared in Fig. 4 with the experimental spectra of mixtures containing 20% and 50% ^{15}N. As pointed out above, only the lowest frequency mode of each chain is found to have a significant intensity. The frequency of this mode rapidly approaches the frequency of the corresponding pure compound as the chain length increases. Finally, it is evident from the spectrum of 50% $K^{15}NO_3$ that the heavier chain is the more intense absorber, in agreement with the theoretical result.

A value of

$$\mathfrak{P} = \left(\frac{\partial \mathfrak{m}}{\partial S} \right)_0 = 1{\cdot}6 \text{ Debye/Å}$$

for the out-of-plane bending mode of NO_3^- was calculated from the experimental value of f_1 using Eqn (8), with $d = 3{\cdot}225$ Å. In principle, $\mathfrak{P}$ can also be obtained from absolute intensity measurements. However, as pointed out in Chapter 5, this procedure involves some experimental difficulty.

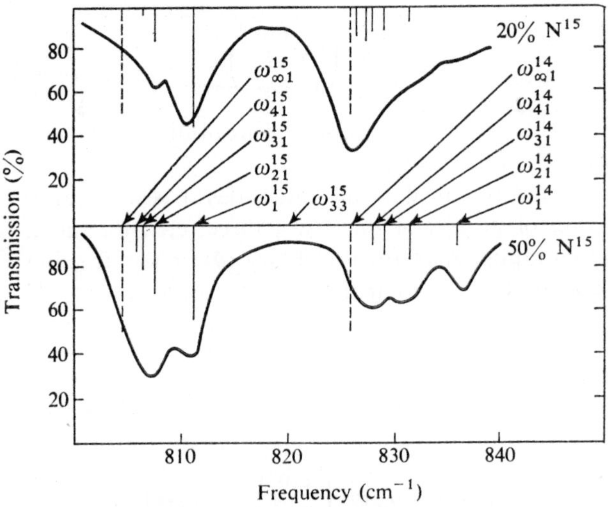

FIG. 4. Infrared absorption of KNO_3 containing 20 per cent and 50 per cent ^{15}N as Nujo emulsions. The solid vertical lines indicate theoretical frequencies and relative intensities. Dotted lines indicate frequencies for isotopically pure crystals [After Decius (**1**)].

This study of coupling in a linear-chain system provides an extremely convincing demonstration of the influence of dipole–dipole interactions on both frequencies and absorption intensities. Furthermore, the spectroscopic effects of the chain length are clearly illustrated. The latter subject will be reconsidered later in this chapter in connection with analyses of the spectra of *n*-paraffins.

III. CHAIN SYMMETRY AND LINE GROUPS: POLYETHYLENE

In the linear-dipole model considered in the previous section the one-dimensional repeat unit or motive is a dipole confined to the chain axis. Thus, aside from the translational symmetry characteristic of an infinite or cyclic chain, the only element of symmetry is the identity, E. The one-dimensional space group, or *line group*, of the infinite chain is then just the infinite translation group, and the factor group is isomorphic with the point group $\mathscr{C}_1$, which consists of the single operation E. It is easily seen that only two line groups can be constructed using one-dimensional repeat units. Their factor groups are isomorphic with the point groups $\mathscr{C}_i$ or $\mathscr{C}_1$ for motives whose symmetries are described, respectively, by operations E and i, or E alone.

It is shown by Bhagavantam and Venkatarayuda [(**5**), p. 17–19] that seven line groups can be formed using two-dimensional motives. In this case each factor group is isomorphic with one of five different point groups. From three-dimensional motives these authors developed 31 line groups, whose factor groups can be classified using eight different point groups. They did not, however, consider motives having higher than two-fold symmetry about the chain axis.

TABLE 2

Factor group order	Symmetry elements of factor group	Isomorphic point group
	(a) Point symmetry operations only	
1	E	$\mathscr{C}_1$
2	E, i	$\mathscr{C}_i$
2	E, σ_h	$\mathscr{C}_s$
2	$E, \sigma_v{}'; E, \sigma_v{}''$	$\mathscr{C}_s$
2	$E, C_2{}'; E, C_2{}''$	$\mathscr{C}_2$
p	$E, (p-1)\, C_p$	$\mathscr{C}_p$
$2p$	$E, (p-1)\, C_p, p\sigma_v$	$\mathscr{C}_{pv}$
4	$E, \sigma_h, \sigma_v{}', C_2{}''; E, \sigma_h, \sigma_v{}'', C_2{}'$	$\mathscr{C}_{2v}$
4	$E, \sigma_v{}', C_2{}', i; E, \sigma_v{}'', C_2{}'', i$	$\mathscr{C}_{2h}$
4	E, σ_h, C_2, i	$\mathscr{C}_{2h}$
4	$E, C_2, C_2{}', C_2{}''$	$\mathscr{D}_2 \equiv \mathscr{V}$
8	$E, \sigma_h, \sigma_v{}', \sigma_v{}'', C_2, C_2{}', C_2{}'', i$	$\mathscr{D}_{2h} \equiv \mathscr{V}_h$
	(b) Point symmetry operations, screw rotations, and glide reflections	
2	$E, \sigma_v{}'^{\kappa}, E, \sigma_v{}''^{g}$	$\mathscr{C}_s$
p	$E, (p-1)\, C_p{}^S$	$\mathscr{C}_p$
4	$E, \sigma_v{}'^{g}, \sigma_v{}'', C_2{}^S; E, \sigma_v{}', \sigma_v{}''^{g}, C_2{}^s$	$\mathscr{C}_{2v}$
4	$E, \sigma_v{}'^{g}, \sigma_v{}''^{g}, C_2$	$\mathscr{C}_{2v}$
4	$E, \sigma_h, \sigma_v{}'^{g}, C_2{}''; E, \sigma_h, \sigma_v{}''^{g}, C_2{}'$	$\mathscr{C}_{2v}$
4	$E, \sigma_v{}'^{g}, C_2{}', i; E, \sigma''^{g}, C_2{}'', i$	$\mathscr{C}_{2h}$
4	$E, \sigma_h, C_2{}^s, i$	$\mathscr{C}_{2h}$
4	$E, C_2{}^S, C_2{}', C_2{}''$	$\mathscr{D}_2 \equiv \mathscr{V}$
8	$E, \sigma_h, \sigma_v{}', \sigma_v{}''^{g}, C_2{}^s, C_2{}', C_2{}'', i; E, \sigma_h, \sigma'^{g}, \sigma_v{}'', C_2{}^s, C_2{}', C_2{}'', i$	$\mathscr{D}_{2h} \equiv \mathscr{V}_h$
8	$E, \sigma_h, \sigma_v{}'^{g}, \sigma_v{}''^{g}, C_2, C_2{}', C_2{}'', i$	$\mathscr{D}_{2h} \equiv \mathscr{V}_h$

The results of Bhagavantam and Venkatarayuda were generalized by Zbinden [(6), p. 47] to include motives possessing p-fold rotation axes parallel to the chain direction. The possible factor-group symmetry operations are given in Table 2. An important distinction is made between factor groups containing only point-symmetry operations and those involving screw rotations and glide reflections. Factor groups of the latter type apply only to infinite or cyclic chains, while those containing only point-symmetry operations apply to finite, noncyclic chains as well.

As an example of the application of line-group analysis, consider the infinite polyethylene chain shown in Fig. 5. The carbon skeleton of this chain is assumed to have a planar zigzag structure. The chain axis is a two-fold screw axis, as well as the glide direction of the operation $\sigma^g(xz)$. The line-group operations are E, $C_2^s(z)$, $C_2(y)$, $C_2(x)$, $i, \sigma(xy)$, $\sigma^g(xz)$, and $\sigma(yz)$, corresponding to the symmetry elements indicated in Fig. 5. Thus, $\mathscr{D}_{2h}$ is the point group which is isomorphic with the factor group, and the 18 degrees of freedom of the six-atom repeat unit can be classified as shown in Table 3 using the method of Chapter 4.

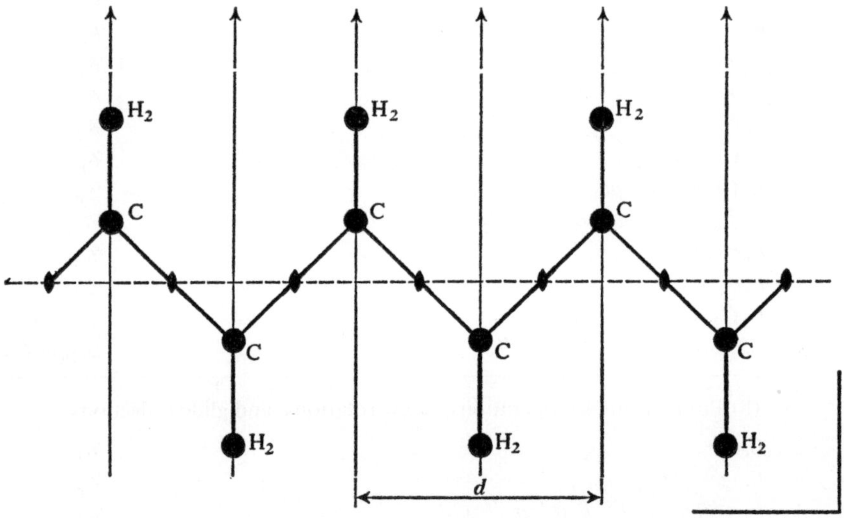

FIG. 5. Symmetry properties of an extended polyethylene chain.

Since the chain has been assumed to be isolated, the external motions of translation in the x or y directions, as well as the rotation about the z axis, have zero frequencies and can be removed, as for a gaseous molecule. These degrees of freedom are sometimes referred to as nongenuinc vibrations. Translation in the z direction is analogous to the acoustical

TABLE 3

$\mathscr{D}_{2h}$	E	$C_2(z)$	$C_2(y)$	$C_2(x)$	i	$\sigma(xy)$	$\sigma(zx)$	$\sigma(yz)$			$n_{\text{tot}}^{(\gamma)}$	$n_{\text{rot}}^{(\gamma)}$	$n_{\text{trans}}^{(\gamma)}$	$n_{\text{vib}}^{(\gamma)}$	
A_g	1	1	1	1	1	1	1	1		$\alpha_{xx}, \alpha_{yy}, \alpha_{zz}$	3	0	0	3	$(\nu, \delta, \delta_{CC})$
B_{1g}	1	1	−1	−1	1	1	−1	−1	R_z	α_{xy}	1	0	0	1	(τ)
B_{2g}	1	−1	1	−1	1	−1	1	−1	R_y	α_{zx}	3	1	0	2	(ν, ρ)
B_{3g}	1	−1	−1	1	1	−1	−1	1	R_x	α_{yz}	2	0	0	2	(ω, ν_{CC})
A_u	1	1	1	1	−1	−1	−1	−1			1	0	0	1	(τ)
B_{1u}	1	1	−1	−1	−1	−1	1	1	T_z		3	0	1	2	(ν, δ)
B_{2u}	1	−1	1	−1	−1	1	−1	1	T_y		2	0	1	1	(ω)
B_{3u}	1	−1	−1	1	−1	1	1	−1	T_x		3	0	1	2	(ν, ρ)
m_j	6	2	0	0	0	6	0	2							
χ_j	18	−2	0	0	0	6	0	2							

vibrations of a crystal lattice. Thus, for in-phase motion of all repeat units, this frequency is also zero.

The distribution of the remaining, genuine vibrations is shown in the last column of Table 3. Approximate descriptions of these various modes are given in Fig. 6. However, it should be recalled that these results are derived from symmetry arguments applied to the internal coordinates. Hence, for species containing more than one genuine vibration the actual normal modes are linear combinations of the symmetry coordinates suggested by Fig. 6. This mixing of internal coordinates is often neglected in the first approximation.

As demonstrated in Chapter 4, selection rules for infrared and Raman fundamentals can be determined on the basis of the factor group. Hence, for the infinite polyethylene chain the results shown in Table 3 predict five infrared-active fundamentals and eight modes active in the Raman effect.

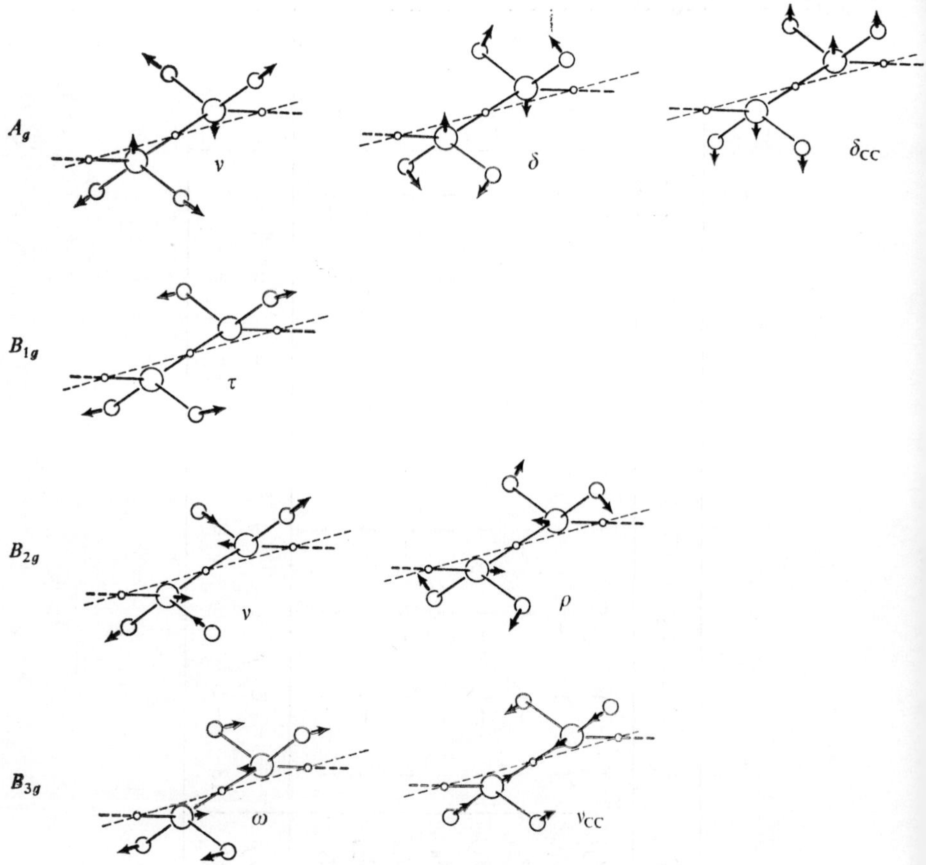

FIG. 6. Approximate forms of the normal modes of vibration of polyethylene.

IV. Finite Chains: n-Paraffins

The vibrations of normal paraffins have been considered by several authors (**7–11**). The present treatment will follow the analysis by Snyder, who developed a simple coupled-oscillator model to account for the vibrations of hydrocarbon chains of finite length.

Hydrocarbon chains having an even or odd number of carbon atoms can be distinguished by their point symmetries. Thus, as shown in Fig. 7, point group $\mathscr{C}_{2v}$ or $\mathscr{C}_{2h}$ serves respectively, as the basis for the vibrational analysis of chains made up of an even or odd number of carbon atoms. Here again, it is assumed that the chains are fully extended, that is, that the carbon skeleton has a planar zigzag configuration. Furthermore, all interaction between chains is neglected.

Since a hydrocarbon chain containing n carbon atoms has $3n + 2$ atoms in total, $9n$ vibrational degrees of freedom are expected. The vibrational analysis can be carried out in the usual manner, as summarized in

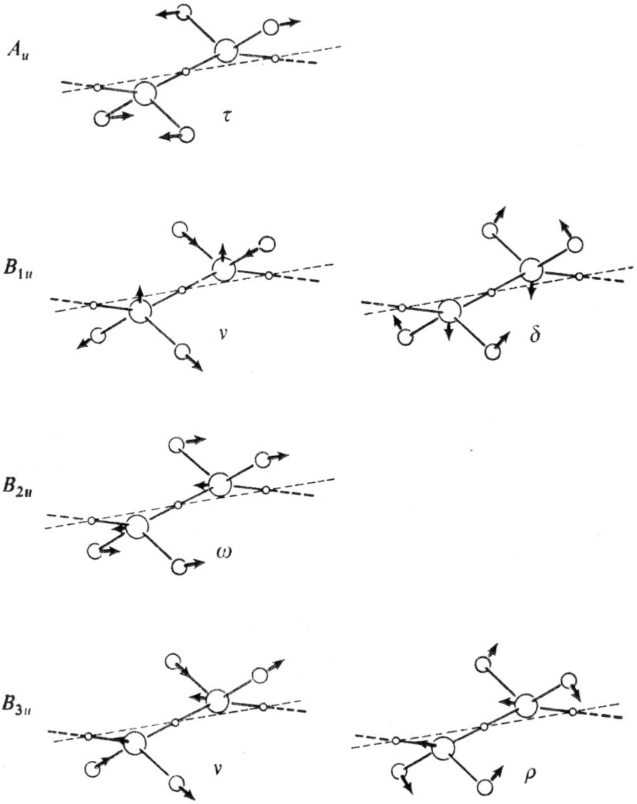

TABLE 4

$\mathscr{C}_{2v}$ (n odd)	E	C_2	$\sigma(zx)$	$\sigma(yz)$		$n_{tot}^{(\gamma)}$	$n_{vib}^{(\gamma)}$
A_1	1	1	1	1	$T_z, \alpha_{xx}, \alpha_{yy}, \alpha_{zz}$	$\frac{5}{2}(n+1)$	$\frac{1}{2}(5n+3)$
A_2	1	1	-1	-1	R_z, α_{xy}	$2n$	$2n-1$
B_1	1	-1	1	-1	$T_x, R_y \alpha_{zx}$	$2(n+1)$	$2n$
B_2	1	-1	-1	1	$T_y, R_x \alpha_{yz}$	$\frac{1}{2}(5n+3)$	$\frac{1}{2}(5n-1)$
χ_j	$9n+6$	-1	3	$n+2$			

TABLE 5

$\mathscr{C}_{2h}$ (n even)	E	C_2	i	σ_v		$n_{tot}^{(\gamma)}$	$n_{vib}^{(\gamma)}$
A_g	1	1	1	1	$R_x, \alpha_{xx}, \alpha_{yy}$	$\frac{1}{2}(5n+4)$	$\frac{1}{2}(5n+2)$
B_g	1	-1	1	-1	$R_y, R_z, \alpha_{zz}, \alpha_{yz}$	$2n+1$	$2n-1$
A_u	1	1	-1	-1	$T_x, \alpha_{xy}, \alpha_{zx}$	$2n+1$	$2n$
B_u	1	-1	-1	1	T_y, T_z	$\frac{1}{2}(5n+4)$	$\frac{5}{2}n$
χ_j	$9n+6$	0	0	$n+2$			

Tables 4 and 5. It is seen that chains having even or odd numbers of carbon atoms have quite different selection rules. For example, from the last column of Table 4 is it found that when n is odd, $7n+1$ vibrational fundamentals should be infrared-active. On the other hand, for even values of n, Table 5 predicts $\frac{9}{2}n$ infrared-active fundamentals.

The internal coordinates used to describe the vibrations of hydrocarbon chains can be classified by symmetry, as shown in Table 6. This procedure allows the internal coordinates of the two types of finite chain to be correlated with those of the infinite chains. Furthermore, the possibility of coordinate mixing in the various normal modes can be easily detected.

In the coupled-oscillator model each type of coordinate (CH_2 rock, CH_2 wag, etc.) of each of the $\mathscr{N} = n-2$ methylene groups is replaced by an oscillator having the vibrational frequency of the corresponding mode of the infinite chain. Each set of such coupled oscillators might then be expected to give an approximate description of the analogous set of normal modes of a finite chain. It is assumed in this first approximation that no mixing occurs between the various sets. Furthermore, all interactions with the methyl end-groups are neglected.

TABLE 6

Group	Mode	No. of coord	$\mathscr{C}_{2v}$ (n odd) $A_1[R(p), IR(z)]$	$A_2[R(d)]$	$B_1[R(d), IR(x)]$	$B_2[R(d), IR(y)]$	$\mathscr{C}_{2h}$ (n even) $A_g[R(p)]$	$B_g[R(d)]$	$A_u[IR(x)]$	$B_u[IR(yz)]$
methyl	stretch	6	2	1	1	2	2	1	1	2
	bend	6	2	1	1	2	2	1	1	2
	rock + wag	4	1	1	1	1	1	1	1	1
	torsion	2	0	1	1	0	0	1	1	0
methylene	stretch (ν)	$2(n-2)$	$\frac{1}{2}(n-1)$	$\frac{1}{2}(n-3)$	$\frac{1}{2}(n-1)$	$\frac{1}{2}(n-3)$	$\frac{1}{2}(n-2)$	$\frac{1}{2}(n-2)$	$\frac{1}{2}(n-2)$	$\frac{1}{2}(n-2)$
	bend (δ)	$n-2$	$\frac{1}{2}(n-1)$	0	0	$\frac{1}{2}(n-3)$	$\frac{1}{2}(n-2)$	0	0	$\frac{1}{2}(n-2)$
	rock (ρ)	$n-2$	0	$\frac{1}{2}(n-3)$	$\frac{1}{2}(n-1)$	0	0	$\frac{1}{2}(n-2)$	$\frac{1}{2}(n-2)$	0
	wag (ω)	$n-2$	$\frac{1}{2}(n-3)$	0	0	$\frac{1}{2}(n-1)$	$\frac{1}{2}(n-2)$	0	0	$\frac{1}{2}(n-2)$
	twist (τ)	$n-2$	0	$\frac{1}{2}(n-1)$	$\frac{1}{2}(n-3)$	0	0	$\frac{1}{2}(n-2)$	$\frac{1}{2}(n-2)$	0
skeleton	stretch	$n-1$	$\frac{1}{2}(n-1)$	0	0	$\frac{1}{2}(n-1)$	$\frac{1}{2}n$	0	0	$\frac{1}{2}(n-2)$
	bend	$n-2$	$\frac{1}{2}(n-1)$	0	0	$\frac{1}{2}(n-3)$	$\frac{1}{2}(n-2)$	0	0	$\frac{1}{2}(n-2)$
	torsion	$n-3$	0	$\frac{1}{2}(n-3)$	$\frac{1}{2}(n-3)$	0	0	$\frac{1}{2}(n-4)$	$\frac{1}{2}(n-2)$	0
	Total	$9n$	$\frac{1}{2}(5n+3)$	$2n-1$	$2n$	$\frac{1}{2}(5n-1)$	$\frac{1}{2}(5n+2)$	$2n-1$	$2n$	$\frac{5}{2}n$

The simple coupled-oscillator model developed in the first section of this chapter can be generalized to include both kinetic and potential coupling between nearest neighbors, next-nearest neighbors, etc. The **F** and **G** matrices are then striped matrices, that is

$$\begin{aligned} f_{i,i} &\equiv f_0, \\ f_{i,i+1} &= f_{i+1,i} \equiv f_1, \\ &\ldots\ldots \end{aligned} \tag{22}$$

or, in general,

$$f_{i,i+l} = f_{i+l,i} \equiv f_l, \qquad l = 0, 1, 2, \ldots, \mathcal{N} - 1. \tag{23}$$

The elements of **G** are analogous. The elements of **F** and **G** account, respectively, for potential and kinetic-energy contributions to coupling between oscillators separated by l unit-cell dimensions. Normally, for

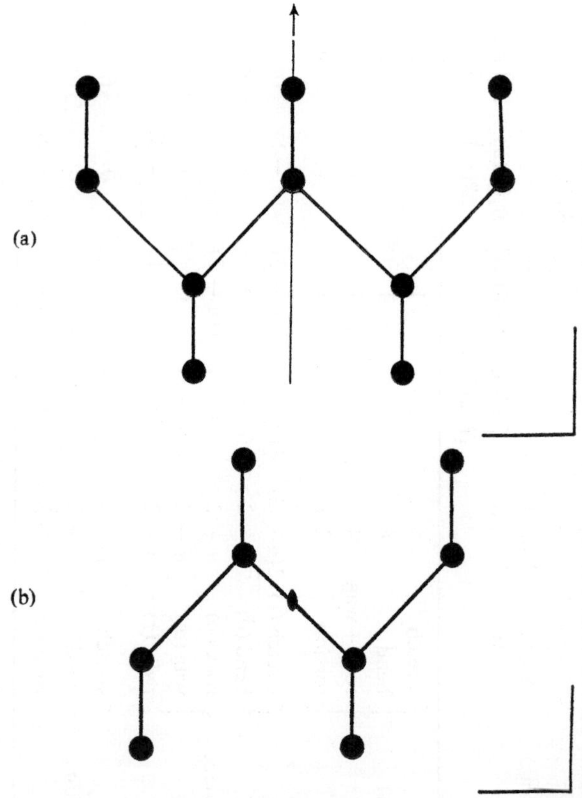

FIG. 7. Symmetry properties of finite hydrocarbon chains (a) n odd (symmetry $\mathscr{C}_{2v}$), (b) n even (symmetry $\mathscr{C}_{2h}$).

TABLE 7

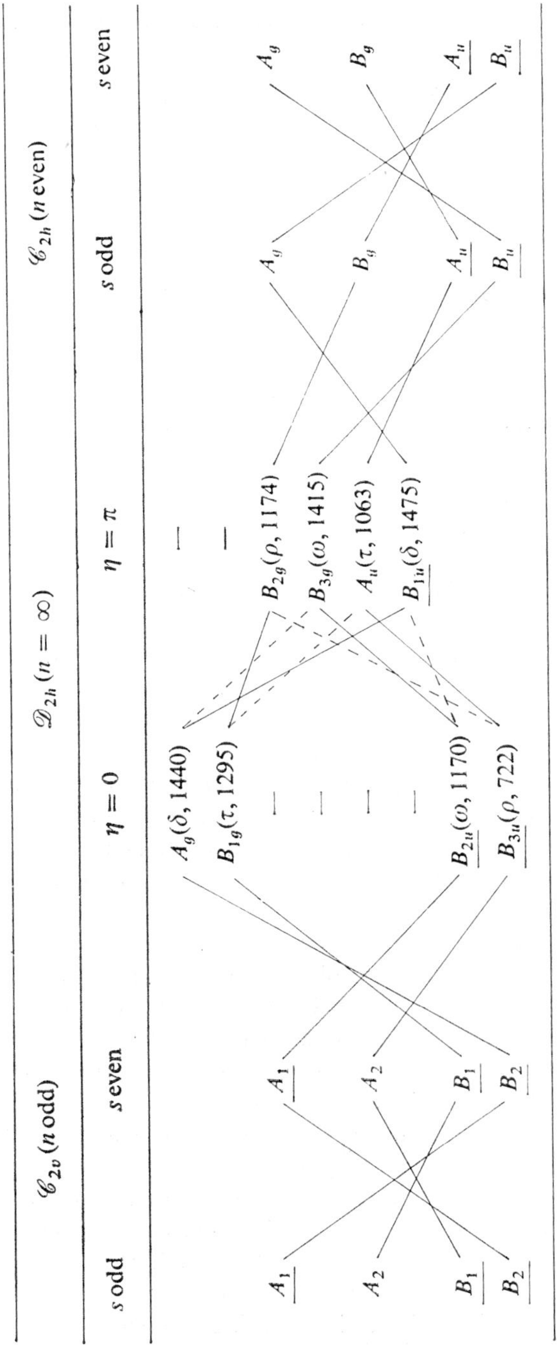
$\mathscr{C}_{2v}$ (n odd)
s odd
s even
$\mathscr{D}_{2h}$ ($n = \infty$)
$\eta = 0$
$\eta = \pi$
$\mathscr{C}_{2h}$ (n even)
s odd
s even
A_1
A_2
B_1
B_2
$A_g(\delta, 1440)$
$B_{1g}(\tau, 1295)$
$B_{2u}(\omega, 1170)$
$B_{3u}(\rho, 722)$
$B_{2g}(\rho, 1174)$
$B_{3g}(\omega, 1415)$
$A_u(\tau, 1063)$
$B_{1u}(\delta, 1475)$
A_g
B_g
A_u
B_u

$l > 3$, the elements of **G** and **F** are very small. In general, the matrix **G F** has elements

$$(\mathbf{G\,F})_j = (\mathbf{G\,F})_{i,i+j} = (\mathbf{G\,F})_{i+j,i} = g_0 f_j + \sum_{l=1} g_l (f_{l+j} + f_{l-j}). \qquad (24)$$

The eigenvalues of **G F** for a chain of $\mathcal{N}$ oscillators are given by

$$\lambda_{\mathcal{N},s} = 4\,\pi^2\,\nu^2_{\mathcal{N},s} = g_0 f_0 + \sum_{j=1} (\mathbf{G\,F})_j \cos j\eta_{\mathcal{N},s}, \qquad (25)$$

where, as in Eqn (5), the phase shift between adjacent oscillators for the sth chain mode is given by

$$\eta_{\mathcal{N},s} = \frac{\pi\,s}{\mathcal{N}+1}, \qquad s = 1, 2, \ldots, \mathcal{N}. \qquad (26)$$

Hence, assignment of an observed absorption band involves specification of the value of s, as well as the type of internal coordinate. Even when mixing of the various coordinates occurs in a set of normal modes (rocking and twisting, for example), the frequencies are still functions only of the phase angle, $\eta_{\mathcal{N},s}$.

The usefulness of Eqn (25) arises from the fact that the matrix elements given by Eqn (24) depend only on the geometry of the paraffin chain and the force constants. They are, therefore, independent of the chain length. Hence, once a suitable set of force constants has been determined, Eqn (25) can be used to predict the normal-mode frequencies of any normal paraffin. Furthermore, in the limit of an infinite chain, optical activity is restricted to in-phase ($\eta = 0$) or out-of-phase ($\eta = \pi$) modes. This last result follows from the fact that the repeat unit of the infinite chain contains two methylene groups. Hence, if an oscillator is associated with each group, both symmetrical and antisymmetrical motions are permissable in the factor-group approximation.

If the symmetric and antisymmetric combinations of the internal coordinates of a repeat unit are considered separately, the symmetry species of the $\mathcal{D}_{2h}$ point group can be divided, as shown in Table 7. Here, the in-phase and out-of-phase combinations are based, rather arbitrarily, on symmetry or antisymmetry with respect to the glide reflection $\sigma^g(xy)$. Indicated also in the central part of this table are some of the internal coordinates illustrated in Fig. 5, along with the vibrational assignments given by Snyder and Schachtschneider (**11**).

The $\eta = 0$ and $\eta = \pi$ species of the infinite chain are correlated through the point groups of both n-odd and n-even finite chains, that is, *via* their common subgroup, $\mathcal{C}_s$ [operations E, $\sigma(yz)$]. Hence, even if the normal modes of the infinite chain were to correspond exactly to internal coordinates, mixing of certain coordinates could occur in finite chains.

Of course the actual form of the modes can only be determined by a normal-coordinate calculation.

Table 7 provides a basis for assignment of the vibrational fundamentals of the finite chains, as well as those of the infinite one. Consider, for example, the vibrational modes which might be expected to correspond approximately to the wagging coordinate. For the infinite chain, only two wagging fundamentals should be infrared-active. They are assigned to bands at 1170 cm^{-1} ($B_{2u}, \eta = 0$) and 1415 cm^{-1} ($B_{3g}, \eta = \pi$). However, in the case of finite chains the modes defined by Eqn (26) can be classified as shown in Table 6 according to the point group of the chain. Furthermore, for each chain symmetry the modes can be further distinguished by the parity of s. These results are also summarized in Table 7.

Continuing with the example of the wagging modes of n-paraffins, it is

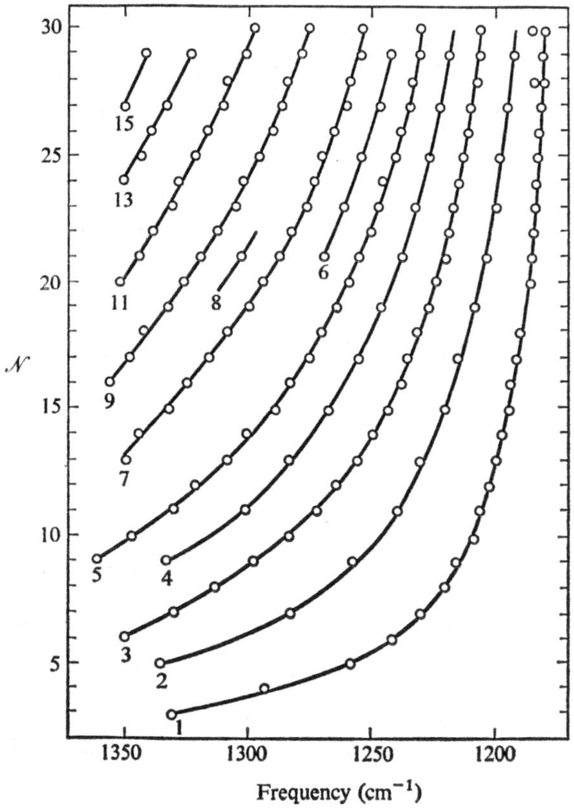

FIG. 8. Methylene wagging-mode array for n-paraffins C_3H_8 through n-$C_{30}H_{62}$ [After Snyder (**10**)].

found from Table 7 that with $\mathcal{N}$ odd (symmetry $\mathscr{C}_{2v}$), the wagging vibrations are of species A_1 when s is even, and of species B_2 when s is odd. By reference to Table 6 it is seen that in the former case $\frac{1}{2}(\mathcal{N} - 3)$ fundamental absorption bands of species A_1, having polarizations perpendicular to the chain axis, should be observed in the region between 1170 cm^{-1} and 1415 cm^{-1}. Similarly, for odd values of s, a second series of $\frac{1}{2}(\mathcal{N} - 1)$ bands (species B_2), in this case polarized parallel to the chain axis, should be active in the same spectral region.

When $\mathcal{N}$ is even, the point symmetry of the chain is $\mathscr{C}_{2h}$ and the wagging motions are of species B_u or A_g as s is, respectively, odd or even. In this case only half of the $\mathcal{N} - 2$ wagging modes, those which are of species B_u (s odd), are infrared-active.

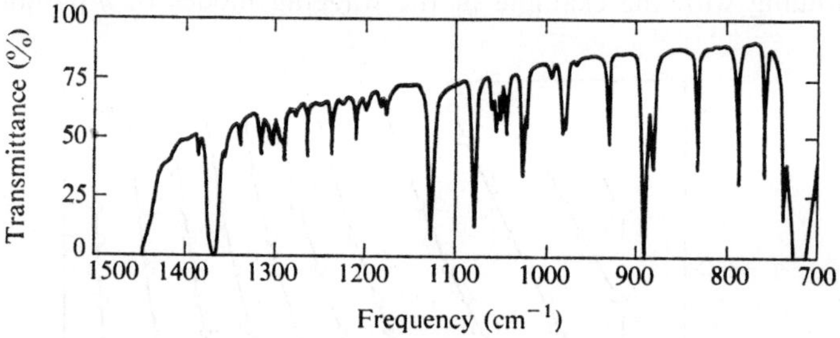

FIG. 9. Infrared absorption spectrum of n-$C_{26}H_{54}$ at −180°C [After Snyder (**10**)].

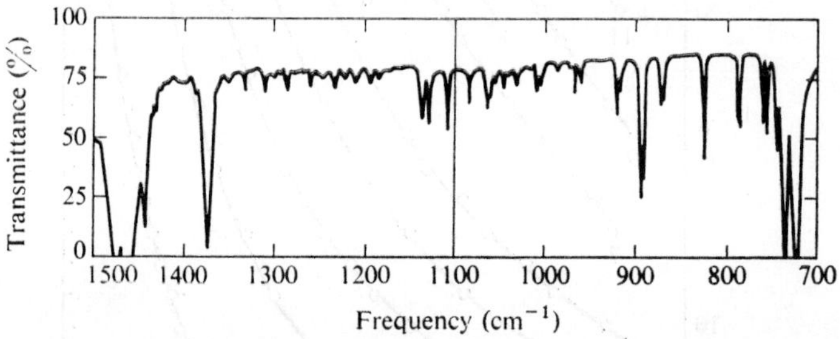

FIG. 10. Infrared absorption spectrum of n-$C_{27}H_{56}$ at −180°C [After Snyder (**10**)].

The results predicted above for the methylene wagging modes are compared in Fig. 8 with the experimental results reported by Snyder and Schachtschneider. The various branches are identified by their values of s following the proposed assignments. The experimental points are obtained by identifying the band series shown for example, in the spectra of Figs. 9† and 10.

† Many of the absorption bands of n-even paraffins exhibit splitting due to interchain coupling. This effect, which is neglected in this treatment of isolated chains, will be considered in the following section.

The absence of points corresponding to even values of both s and $\mathcal{N}$ provides firm support for the qualitative treatment presented above. Furthermore, using Eqn (26), the data lead to the dispersion curve shown in Fig. 11. The points fall very close to a common curve, even for the shortest chains, indicating relatively little interaction between these vibrations and those of the methyl end-groups. The calculation of the vibrational frequencies from a suitable force field will be discussed later in this chapter.

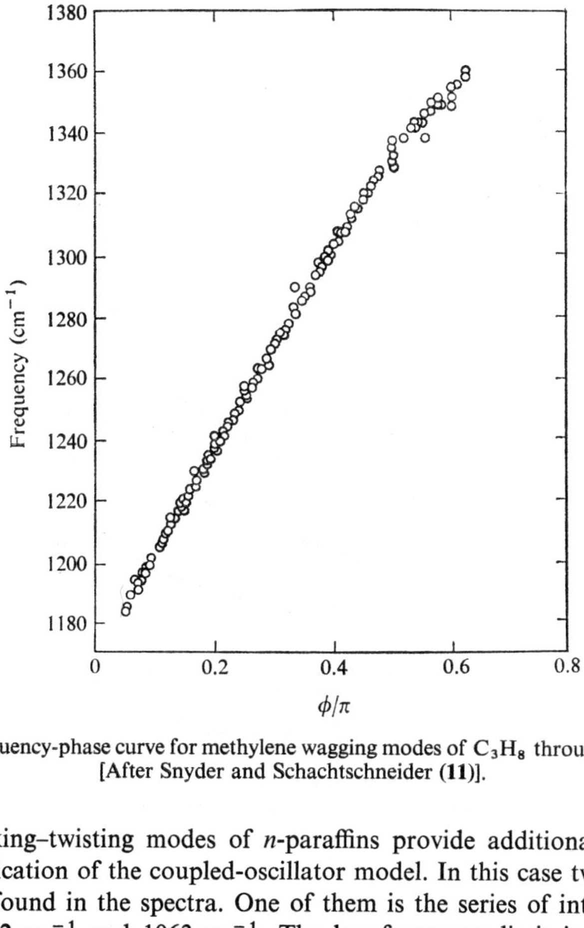

FIG. 11. Frequency-phase curve for methylene wagging modes of C_3H_8 through n–$C_{30}H_{62}$ [After Snyder and Schachtschneider (**11**)].

The rocking–twisting modes of *n*-paraffins provide additional examples of the application of the coupled-oscillator model. In this case two series of bands are found in the spectra. One of them is the series of intense bands between 722 cm^{-1} and 1063 cm^{-1}. The low-frequency limit is formed by the very strong and characteristic band of polyethylene, which is assigned to a methylene rocking motion of species B_{3u}. Following Snyder and Schactschneider's assignments, the high-frequency limit to this series results from a CH_2 twisting motion of species A_u. Hence, their assignments favor the correlation $B_{3u} \Leftrightarrow A_u$, shown by a heavy line in Table 7, rather than the alternative possibility, $B_{3u} \Leftrightarrow B_{2g}$. Thus, rocking and twisting coordinates are mixed except in the infinite chain at the limits $\eta = 0$ and $\eta = \pi$. It will be shown later that normal-coordinate calculations support this conclusion.

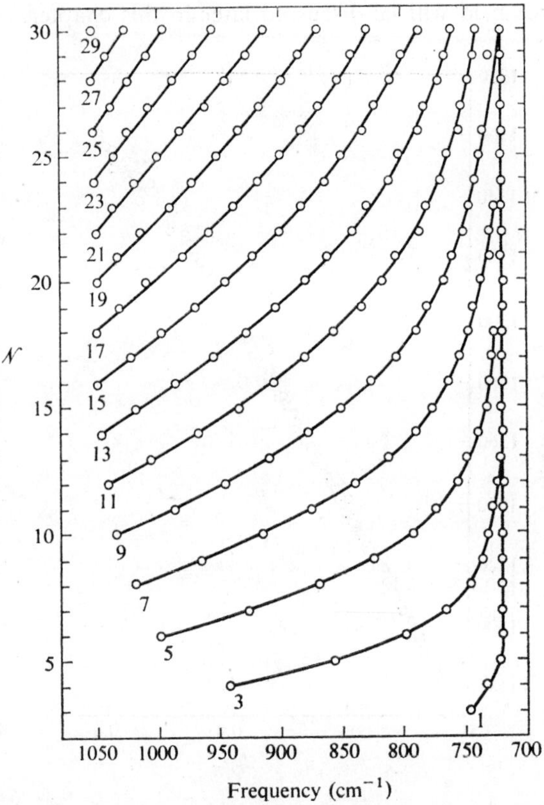

FIG. 12. Methylene rocking–twisting-mode array for *n*-paraffins of C_3H_8 through *n*-$C_{30}H_{62}$ [After Snyder and Schachtschneider (**11**)].

The array of experimental points shown in Fig. 12 summarizes the experimental results for this rocking–twisting series. As predicted by Table 7, only odd *s*-modes are infrared-active, regardless of the parity of *n*. However, for very short chains interaction of these modes with the methyl end-groups results in some frequency perturbation, as well as the appearance of several very weak bands for *s*-even modes.

A similar, but much weaker series of rocking–twisting modes is assigned in the region 1174 cm^{-1} to 1295 cm^{-1}. This series arises from the other rocking–twisting correlation ($B_{1g} \Leftrightarrow B_{2g}$) shown in Table 7.

V. Crystalline Polymers and Chain Interactions

In the previous section a simple coupled-oscillator model was used to interpret the vibrational spectra of *n*-paraffins. This model has been applied with considerable success to a number of polymer systems. However, the isolated-chain approximation necessarily ignores interchain forces, which, depending on the crystal structure, might be expected to have an effect on the spectra of crystalline polymers.

It has been shown by Snyder (**10**) that many of the absorption bands of crystalline *n*-paraffins having orthorhombic or monoclinic structures are split into doublets. This effect, which is easily seen by comparison of Figs.

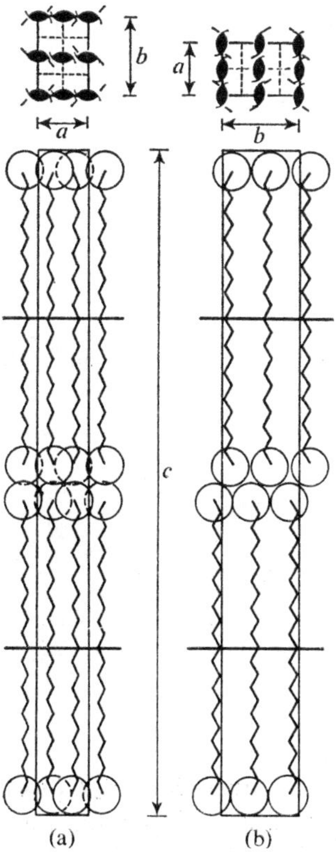

Fig. 13. Structure of the unit cell of crystalline *n*-tricosane, (a) *ac* projection: (b) *bc* projection [After Smith (**12**).]

9 and 10, can be accounted for using the known crystal structures by applying the factor-group method. The crystal structures of the *n*-paraffins are summarized in Table 8.

TABLE 8

No. of carbon atoms	Crystal system	Space group	Molecules per unit cell (Z)
21–29 (*n* odd)	orthorhombic	$Pbcm = \mathscr{D}_{2h}^{11}$	4
26 (*n* even)	monoclinic	$P2_1/a = \mathscr{C}_{2h}^{5}$	2
26 (*n* even)	triclinic	?	1
∞	orthorhombic	$Pnam = \mathscr{D}_{2h}^{16}$	2

Consider first, *n*-paraffins having the orthorhombic structure. These molecules, which have an odd number of carbon atoms, crystallize in sheets perpendicular to the chain axis (y), with four molecules per primitive cell (space group $Pbcm \equiv \mathscr{D}_{2h}^{11}$), as shown in Fig. 13. The symmetry operations of the factor group are E, $C_2^s(Z)$, $C_2^s(Y)$, $C_2^s(Z)$, $i, \sigma(XY)$, $\sigma^{g(Z)}(YZ)$, and $\sigma^{g(X)}(XZ)$. Of these operations, only E and $\sigma(XY)$ leave a given molecule fixed. Thus, the site group is $\mathscr{C}_s$ and the correlations between the molecular point group and the unit-cell group can be easily established, as shown in Table 9.

TABLE 9

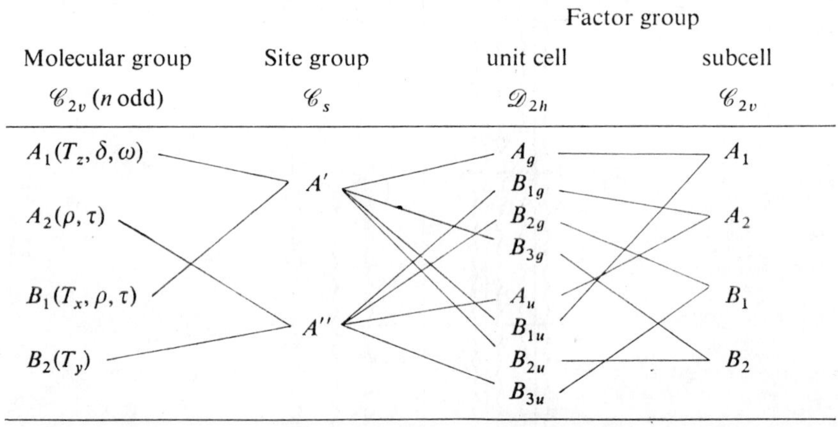

Factor group
Molecular group
Site group
unit cell
subcell
$\mathscr{C}_{2v}$ (n odd)
$\mathscr{C}_s$
$\mathscr{D}_{2h}$
$\mathscr{C}_{2v}$
$A_1(T_z, \delta, \omega)$
$A_2(\rho, \tau)$
$B_1(T_x, \rho, \tau)$
$B_2(T_y)$
A'
A''
A_g
B_{1g}
B_{2g}
B_{3g}
A_u
B_{1u}
B_{2u}
B_{3u}
A_1
A_2
B_1
B_2

It should be noted that the operations E, $C_2^s(X)$, $\sigma(XY)$, and $\sigma^{g(X)}(XY)$, which confine molecules to a given layer, form a subgroup of the unit-cell group. This subgroup, which is isomorphic with point group $\mathscr{C}_{2v}$, describes the symmetry of a subcell containing two molecules. Correlation of the vibrations of the subcell with those of the entire unit cell are included in Table 9.

Selection rules for the infrared and Raman fundamentals of isolated chains were summarized in Table 6. From the correlation diagram of Table 9 it is seen that the lower symmetry imposed by the site can, in principle, allow infrared activity of the A_2-species vibrations of the isolated chain.

The correlation between the factor group and the site group indicates that each molecular vibration has four components in the crystal. In the infrared spectrum one band should be observed for each vibration parallel to the chain axis (y), while a doublet would be expected for each perpendicular vibration. These selection rules are the same as found above for polyethylene.

As the interaction between molecules in the same sheet would be expected to be stronger than the interaction between molecules of different sheets, the factor-group analysis based on the two-molecule subcell might provide a better description of the crystal vibrations. The correlation diagram (Table 9) shows that the effect of choosing the smaller cell is to remove the distinction between g and u phases of the relative molecular motion in adjacent sheets.

Turning now to the series of n-paraffins having an even number of carbon atoms, Table 8 indicates that for $n \geqslant 26$ the crystal structure is monoclinic, of space group $P2_1/a \equiv \mathscr{C}_{2h}{}^5$, with two molecules per unit cell. This structure is very similar to the orthorhombic structure discussed above, but in this case the chains are tilted with respect to the XY plane of Fig. 13. The site symmetry of a molecule is $\mathscr{C}_i$. The symmetry species of the molecular point group are correlated with those of the site group and the unit-cell group in Table 10. It is apparent that the presence of two molecules per unit cell can lead to doubling of fundamental absorption bands.

TABLE 10

Molecular group $\mathscr{C}_{2h}$	Site group $\mathscr{C}_i$	Factor group (unit cell) $\mathscr{C}_{2h}$
A_g		A_g
	A_g	
B_g		B_g
$A_u(T_x)$		$A_u(T_Z)$
	A_u	
$B_u(T_y, T_z)$		$B_u(T_X, T_Y)$

For $n \leqslant 26$, *n*-paraffins having an even number of carbon atoms are reported to crystallize in the triclinic system. Although the exact crystal structure does not appear to have been determined, the presence of but one molecule per unit cell precludes the possibility of correlation coupling. Hence, doubling of vibrational fundamentals should not be observed.

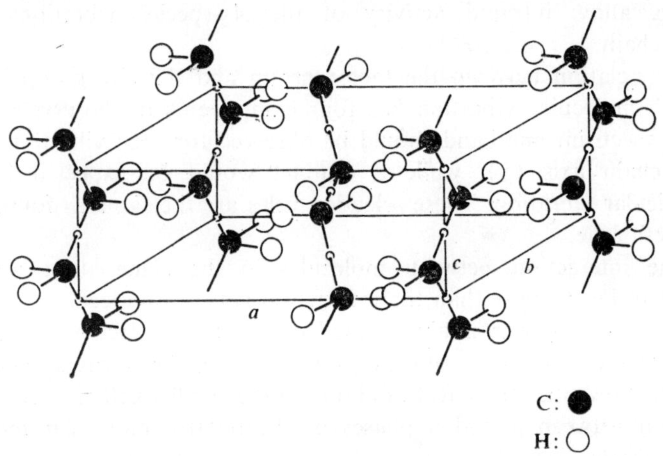

FIG. 14. Unit cell of crystalline polyethylene.

The group-theoretical analysis of the vibrations of crystalline polyethylene was developed by Krimm, *et al.* (**13**), using the crystallographic data of Bunn (**14**). The orthorhombic unit cell (space group $Pnma \equiv \mathscr{D}_{2h}^{16}$) is shown in Fig. 14. Each of the two chains passing through the unit cell contributes two methylene groups to it. Each of the four equivalent $\mathscr{C}_s$

TABLE 11

Single infinite chain $\mathscr{D}_{2h}$	Chain segment $\mathscr{C}_{2h}$	Unit cell $\mathscr{D}_{2h}^{16}$
A_g		A_g
B_{1g}	A_g	B_{1g}
B_{2g}	B_g	B_{2g}
B_{3g}		B_{3g}
A_u		A_u
$B_{1u}(T_z)$	A_u	$B_{1u}(T_Z)$
$B_{2u}(T_y)$	B_u	$B_{2u}(T_Y)$
$B_{3u}(T_x)$		$B_{3u}(T_X)$

sites in the space group *Pnma* is thus occupied by a CH_2 group. However, as the coupling between the two adjacent methylene groups of a given chain is much stronger than the coupling between those of different chains, a more meaningful site-group type of analysis can be made on the basis of a chain segment containing two methylene groups and having effective symmetry $\mathscr{C}_{2h}$. The resulting correlation diagram is shown in Table 11.

The factor-group analysis of the $\mathscr{D}_{2h}^{16}$ unit cell can be easily carried out using the character table for the isomorphic point group $\mathscr{D}_{2h}$ (See Table 3). From Fig. 14 it is found that only the operations E and $\sigma(XY)$ leave atoms unmoved. The resulting distribution of the vibrational modes is given by the quantities $n^{(\gamma)}$ in Table 12, where internal, translatory, and rotatory vibrations have been distinguished. The symmetry species of the internal coordinates are also indicated.

TABLE 12

			Lattice vibs.			
	n_{tot}	n_{acous}	n_{trans}	n_{rot}	n_{int}	Internal coordinates
A_g	6	0	0	1	5	$2\nu, \delta, \rho, \beta$
B_{1g}	6	0	0	1	5	$2\nu, \delta, \rho, \beta$
B_{2g}	3	0	0	0	3	τ, ω, σ
B_{3g}	3	0	0	0	3	τ, ω, σ
A_u	3	0	1	0	2	τ, ω
B_{1g}	3	1	0	0	2	τ, ω
B_{2g}	6	1	1	0	4	$2\nu, \delta, \rho$
B_{3g}	6	1	1	0	4	$2\nu, \delta, \rho$

Using the results of the Table 12 in conjunction with the correlation diagram of Table 11, the following comparison can be made between the predicted spectra of isolated polyethylene chains and crystals. All of the single-chain modes which are polarized perpendicular to the chain axis should be split into two components in infrared spectra of crystalline samples. Of these components, those of species B_{2u} should be polarized along the Y axis, while B_{3u}-species components should be X-polarized. On the other hand, parallel chain vibrations (twisting–rocking and wagging motions) would each be expected to have but one infrared-active component (species B_{1u}) in the crystal. The inactive A_u-species vibration of the isolated chain has the possibility of becoming infrared-active as a result of the reduced symmetry imposed by the crystal field. Finally, two translatory lattice vibrations should be active in the far-infrared spectrum.

As pointed out in the previous section, the strong absorption band at approximately 722 cm^{-1} in the spectrum of polyethylene develops into a series of bands extending upwards to 1063 cm^{-1} in the spectra of *n*-paraffins of finite length. This rocking–twisting series is particularly sensitive to crystalline effects and, hence, serves to illustrate the factor-group selection rules developed in this section for *n*-paraffins of various crystal structures.

Representative spectra of the appropriate region are shown in Figs 15–17. In these spectra, which were obtained from very pure crystalline samples at liquid-nitrogen temperatures, the series of sharp bands is easily identified. The first member of the series, which corresponds to a nearly in-phase rocking motion of adjacent CH_2 groups, is the very strong band near 722 cm^{-1}. In accordance with the selection rules developed above, this band is split into a doublet in the spectra of all paraffins having either the orthorhombic or the monoclinic structures, as well as in the spectrum of polyethylene. In the spectrum of *n*–$C_{24}H_{50}$ (Fig. 17), which is representative of the spectra of triclinic *n*-paraffins, this band does not exhibit splitting.

In both Figs 15 and 16 some higher members of the series are also doublets. In fact, careful analysis of the regularities of the splittings indicates (**10**, **15**, **16**) that the separation between components is a continuous function of the phase shift, η. Furthermore, the splitting increases with decreasing sample temperature. Similar regularities in the relative intensities

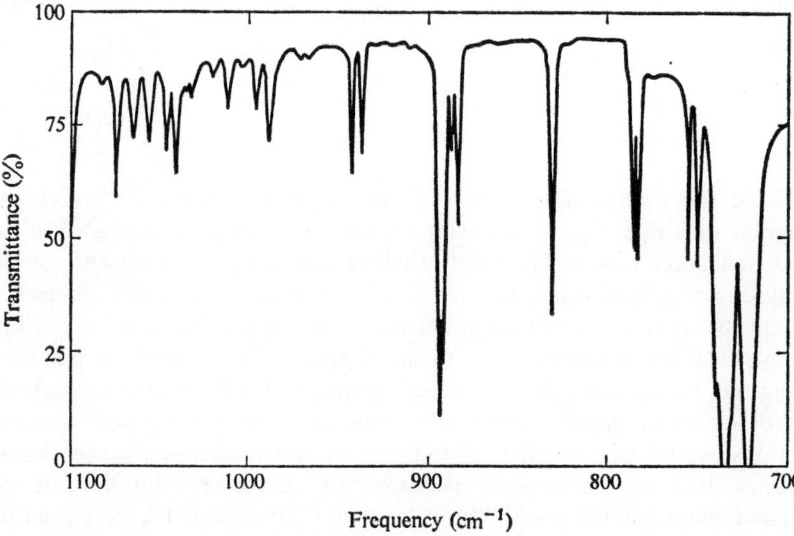

FIG. 15. Infrared spectrum of an orthorhombic *n*-paraffin, *n*-$C_{23}H_{48}$, at −180°C [After Snyder (**10**)].

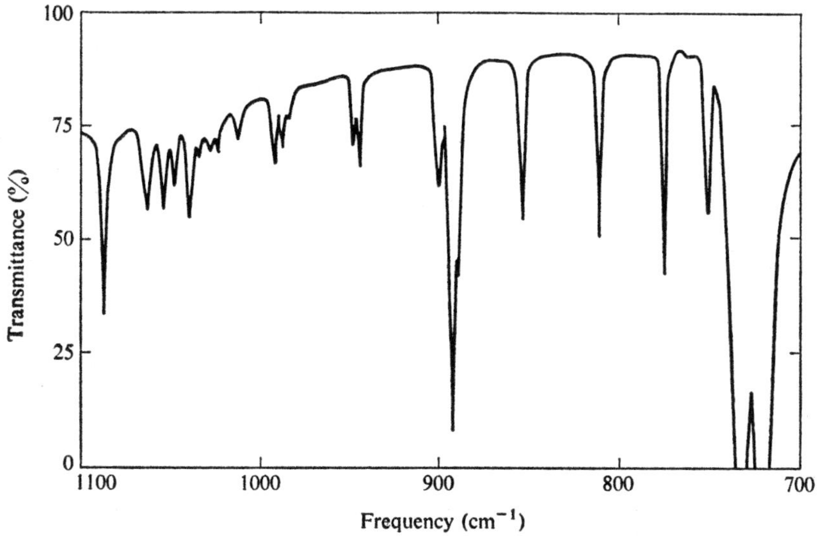

FIG. 16. Infrared spectrum of a monoclinic *n*-paraffin, *n*-$C_{28}H_{58}$, at −180°C [After Snyder (**10**)].

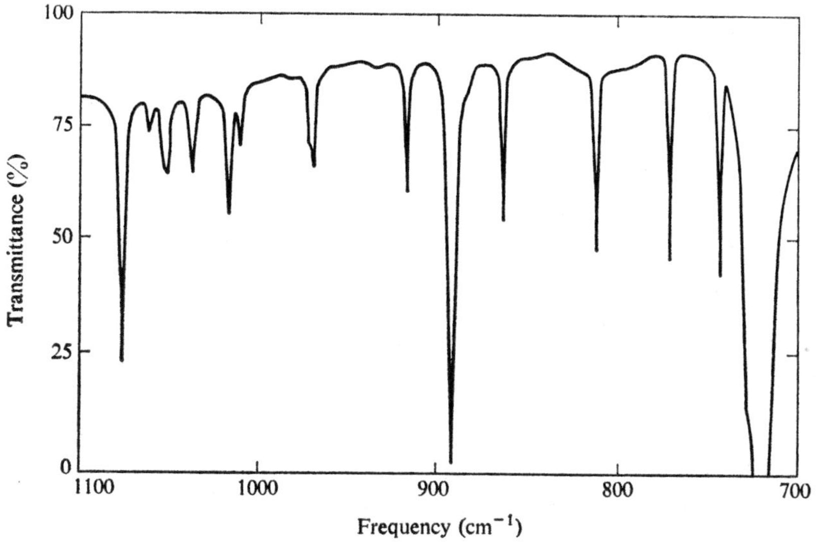

FIG. 17. Infrared spectrum of a triclinic *n*-paraffin, *n*-$C_{24}H_{50}$, at −180°C [After Snyder (**10**)].

of the two components have also been observed. These results will be considered quantitatively later in this chapter, where the calculation of intermolecular forces in polymer crystals will be discussed.

It has been shown in the qualitative discussion presented in this section, that the vibrational spectrum of a polymer depends on its crystalline environment. Thus, the effects of local symmetry and interchain coupling are to produce modifications in selection rules and, in some cases, splitting of vibrational fundamentals. However, not all absorption features which disappear when a sample is melted can be correctly associated with crystallinity, as some bands may arise simply from the extended structure of a given chain. These so-called regularity bands disappear as a result of chain folding or twisting. This question, which is extremely important from a practical point of view, has been reviewed by Zerbi, *et al.* (**17**).

VI. Intrachain Forces in *n*-Paraffins

An effective force field for extended *n*-paraffins and polyethylene has been calculated by Schachtschneider and Snyder (**18–21**). A set of 31 general valence force constants was used to give a least-squares fit to 270 observed vibrational frequencies of the *n*-paraffins propane through *n*-decane, and polyethylene. An additional constant, which was estimated from the torsional frequency of ethane (280 cm^{-1}), was used to account for twisting about the C—C bonds. The complete set of constants and their standard deviations are given in Table 13. The internal coordinates are defined in Fig. 18.

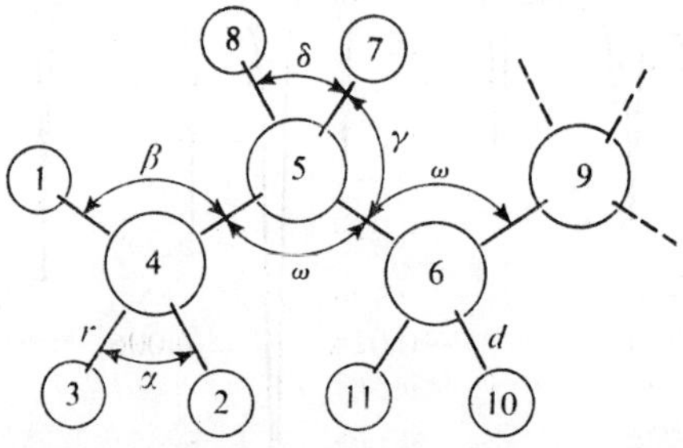

FIG. 18. Internal coordinates for normal paraffins [After Schachtschneider and Snyder (**18**)].

The general valence force field represented in Table 13 contains 11 diagonal force constants including the torsional constant, H_τ. The methyl and methylene groups were given different diagonal force constants, and a number of important interaction constants were included. Notable off-diagonal constants are $F_{R\gamma} = F_{R\beta}$, an interaction between a C—C stretch and an HCC bend with a C—C bond in common, and $F_{R\omega}$, a CC-

TABLE 13 [From reference (**18**)]

No.	Symbol	Force constant[a]	Standard deviation	Example of coordinate or interaction (See Fig. 18)
1	$(CH_3)\ K_r$	4·704	0·003	1-4
2	$(CH_3)\ F_r$	0·039	0·002	1-4 with 2-4
3	$(CH_2)\ K_d$	4·546	0·004	7-5
4	F_d	0·016	0·004	7-5 with 8-5
5	$(CH_3—CH_3)\ K_R$	4·423	0·059	Ethane
6	$(CH_3—CH_2)\ K_R{}^{I}$	4·329	0·026	4-5
7	$(CH_2—CH_2)\ K_R{}^{II}$	4·427	0·042	5-6
8	F_R	0·064	0·030	4-5 with 5-6
9	$F_{R\gamma} = F_{R\beta}$	0·261	0·031	4-5 with 1-4-5
10	$F_{R\gamma}{}'$	−0.004	0·031	4-5 with 7-5-6
11	$F_{R\omega}$	0·351	0·022	4-5 with 4-5-6
12	$(CH_3)\ H_\alpha$	0·540	0·001	1-4-3
13	$(CH_2)\ H_\delta$	0·550	0·003	7-5-8
14	$(CH_3)\ H_\beta$	0·637	0·005	1-4-5
15	$(CH_3)\ F_\beta$	−0·017	0·005	1-4-5 with 2-4-5
16	$(CH_2)\ H_\gamma$	0·666	0·004	7-5-4
17	$(CH_2)\ F_\gamma$	−0·016	0·004	7-5-4 with 8-5-4
18	$(CH_2)\ F_\gamma{}'$	0·023	0·003	7-5-4 with 7-5-6
19	$(CH_2)\ F_{\gamma\omega}$	−0·124	0·039	7-5-4 with 4-5-6
20	H_ω	0·901	0·059	4-5-6
21	$f_\omega{}^t$	0·093	0·021	4-5-6 with 5-6-9
22	$f_{\beta\omega}{}^t = f_{\gamma\omega}{}^t$	0·072	0·011	1-4-5 with 4-5-6
23	$f_{\beta\omega}{}^g = f_{\gamma\omega}{}^g$	−0·058	0·008	7-5-6 with 5-6-9
24	$f_\beta{}^t = f_{\beta\gamma}{}^t = f_\gamma{}^t$	0·106	0·006	3-4-5 wfth 4-5-7
25	$f_\beta{}^g = f_{\beta\gamma}{}^g = f_\gamma{}^g$	−0·024	0·006	3-4-5 with 4-5-8
26	$f_{\beta\gamma}{}'^t = f_\gamma{}'^t$	−0·002	0·001	2-4-5 with 7-5-6
27	$f_{\beta\gamma}{}'^g = f_\gamma{}'^g$	0·002	0·001	2-4-5 with 8-5-6
28	$f_\gamma{}''^t$	−0·001	0·006	8-5-4 with 10-6-9
29	$f_\gamma{}''^g$	0·001	0·006	8-5-4 with 11-6-9
30	$h_{\beta\gamma}{}^c = h_\gamma{}^c$	−0·002	0·003	2-4-5 with 10-6-5
31	$h_{\beta\gamma}{}^c = h_\gamma{}^c$	−0·003	0·002	2-4-5 with 11-6-5
32	H_τ	0·008	—	Torsion about 4-5

[a] Stretching constants in units of mdyne/Å, stretch–bend interaction constants in units of mdyne/rad, and bending constants in units of mdyne-Å/rad^2.

stretch–CCC-bend interaction with a C—C bond in common. This force field was used to calculate vibrational frequencies and the eigenvectors of all members of the series. It was then extended to include higher members of the *n*-paraffin series, and with slight modification to branched hydrocarbon chains.

The results of this normal coordinate treatment of *n*-paraffins are illustrated by the frequency array for the rocking–twisting series shown in Fig. 19.

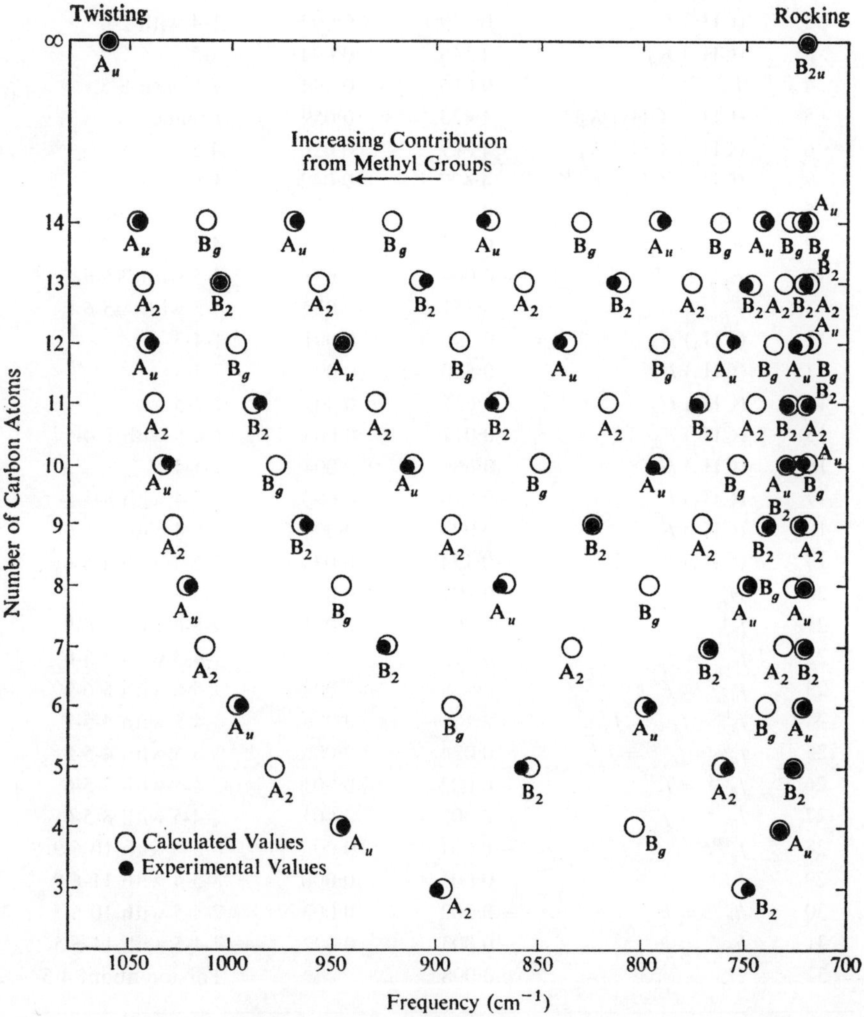

FIG. 19. Calculated and observed frequencies of the methylene twisting–rocking region (1060–720 cm^{-1}) [After Schachtschneider and Snyder (**18**)].

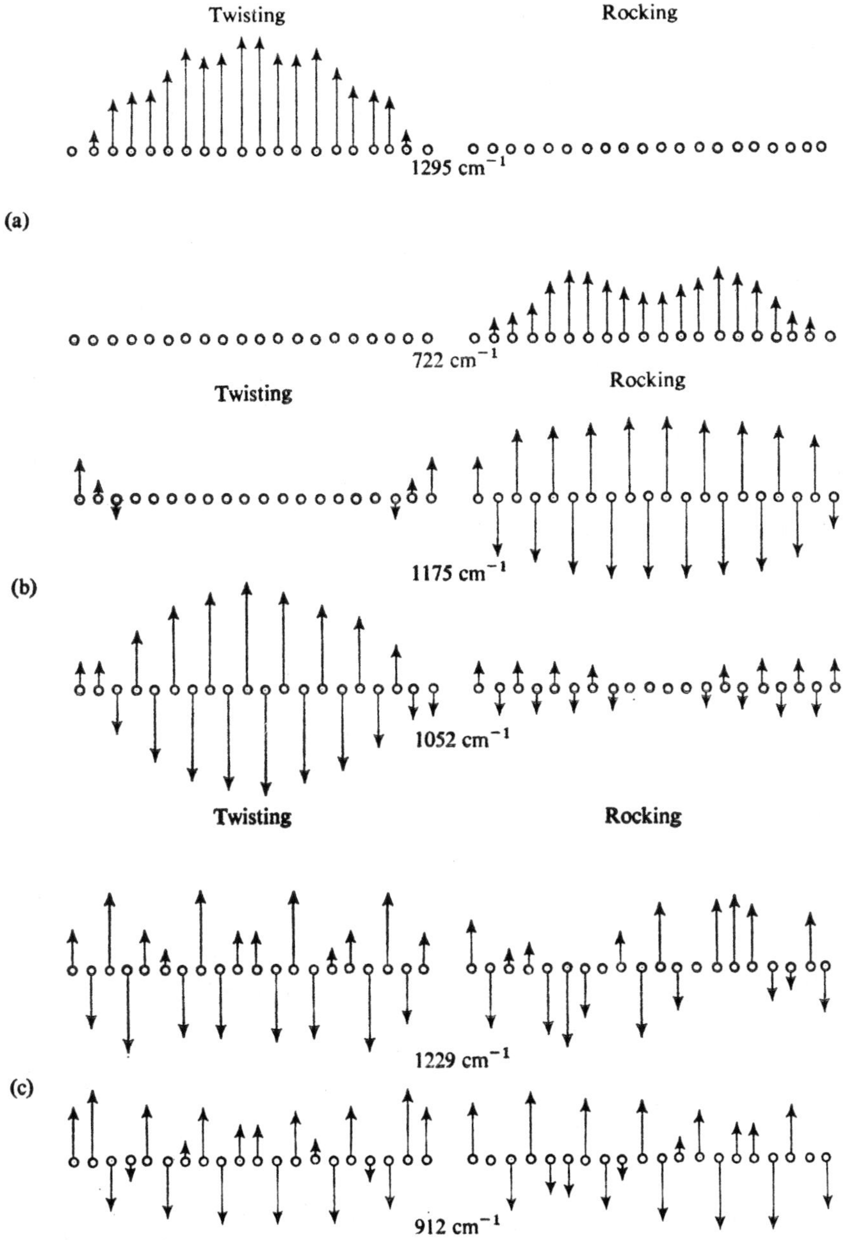

FIG. 20. Form of normal modes of n–$C_{20}H_{42}$. Lengths of arrows are proportional to elements of **L**. (a) In-phase ($\eta = 0$) twisting and rocking modes, (b) Out-of-phase ($\eta = \pi$) twisting and rocking modes, (c) Mixed twisting–rocking modes ($\eta = \pi/2$) [After Schachtschneider and Snyder (**18**)].

The calculated frequencies are in very good agreement with the observed fundamentals. Furthermore, the frequencies of the infrared-active vibrations can be determined. The calculated **L** matrix allows the forms of the various normal modes to be found, as shown in Fig. 20 for n-$C_{20}H_{42}$. Vibrations at frequencies near 1295 and 1061 cm^{-1} are almost pure CH_2 twisting modes, while those near 1170 and 721 cm^{-1} are essentially rocking modes. At intermediate frequencies these two types of internal coordinates are thoroughly mixed, as shown in Fig. 20(c). This result is in agreement with the qualitative predictions made earlier in this chapter.

The normal coordinate treatment also allows the role of the end groups to be evaluated. In general, the effect of the methyl groups on the vibrational modes increases with decreasing chain length. However, even with $n = 20$, a noticeable contribution from the CH_3 groups is seen (Fig. 20). Since coupling between the CH_3-in-phase rocking mode and the C—C stretching modes is quite strong, the observed methyl rocking frequency varies with chain length. Calculated and observed values of the frequency of this in-plane mode are compared in Fig. 21.

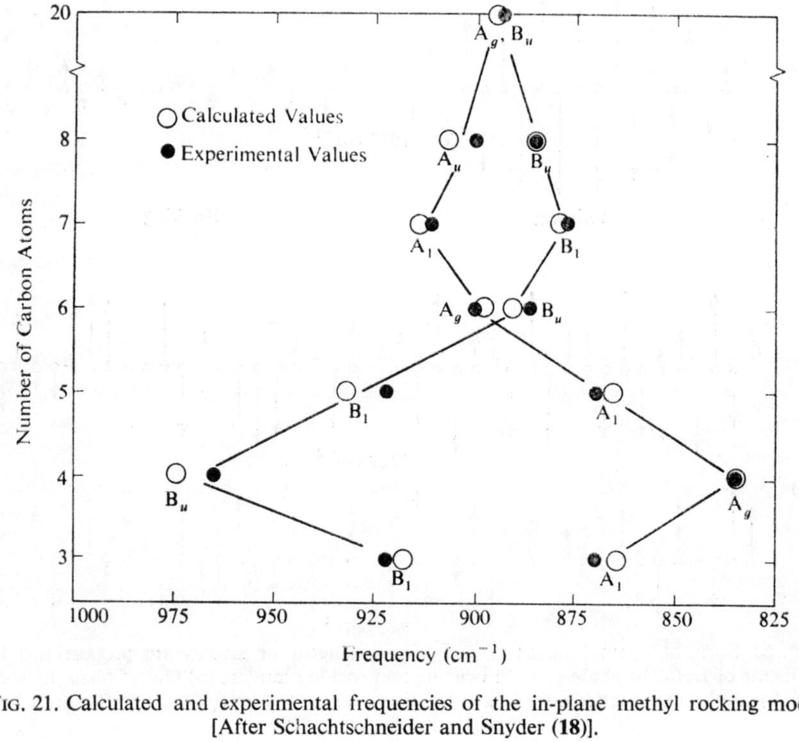

FIG. 21. Calculated and experimental frequencies of the in-plane methyl rocking mode [After Schachtschneider and Snyder (**18**)].

This study of the vibrational spectra of *n*-paraffins illustrates the application of quantitative normal-coordinate analyses. In particular, such calculations can be used to predict with high accuracy the vibrational frequencies of other members of the series using the vibrational assignments for lower homologs. The same authors have applied this method to isotactic and two forms of syndiotactic polypropylene (**22**).

VII. Interchain Coupling in Crystalline Polyethylene

The factor-group analysis of crystalline polyethylene was discussed earlier in this chapter. It was shown that correlation coupling between the two chains in the orthorhombic primitive cell resulted in splitting of vibrations into two components. In the case of vibrations perpendicular to the chain axes (in the X, Y plane of the crystal) both components are infrared-active. Furthermore, from the analysis of the spectra of crystalline *n*-paraffins it appears that the frequency separation between components is a continuous function of the phase shift, η, between adjacent repeat units in a given chain.

A quantitative treatment of the coupling in crystalline polyethylene was made by Tasumi and Shimanouchi (**15**) by calculating a set of intermolecular force constants using the observed splitting of the methylene bending and rocking vibrations at 1475 cm^{-1} ($B_{1u} \Rightarrow B_{2u} + B_{3u}$) and 722 cm^{-1} ($B_{3u} \Rightarrow B_{2u} + B_{3u}$), respectively. An additional value was obtained by extrapolation to infinite chain length of the splitting of the CH_2 twisting mode of *n*-paraffins at 1050 cm^{-1} ($A_u \Rightarrow A_u + B_{1u}$).

The frequency splitting can be directly related to the interchain force field using the method of Chapter 6. There it was pointed out that by application of the approximate separation of high and low frequencies, the internal vibrations are determined by the first three terms of Eqn (58), Chap. 6. In the present example there are but two molecules per unit cell. Hence, from Eqn (58), Chap. 6, the potential energy of the internal vibrations becomes

$$2V \approx \sum_{i=1}^{2} \mathbf{Q}^{0(\text{in})\dagger} \mathbf{\Lambda}_i^{0(\text{in})} \mathbf{Q}_i^{0(\text{in})} + \sum_{i=1}^{2} \mathbf{Q}_i^{0(\text{in})\dagger} \mathscr{L}_i^{0(\text{in})\dagger} \mathbf{\Phi}_i^{\dagger} \mathbf{F}_{ii}' \mathbf{\Phi}_i \mathscr{L}_i^{0(\text{in})} \mathbf{Q}_i^{0(\text{in})}$$
$$+ \mathbf{Q}_1^{0(\text{in})\dagger} \mathscr{L}_1^{0(\text{in})\dagger} \mathbf{\Phi}_1^{\dagger} \mathbf{F}_{12}' \mathbf{\Phi}_2 \mathscr{L}_2^{0(\text{in})} \mathbf{Q}_2^{0(\text{in})}$$
$$+ \mathbf{Q}_2^{0(\text{in})\dagger} \mathscr{L}_2^{0(\text{in})\dagger} \mathbf{\Phi}_2^{\dagger} \mathbf{F}_{21}' \mathbf{\Phi}_1 \mathscr{L}_1^{0(\text{in})} \mathbf{Q}_1^{(\text{in})}, \tag{27}$$

and the internal frequencies are given approximately by the eigenvalues of the matrix

$$\mathbf{D} = \left(\begin{array}{c|c} \mathbf{\Lambda}^0 + \mathscr{L}_1^{0(\text{in})\dagger}\mathbf{\Phi}_1^{\dagger}\mathbf{F}_{11}'\mathbf{\Phi}_1\mathscr{L}_1^{0(\text{in})} & \mathscr{L}_1^{0(\text{in})\dagger}\mathbf{\Phi}_1^{\dagger}\mathbf{F}_{12}'\mathbf{\Phi}_2\mathscr{L}_2^{0(\text{in})} \\ \hline \mathscr{L}_2^{0(\text{in})\dagger}\mathbf{\Phi}_2^{\dagger}\mathbf{F}_{21}'\mathbf{\Phi}_1\mathscr{L}_1^{0(\text{in}))} & \mathbf{\Lambda}^0 + \mathscr{L}_2^{0(\text{in})\dagger}\mathbf{\Phi}_2^{\dagger}\mathbf{F}_{22}'\mathbf{\Phi}_2\mathscr{L}_2^{0(\text{in})} \end{array}\right). \tag{28}$$

Thus, the splittings are determined entirely by the submatrices containing $\mathbf{F}_{12}' = \mathbf{F}_{21}'$, while frequency shifts produced by the crystalline field result from the elements in $\mathbf{F}_{11}' = \mathbf{F}_{22}'$. The separations between the components of the split bands are given approximately by

$$\Delta\mathbf{\Lambda} \approx 2\mathscr{L}_1^{0(\text{in})\dagger}\mathbf{\Phi}_1^{\dagger}\mathbf{F}_{12}'\mathbf{\Phi}_2\mathscr{L}_2^{0(\text{in})}. \tag{29}$$

The eignevectors $\mathscr{L}_1^{0(\text{in})} = \mathscr{L}_2^{0(\text{in})}$ in Eqn (29) are known as functions of the phase shift, η, from the normal coordinate treatment of an isolated chain, as discussed, for example, in the previous section. The matrices $\mathbf{\Phi}_1$ and $\mathbf{\Phi}_2$ are obtainable from a knowledge of the crystal structure. Finally, the force-constant matrix, $\mathbf{F}_{12}'$, can be set up using a simple set of intermolecular coordinates such as q_1 through q_4 shown in Fig. 22. The

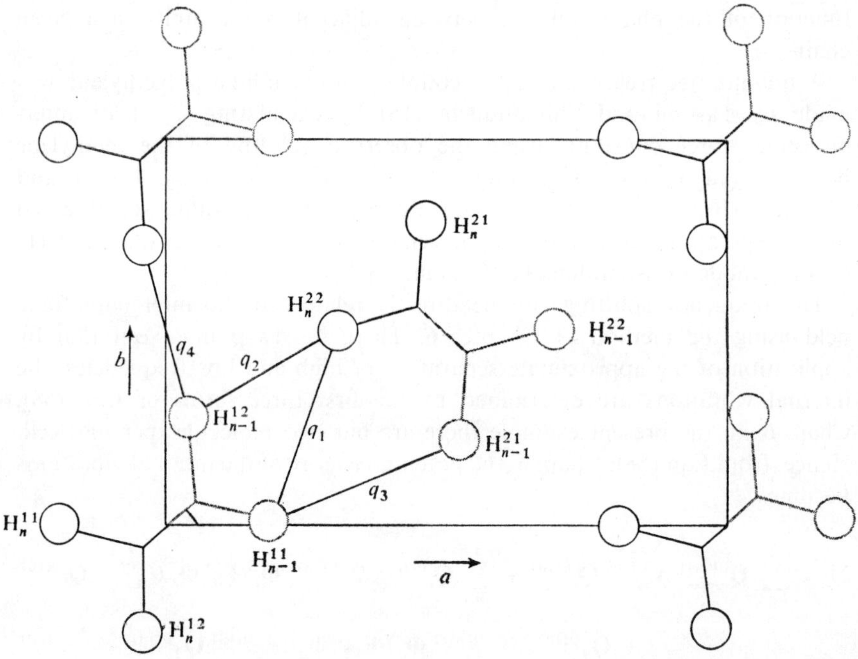

FIG. 22. Interchain coordinates for crystalline polythylene [From Tasumi and Shimanoucih (15)].

coordinates q_1 and q_2 each occur eight times in the unit cell, while q_3 and q_4 occur four times each.

The **F** matrix based on this system of 24 coordinates is transformed to crystal-fixed, optical coordinates using the known atomic positions. The resulting matrix, $\mathbf{F}_{12}'$, thus contains four unknown force constants, which in principle, require at least four observed splittings for their evaluation. Since optical coordinates form the basis of $\mathbf{F}_{12}'$, this matrix is a function of η, the phase shift along the chain. Phase shifts in the X and Y directions are set equal to zero, as only optically active lattice modes are of spectroscopic importance.

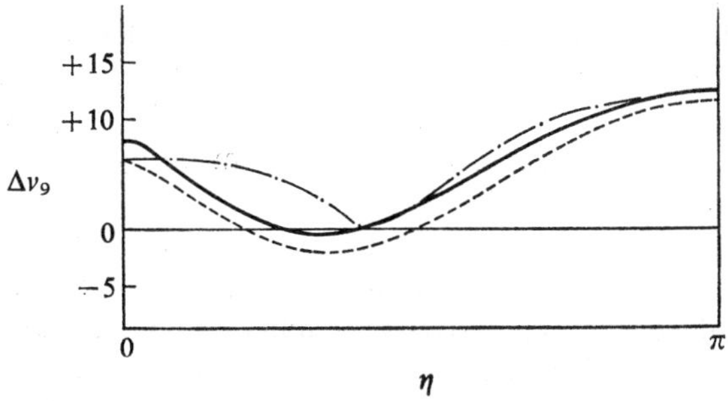

FIG. 23. Calculated and observed splitting of the antisymmetric mode ω_9. Curves —— and - - - - - were calculated by Tasumi and Shimanouchi. Intermediate values of η were used in calculating the former curve. Curve —·—·— is from observed data. [From reference (**15**)].

In the calculation by Tasumi and Shimanouchi one force constant was assumed and the other three calculated from splitting, as indicated in Table 14. These values are, in general, in agreement with H – H repulsion constants obtained from the spectra of other molecules (See Fig. 6 of Chapter 6). Furthermore, as $\mathscr{L}_1^{0(\text{in})} = \mathscr{L}_2^{0(\text{in})}$ is in this case a known function of η, the splitting can be calculated for various values of the phase shifts and compared with results obtained from spectra of n-paraffins. Figure 23 shows the experimental splitting of the series of rocking–twisting bands between 722 cm^{-1} and 1050 cm^{-1} and the calculated results using the force constants of Table 14. A least-squares fit using experimental

TABLE 14

H ··· H distance Å	Force constant[a] (mdyn/Å)
$q_1 = 2{\cdot}945$	$f_1 = 0{\cdot}0045$
$q_2 = 2{\cdot}743$	$f_2 = 0{\cdot}0133$
$q_3 = 2{\cdot}755$	$f_3 = 0{\cdot}0080$
$q_4 = 2{\cdot}575$	$f_4 = 0{\cdot}02$ (assumed)

[a] From reference (**15**).

points for intermediate values of η, obtained from the spectra of n-paraffins, provides somewhat better agreement between the two curves of Fig. 23 and some modification of the intermolecular force constants.

Correlation splitting has recently been observed in the Raman spectra of crystalline polyethylene at liquid-nitrogen temperatures (**16**). The results were found to be in good agreement with the splitting calculated from the force field and geometry used by Tasumi and Krimm (**23**).

VIII. Vibrations of Helical Chains

As many high polymers are known to have helical structures, it is useful to extend the line-group method presented earlier in this chapter to include such configurations. The application of group theory to the vibrational analysis of helices was originally made by Higgs (**24**) and later generalized by Tadokoro (**25**).

Consider an isolated infinite helical molecule in which the crystallographic repeat unit contains n chemical units and t turns. For example, crystalline polyoxymethylene has nine $-CH_2O-$ units and makes five turns in the unit-cell length of 17·3 Å. Thus, one has $n = 9$ and $t = 5$ for this structure. The planar zigzag chain of polyethylene considered earlier becomes a special case of a helix in which $n = 2$ and $t = 1$.

As in the simple line-group analysis, the factor group can be used to classify the vibrations of helical molecules and to determine infrared and Raman selection rules. The factor group, which is sometimes represented by the symbol $\mathscr{C}(2\pi\, t/n)$ (**26**), is isomorphic with the point group $\mathscr{C}_n$. The

TABLE 15

$\mathscr{C}(2\pi t/n)$	E	C^1	C^2	...	C^{n-1}	$n^{(\gamma)}$	
A	1	1	1	...	1	$3m$	$T_{\parallel}, R_{\parallel}, \alpha_{xx}+\alpha_{yy}, \alpha_{zz}$
E_1	$\begin{cases}1\\1\end{cases}$	ε ε^*	ε^2 ε^{2*}	... ...	$\varepsilon^{(n-1)}$ $\varepsilon^{(n-1)*}$	$3m$	$(T_{\perp}, T_{\perp})\ (\alpha_{yz}, \alpha_{zx})$
E_2	$\begin{cases}1\\1\end{cases}$	ε^2 ε^{2*}	ε^4 ε^{4*}	... ...	$\varepsilon^{2(n-1)}$ $\varepsilon^{2(n-1)*}$	$3m$	$(\alpha_{xx}-\alpha_{yy}, \alpha_{xy})$
E_3	$\begin{cases}1\\1\end{cases}$	ε^3 ε^{3*}	ε^6 ε^{6*}	... ...	$\varepsilon^{3(n-1)}$ $\varepsilon^{3(n-1)*}$	$3m$	
⋮							
$E_{(n-1)/2}$	$\begin{cases}1\\1\end{cases}$	$\varepsilon^{(n-1)/2}$ $\varepsilon^{(n-1)/2*}$	ε^{n-1} ε^{n-1*}	... ...	$\varepsilon^{(n-1)^2/2}$ $\varepsilon^{(n-1)^2/2*}$	$3m$	

* $\varepsilon \equiv \exp(2\pi i t/n)$.

character tables, which can be easily developed by standard methods [(**3**), p. 312], [(**27**), Chap. 4] are shown in Tables 15 and 16 for odd and even values of n, respectively. The symmetry operation C^1 is a rotation by an angle of $2\pi t/n$ about the axis of the helix followed by a translation along the axis by $1/n$ of the unit cell dimension. The symbol C^k represents this symmetry operation performed k times in succession.

TABLE 16

$\mathscr{C}(2\pi t/n)$	E	C^1	C^2	...	C^{n-1}	$n^{(\gamma)}$	
A	1	1	1	...	1	$3m$	$T_{\parallel}, R_{\parallel}, \alpha_{xx}+\alpha_{yy}, \alpha_{zz}$
B^a	1	-1	1	...	1	$3m$	
E_1	1	ε	ε^2	...	ε^{n-1}	$3m$	$(T_{\perp}, T_{\perp})\,(\alpha_{yz}, \alpha_{zx})$
	1	ε^*	ε^{2*}	...	ε^{n-1*}		
$E_2{}^a$	1	2	4	...	$\varepsilon^{2(n-1)}$	$3m$	$(\alpha_{xx}-\alpha_{yy}, \alpha_{xy})$
	1	2	4	...	$\varepsilon^{2(n-1)*}$		
E_3	1	3	6	...	$\varepsilon^{3(n-1)}$	$3m$	
	1	3	6	...	$\varepsilon^{3(n-1)*}$		
⋮							
$E_{\frac{1}{2}n-1}$	1	$\varepsilon^{\frac{1}{2}n-1}$	ε^{n-2}	...	$\varepsilon^{(n-1)(\frac{1}{2}n-1)}$	$3m$	
	1	$\varepsilon^{\frac{1}{2}n-1*}$	ε^{n-2*}	...	$\varepsilon^{(n-1)(\frac{1}{2}n-1)*}$		

* $\varepsilon \equiv \exp(2\pi i t/n)$
[a] When $\frac{1}{2}n \leqslant 2$, species B, rather than E_2, contains two components of $\boldsymbol{\alpha}$.

The symmetry species of the various components of the dipole moment vector and the polarizability tensor are found using the method of Chapter 2, that is, by application of Eqns (2) and (73) of Chap. 2, and the magic formula. These results are shown in Tables 15 and 16, along with the classification of the internal and external degrees of freedom. If each chemical unit contains m atoms, the unit cell is made up of mn atoms having $3mn$ degrees of freedom. The external coordinates of the system are designated $T_{\parallel}$ and $T_{\perp}$ for translations parallel and perpendicular to the axis, respectively, and $R_{\parallel}$ for rotation about the axis.

In the event the chemical unit has local symmetry its symmetry operations must be combined with those of the factor group $\mathscr{C}(2\pi t/n)$. Thus, for the polyoxymethylene chain shown in Fig. 24 there are nine C_2 operations about axes perpendicular to the axis of the helix. Furthermore, the C^k operations form pairs belonging to the same class, as shown in Table 17. The resulting group is isomorphic with the point group $\mathscr{D}_n$, whose character table is constructed following the general rules for dihedral groups [(**3**), p. 318], [(**27**), p. 152].

TABLE 17

$\mathscr{D}\,(10\pi/9)$	E	$2C^1$	$2C^2$	$2C^3$	$2C^4$	$9C_2$		$n_{\mathrm{tot}}^{(\gamma)}$	$n_{\mathrm{skel}}^{(\gamma)}$
A_1	1	1	1	1	1	1	$\alpha_{xx} + \alpha_{yy}, \alpha_{zz}$	5	2
A_2	1	1	1	1	1	-1	$T_{\parallel}, R_{\parallel}$	7	4
E_1	2	$2\cos\phi$	$2\cos 2\phi$	$2\cos 3\phi$	$2\cos 4\phi$	0	$(T_{\perp}, T_{\perp})\,(\alpha_{yz}, \alpha_{zx})$	12	6
E_2	2	$2\cos 2\phi$	$2\cos 4\phi$	$2\cos 6\phi$	$2\cos 8\phi$	0	$(\alpha_{xx} - \alpha_{yy}, \alpha_{xy})$	12	6
E_3	2	$2\cos 3\phi$	$2\cos 6\phi$	$2\cos 9\phi$	$2\cos 12\phi$	0		12	6
E_4	2	$2\cos 4\phi$	$2\cos 8\phi$	$2\cos 12\phi$	$2\cos 16\phi$	0		12	6
χ (tot)	54	0	0	0	0	-2			
χ (skel)	108	0	0	0	0	-2			

$\phi = 10\pi/9$.

The results of the factor-group analysis of the helical polyoxymethylene chain are summarized in Table 17. The column headed $n_{tot}^{(\gamma)}$ gives the distribution of normal modes among the symmetry species. The external vibrations of rotation and translation should of course be subtracted from the entries in this column. These results predict 16 vibrational fundamentals $(5A_2 + 11E_1)$ active in infrared absorption and 28 Raman-active vibrations $(5A_1 + 11E_1 + 12E_2)$.

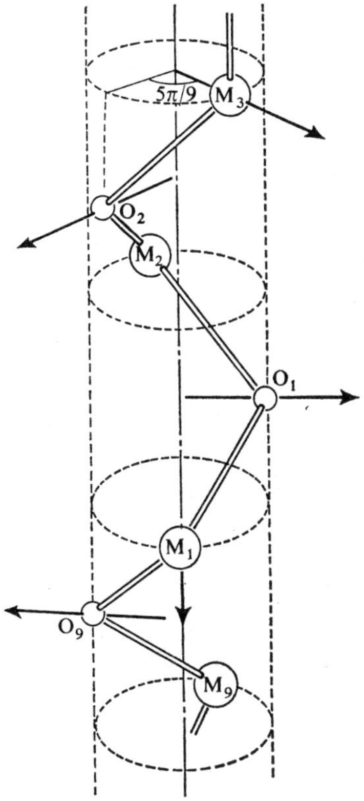

FIG. 24. Symmetry properties of the polyoxymethylene chain. Arrows indicate two-fold symmetry axes. Each methylene group is represented by M.

The skeletal vibrations of polyoxymethylene have been estimated from a simple force field by Tadokoro **(25)**, **(28)** using the **F** – **G** method. The calculation follows the same general procedure outlined earlier in the discussion of polyethylene. However, the presence of complex entries in the character table for $\mathscr{C}(2\pi t/n)$ results in some complication of the problem if one of these groups is used as the basis of the calculation. For details, the reader is referred to the original articles.

References

1. Decius, J. C. *J. Chem. Phys.* **23,** 1290 (1955).
2. Hirschfelder, J. O., Curtiss, C. F. and Bird, R. B. "Molecular Theory of Gases and Liquids", Wiley, New York (1954).
3. Wilson, E. B., Jr., Decius, J. C. and Cross, P. C. "Molecular Vibrations", McGraw-Hill, New York (1955).
4. Flory, P. J. *J. Am. Chem. Soc.* **58,** 1877 (1936).
5. Bhagavantam, S. and Venkatarayudu, T. "Theory of Groups and its Application to Physical Problems", Third Edition, Andhra University, Waltair (1962).
6. Zbinden, R. "Infrared Spectroscopy of High Polymers", Academic Press, New York (1964).
7. Primas, H. and Günthard, H. H. *Helv. Chim. Acta* **36,** 1659, 1791 (1953).
8. Tschamler, H. *J. Chem. Phys.* **22,** 1845 (1954).
9. Theimer, O. *J. Chem. Phys.* **27,** 406 (1957).
10. Snyder, R. G. *J. Mol. Spectry.* **4,** 411 (1960); **7,** 116 (1961).
11. Snyder, R. G. and Schachtschneider, J. H. *Spectrochim. Acta* **19,** 85 (1963).
12. Smith, A. E. *J. Chem. Phys.* **21,** 2229 (1953).
13. Krimm, S., Liang, C. Y. and Sutherland, G. B. B. M. *J. Chem. Phys.* **25,** 549 (1956).
14. Bunn, C. W. *Trans. Faraday Soc.* **35,** 482 (1939).
15. Tasumi, M. and Shimanoushi, T. *J. Chem. Phys.* **43,** 1245 (1965).
16. Boerio, F. J. and Koenig, J. L. *J. Chem. Phys.* **52,** 3425 (1970).
17. Zerbi, G., Ciampelli, F. and Zamboni, V. *J. Polymer Sci.* **7C,** 141 (1963).
18. Schachtschneider, J. H. and Snyder, R. G. *J. Polymer Sci.* **7C,** 99 (1963).
19. Schachtschneider, J. H. and Snyder, R. G. *Spectrochim. Acta* **19,** 117 (1963).
20. Snyder, R. G. *J. Chem. Phys.* **47,** 1316 (1967).
21. Tasumi, M., Shimanouchi, T. and Miyazawa, T. *J. Mol. Spectry.* **9,** 261 (1962).
22. Schachtschneider, J. H. and Snyder, R. G. *Spectrochim. Acta* **20,** 853 (1964).
23. Tasumi, M. and Krimm, S. *J. Chem. Phys.* **46,** 755 (1967).
24. Higgs, P. W. *Proc. Roy. Soc. (London)* **A220,** 470 (1953).
25. Tadokoro, H. *J. Chem. Phys.* **33,** 1558 (1960).
26. Liang, C. Y. and Krimm, S. *J. Chem. Phys.* **25,** 563 (1956).
27. Barchewitz, M. P. "Spectroscopic infrarouge I. Vibrations moleculaires", Gauthier-Villars, Paris (1961).
28. Miyazawa, T., Ideguchi, Y. and Fukushima, K. *J. Chem. Phys.* **38,** 2709 (1963).

Chapter 8

Spectra of Impure Crystals

The vibrational spectra of crystals containing low concentrations of impurities are of special interest for two reasons. If attention is focussed on the spectra of polyatomic impurities, these species can be studied in the absence of correlation coupling. Ordinarily, their rotational degrees of freedom are "frozen out" and the spectra consist of the pure-vibrational transitions of the essentially isolated impurity. On the other hand, if the lattice vibrations of an impurity system are considered, it is seen that the presence of an impurity results in the modification of the lattice frequencies and the appearance of so-called localized modes, which correspond roughly to the translational and rotational motions of the impurity in its surrounding cage.

Two different types of impurity system have been extensively studied: (1) ionic crystals (primarily alkali halides) "doped" with a low concentration of small impurity ions, and (2) solid matrices of inert molecules (usually the rare gases) in which have been trapped a small percentage of polyatomic molecules. Although the methods of sample preparation are very different in these cases, the two systems are quite analogous from the theoretical point of view. The term *matrix isolation*, which was suggested by Pimentel (**1**), is appropriate to both types of system, although it is usually used to describe molecules or free radicals in rare-gas matrices.

I. Infrared Spectra and Anharmonicity of the Cyanate Ion

The study of the vibrational spectrum of a polyatomic ion is greatly facilitated by substituting it in the crystal lattice of another compound. This technique was apparently first reported in 1928 by Maslakowez (**2**), who observed the infrared absorptions of the nitrate ion in solution in KCl. The infrared spectra of numerous polyatomic ions in ionic matrices have been reported by van der Elsken (**3**), Maki and Decius (**4**), Price *et al.* (**5**), Bryant and Turrell (**6**), and others. Alkali halides were used as matrix materials because of their transparence in the infrared region and the comparative ease of sample preparation.

The infrared absorption spectra of dilute solid solutions of NCO^- in various alkali halides were studied in detail by Maki and Decius (**4**). Single-crystal samples were prepared by modification of the Kyropoulos method (7), in which a seed crystal is slowly pulled from the fused alkali halide containing approximately 1% of the corresponding alkali cyanate.

The infrared spectra of these crystals were recorded from 600 to 5000 cm^{-1} at temperatures ranging from 50° to 480°K. Over seventy distinct absorption features were observed and assigned in the spectra of concentrated samples of the various isotopic species of NCO^- in KBr and KI matrices. These observations allowed the harmonic force constants and most of the anharmonicity constants to be determined for NCO^- in these two host lattices.

The vibrational energy (expressed in cm^{-1}) of a free polyatomic molecule having doubly degenerate vibrations is usually written in the form [(**8**), p. 210]

$$E_{\text{vib}} = \sum_{k} \omega_k \left(v_k + \frac{d_k}{2} \right) + \sum_{k} \sum_{k' \geqslant k} X_{kk'} \left(v_k + \frac{d_k}{2} \right) \left(v_{k'} + \frac{d_{k'}}{2} \right) + \sum_{k} \sum_{k' \geqslant k} g_{kk'} l_k l_{k'}, \tag{1}$$

where $d_k = 1$ or 2 is the degeneracy of the kth normal mode, $X_{kk'} = X_{k'k}$ is the anharmonicity constant which couples the normal modes k and k', and $g_{kk'}$ is a constant which takes into account the effect of rotation about the molecular axis. The quantum number l_k is given by $l_k = v_k, v_k - 2, v_k - 4 \ldots 1$ or 0. In spectroscopic applications the zero-point energy is not involved and the vibrational energy can be referred to the lowest vibrational level; then,

$$E_{\text{vib}} - E_{\text{vib}}^0 = \sum_{k} \omega_k{}^0 v_k + \sum_{k} \sum_{k' \geqslant k} X_{kk'}^0 v_k v_{k'} + \sum_{k} \sum_{k' \geqslant k} g_{kk'} l_k l_{k'} + \ldots, \tag{2}$$

where

$$\omega_k{}^0 = \omega_k + X_{kk'} d_k + \tfrac{1}{2} \sum_{k' \neq k} X_{kk'} d_{k'} + \ldots \tag{3}$$

and $X_{kk'}^0 = X_{kk'}$ if higher terms are neglected. The frequencies of the vibrational fundamentals are then given by

$$v_k = \omega_k{}^0 + X_{kk} + g_{kk} = \omega_k + X_{kk}(1 + d_k) + \tfrac{1}{2} \sum_{k' \neq k} X_{kk'} d_{k'} + g_{kk'}. \tag{4}$$

For a linear triatomic molecule Eqns (1) and (2) become, respectively,

$$E_{\text{vib}} = \omega_1(v_1 + \tfrac{1}{2}) + \omega_2(v_2 + 1) + \omega_3(v_3 + \tfrac{1}{2}) + X_{11}(v_1 + \tfrac{1}{2})^2 + X_{22}(v_2 + 1)^2 + X_{33}(v_3 + \tfrac{1}{2})^2 + X_{12}(v_1 + \tfrac{1}{2})(v_2 + 1) + X_{13}(v_1 + \tfrac{1}{2})(v_3 + \tfrac{1}{2}) + X_{23}(v_2 + 1)(v_3 + \tfrac{1}{2}) + g_{22}\, l_2^{\,2} + \ldots, \quad (5)$$

and

$$E_{\text{vib}} - E^0_{\text{vib}} = \sum_{k=1}^{3} \omega_k^{\,0}\, v_k + \sum_{k' \leqslant k = 1}^{3} X^0_{kk'}\, v_k v_{k'} + g_{22}\, l_2^{\,2}, \quad (6)$$

where $k = 1, 3$ for the stretching modes and $k = 2$ for the doubly degenerate bending vibration. A given vibrational state of the molecule is thus specified by the set of four quantum numbers $v_1\, v_2^{\,l_2}\, v_3$, where the values of $l_2 = 0, 1, 2, \ldots$, which correspond to species Σ, Π, $\Delta \ldots$, represent the vibrational angular momentum about the molecular axis.

For a molecule or complex ion in the solid state, Eqns (5) and (6) should be augmented by terms representing overall rotation and translation. However, as the energies of these external motions are quite small compared to the internal vibrational energy, it is possible to fit the observed infrared spectrum of NCO^- by including only the internal degrees of freedom, provided that the strong Fermi resonance between ν_1 and $2\nu_2$ is considered.

In general, the energies of Fermi multiplets can be calculated from the determinantal equation

$$\begin{vmatrix} W_1^{\,0} - W & W_{12} & \cdots & W_{1n} \\ W_{21} & W_2^{\,0} - W & & \\ \vdots & & & \\ W_{n1} & & & W_n^{\,0} - W \end{vmatrix} = 0, \quad (7)$$

where the $W_n^{\,0}$'s are the unperturbed energies, W is the perturbed energy, and

$$W_{nn'} = g_{122} \int \psi_n^{*}\, Q_1\, Q_2^{\,2}\, \psi_{n'}\, d\tau. \quad (8)$$

The factor g_{122} which is the coefficient of $Q_1\, Q_2^{\,2}$ in the anharmonic potential expression, is neglected in the harmonic approximation [See Eqn (23), Chap. 1]. For Fermi doublets of the type ν_1, $2\nu_2$, (**8**)

$$W_{v_1,\, v_2,\, v_3;\, v_1 - 1,\, v_2 + 2,\, v_3} = -\,2^{-3/2} b\, v_1^{\,1/2} [(v_2 + 2)^2 - l_2^{\,2}]^{1/2}. \quad (9)$$

Thus, with v_1, $v_2^{\,l_2}$, $v_3 = 1\,0^0\,0$ one finds

$$W_{10^00,\, 02^00} = -\, b/\sqrt{2}. \quad (10)$$

TABLE 1 [From reference (**4**)].

Frequency obs (cm^{-1})	Lower state $v_1v_2{}^lv_3$	Upper state $v_1v_2{}^lv_3$	Remarks	$\nu_{obs} - \nu_{calc}$ (cm^{-1})
612·0	$0\,0^0\,0$	$0\,1^1\,0$	C^{13}	0·0[c]
629·4	$0\,0^0\,0$	$0\,1^1\,0$	...	0·0[c]
1175·8[a]	$0\,0^0\,0$	$1\,0^0\,0$	O^{18}	...
...	$0\,0^0\,0$	$0\,2^0\,0$	O^{18}	...
1188·9[a]	$0\,0^0\,0$	$1\,0^0\,0$	N^{15}	...
1282·3[a]	$0\,0^0\,0$	$0\,2^0\,0$	N^{15}	...
1191·0[a]	$0\,0^0\,0$	$1\,0^0\,0$	C^{13}	0·0[c]
1272·5	$0\,0^0\,0$	$0\,2^0\,0$	C^{13}	0·0[c]
1205·5	$0\,0^0\,0$	$1\,0^0\,0$	...	0·0[c]
1292·6	$0\,0^0\,0$	$0\,2^0\,0$	...	0·0[c]
2094·6[b]	$0\,0^0\,0$	$0\,0^0\,1$	$N^{15}C^{13}$	...
2104·8[b]	$0\,0^0\,0$	$0\,0^0\,1$	$C^{13}O^{18}$	...
2112·8	$0\,0^0\,0$	$0\,0^0\,1$	C^{13}	0·0[c]
2152·5[a]	$0\,0^0\,0$	$0\,0^0\,1$	N^{15}	...
2161·5[a]	$0\,0^0\,0$	$0\,0^0\,1$	O^{18}	...
2169·6	$0\,0^0\,0$	$0\,0^0\,1$	vs[e]	0·0[c]
2360·3[b]	$0\,0^0\,0$	$2\,0^0\,0$	C^{13}	3·3[d]
2462·6	$0\,0^0\,0$	$1\,2^0\,0$	C^{13}	−2·1[d]
2554·2	$0\,0^0\,0$	$0\,4^0\,0$	C^{13}	−3·5[d]
2392·9	$0\,0^0\,0$	$2\,0^0\,0$	...	2·3[d]
2487·3	$0\,0^0\,0$	$1\,2^0\,0$	...	2·6[d]
2602·5	$0\,0^0\,0$	$0\,4^0\,0$	...	−4·7[d]
2714·0[b]	$0\,0^0\,0$	$0\,1^1\,1$	C^{13}	−0·3
2788·1	$0\,0^0\,0$	$0\,1^1\,1$	...	0·2
3284·1	$0\,0^0\,0$	$1\,0^0\,1$	C^{13}	0·0[c]
3366·0[a]	$0\,0^0\,0$	$0\,2^0\,1$	C^{13}	0·0[c]
3322·0[a]	$0\,0^0\,0$	$1\,0^0\,1$	N^{15}	...
3414·7	$0\,0^0\,0$	$0\,2^0\,1$	N^{15}	...
3355·0	$0\,0^0\,0$	$1\,0^0\,1$	...	0·0[c]
3442·2	$0\,0^0\,0$	$0\,2^0\,1$	...	0·0[c]
4524·7	$0\,0^0\,0$	$2\,0^0\,1$	...	5·6
4623·0	$0\,0^0\,0$	$1\,2^0\,1$	...	7·5
4737·5	$0\,0^0\,0$	$0\,4^0\,1$	...	1·1
574·4	$0\,1^1\,0$	$1\,0^0\,0$	...	−1·7
663·2	$0\,1^1\,0$	$0\,2^0\,0$	...	0·0
1174·4[b]	$0\,1^1\,0$	$1\,1^1\,0$	C^{13}	...
1285·7[b]	$0\,1^1\,0$	$0\,3^1\,0$	C^{13}	...
1188·1	$0\,1^1\,0$	$1\,1^1\,0$	...	1·9[d]
1309·1	$0\,1^1\,0$	$0\,3^1\,0$	...	−1·8[d]
2102·2	$0\,1^1\,0$	$0\,1^1\,1$	C^{13}	−0·1[c]
2158·6	$0\,1^1\,0$	$0\,1^1\,1$	...	0·0[c]
...	$0\,1^1\,0$	$2\,1^1\,0$	...	...
...	$0\,1^1\,0$	$1\,3^1\,0$	...	...
2628·3	$0\,1^1\,0$	$0\,5^1\,0$	...	−11·1
2725	$0\,1^1\,0$	$1\,0^0\,1$	vw[e]	−0·6
2813	$0\,1^1\,0$	$0\,2^0\,1$	vw[e]	0·2
2776·6	$0\,1^1\,0$	$0\,2^2\,1$	...	0·0[c]
3257·2[b]	$0\,1^1\,0$	$1\,1^1\,1$	C^{13}	...
...	$0\,1^1\,0$	$0\,3^1\,1$	C^{13}	...
3325·7	$0\,1^1\,0$	$1\,1^1\,1$	...	1·4
...	$0\,1^1\,0$	$0\,3^1\,1$	...	...
1175·5	$0\,2^2\,0$	$1\,2^2\,0$	...	5·4
1320·4	$0\,2^2\,0$	$0\,4^2\,0$	...	−5·7
2147·9	$0\,2^2\,0$	$0\,2^2\,1$	...	0·4

[a] Observable only at low temperatures.
[b] Observed only in 60% C^{13} sample.
[c] Used in evaluating energy constants.
[d] Sum of this multiplet used in evaluating energy constants.
[e] Very strong, vs; very weak, vw.

The constant b, which is proportional to g_{122}, can be considered as a parameter which, in addition to the ten coefficients in Eqn (6), is to be used in fitting the observed frequencies.

TABLE 2.

	KBr solid solution		KI solid solution		
	$N^{14}C^{12}O^{16}$	$N^{14}C^{13}O^{16}$	$N^{14}C^{12}O^{16}$	CO_2[a]	N_2O[b]
ω_1^0	1254·03	1253·96	1248·2	1345·06	1282·11
ω_2^0	629·64	612·1	627·9	667·02	588·43
ω_3^0	2182·60	2125·8	2168·8	2361·8	2238·85
X_{11}	−4·92	−4·91	−4·35	−2·20	−5·21
X_{22}	−2·58	−2·44	−2·7	−0·75	−0·17
X_{33}	−13·0[c]	−12·3[c]	−13·0[c]	−12·50	−15·10
X_{12}	9·40	9·16	9·20	3·76	0·52
X_{23}	−11·05	−10·5	−11·10	−11·58	−14·22
X_{31}	−18·0	−17·6	−19·80	−21·84	−27·26
g	2·33	2·3	2·8	1·03	0·52
b	61·59	52·17	61·65	72·1	40·0
ω_1	1258·55	1256·88	1253·2	1354·42	1300·43
ω_2	635·62	617·8	634·25	672·43	595·62
ω_3	2215·7	2156·65	2202·80	2396·8	2281·80

[a] See reference (**9**).
[b] See reference (**10**).
[c] Assumed value (See text).

Infrared absorptions and their assignments, as reported by Maki and Decius (**4**), for NCO^- in KBr are shown in Table 1. The harmonic frequencies, anharmonicity constants, and Fermi resonance parameters calculated from these data are given in Table 2. As no overtones of ν_3 were observed, the quantity X_{33} could not be evaluated and was inferred from corresponding data for the isoelectronic molecules CO_2 and N_2O. In Table 1 the observed frequencies are compared with those calculated from Eqn (6) and the entries in Table 2. The harmonic frequencies ω_1, ω_2, and ω_3 of

Table 2 can be used to calculate the force constants. The secular equations in this case lead to

$$|\mathbf{GF} - \mathbf{E}\lambda_k| =$$

$$\left| \begin{pmatrix} \mu_N + \mu_C & -\mu_C & 0 \\ -\mu_C & \mu_C + \mu_O & 0 \\ 0 & 0 & \dfrac{\mu_N}{r_{NC}^2} + \dfrac{\mu_O}{r_{CO}^2} + \left(\dfrac{1}{r_{NC}} + \dfrac{1}{r_{CO}}\right)^2 \mu_C \end{pmatrix} \begin{pmatrix} f_{NC} & f' & 0 \\ f' & f_{CO} & 0 \\ 0 & 0 & f_\alpha \end{pmatrix} - \mathbf{E}\lambda_k \right| = 0, \quad (11)$$

or,

$$\mu_N f_{NC} + \mu_O f_{CO} + \mu_C (f_{NC} + f_{CO} - f') = \lambda_1 + \lambda_3 = 4\pi^2 (\omega_1{}^2 + \omega_3{}^2), \quad (12)$$

$$(\mu_N \mu_C + \mu_C \mu_O + \mu_O \mu_N)(f_{NC} f_{CO} - f'^2) = \lambda_1 \lambda_3 = 16\pi^4 \omega_1{}^2 \omega_3{}^2, \quad (13)$$

and

$$\left[\frac{\mu_N}{r_{NC}^2} + \frac{\mu_O}{r_{CO}^2} + \left(\frac{1}{r_{NC}} + \frac{1}{r_{CO}}\right)^2 \mu_C\right] f_\alpha = \lambda_2 = 4\pi^2 \omega_2{}^2. \quad (14)$$

It is important to note that in this work the sum rule (**11**) was applied to the frequencies of the various isotopic species allowing a unique set of force constants to be chosen.

The spectroscopic results reported for the KBr : NCO^- system (**4**), (**5**), (**12**) are summarized in Fig. 1. The number of observed spectroscopic transitions is striking in view of the relative simplicity of this triatomic system. As a result, the vibrational energy-level diagram of NCO^- in KBr appears to be as well established as those of the classical gaseous molecules CO_2 and N_2O.

II. Impurity–Lattice Interaction

In the preceding section the intra-ionic potential function for NCO^- was evaluated from the observed vibrational frequencies. The effect of the environment of the NCO^- impurtiy was not considered, although it was found that the potential constants varied somewhat as functions of the nature of the host lattice. This perturbation by the lattice will now be considered more specifically.

If a polyatomic species is introduced as a substitutional impurity in a crystal lattice, its vibrational frequencies are dependent upon the potential function which governs its interaction with the lattice. If a satisfactory model of impurity–lattice interaction is available, a quantitative method of determining intermolecular or interionic forces is offered by measuring the vibrational frequencies of the impurity as functions of temperature, pressure, and the nature of the lattice. Thus, the introduction of an impurity as a "probe" in a crystal provides a direct technique for investigating lattice forces.

The classic model of the interaction of a solute molecule with its solvent was derived by Bauer and Magat (**13**). The more general form of their de-

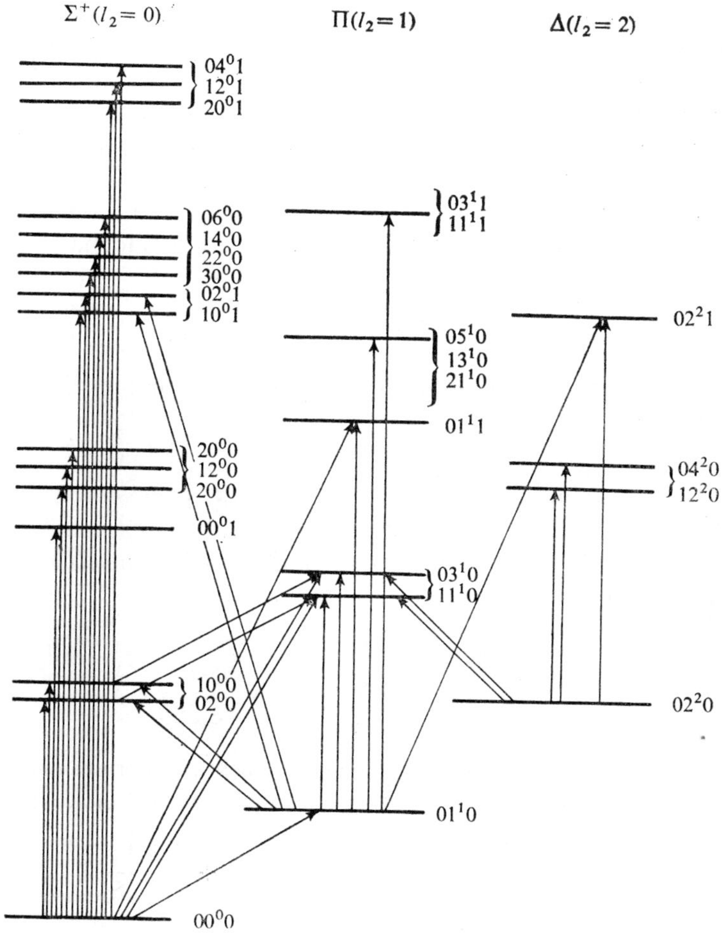

FIG. 1. Observed vibrational transitions in the cyanate ion, NCO^-, in a KBr lattice.

velopment assumes that the vibrational energy of a dissolved molecule can be written as

$$V = V_{in} + V' = \tfrac{1}{2} \sum_{jj'} \mathsf{S}_j \mathsf{F}_{jj'} \mathsf{S}_{j'} + \sum_{jj'j''} \beta_{jj'j''} \mathsf{S}_j \mathsf{S}_{j'} \mathsf{S}_{j''} + \ldots + V', \quad (15)$$

where V' represents the external forces acting on the solute molecule or ion and the potential function for the internal motions has been augmented to include anharmonic terms. This model of solute–solvent interaction is often used to describe solvent effects in liquid solutions (**14**). However, it is equally useful in the analysis of the spectra of impurities or "guests" in solid host lattices (**6**).

In Eqn (15) symmetry coordinates were used as the basis, although any convenient set of internal or external coordinates can be used. The use of symmetry coordinates, however, allows the molecular symmetry to determine restrictions on the elements of $\mathbf{F}$ and $\boldsymbol{\beta}$. The arguments of Chapter 2 which showed that certain elements of $\mathbf{F}$ may vanish by symmetry, can be extended to determine the form of $\boldsymbol{\beta}$ [(**15**), p. 193].

By taking the derivative of Eqn (15) with respect to a given symmetry coordinate S_s, and dropping terms higher than cubic, it is found that

$$\frac{\partial V}{\partial \mathsf{S}_s} = \sum_t \mathsf{F}_{ts} \mathsf{S}_t + 3 \sum_{tt'} \beta_{tt's} \mathsf{S}_t \mathsf{S}_{t'} + \frac{\partial V'}{\partial \mathsf{S}_s}. \quad (16)$$

The condition for equilibrium of the molecular configuration requires that Eqn (16) vanish, viz.,

$$\sum_t \mathsf{F}_{ts} \mathsf{S}_t^0 + 3 \sum_{tt'} \beta_{tt's} \mathsf{S}_t^0 \mathsf{S}_{t'}^0 + \left(\frac{\partial V'}{\partial \mathsf{S}_s}\right)_0 = 0, \quad (17)$$

where the vector $\mathbf{S}^0$ describes the distortion of the solute molecule resulting from perturbation by the solvent. The second derivatives of Eqn (15) are given by

$$\frac{\partial^2 V}{\partial \mathsf{S}_s \partial \mathsf{S}_{s'}} = \mathsf{F}_{ss'} + 6 \sum_t \beta_{tss'} \mathsf{S}_t + \frac{\partial^2 V'}{\partial \mathsf{S}_s \partial \mathsf{S}_{s'}}. \quad (18)$$

These second derivatives, evaluated under the new equilibrium conditions, can be identified with the force constants of the perturbed molecule. Thus,

$$\mathsf{F}_{ss'}' \equiv \left(\frac{\partial^2 V}{\partial \mathsf{S}_s \partial \mathsf{S}_{s'}}\right) = \mathsf{F}_{ss'} + 6 \sum_t \beta_{tss'} \mathsf{S}_t^0 + \left(\frac{\partial^2 V'}{\partial \mathsf{S}_s \partial \mathsf{S}_{s'}}\right)_0. \quad (19)$$

It is apparent from Eqn (19) that, if the potential function governing the vibrations of a polyatomic impurity can be considered to be that of the free molecule or ion, plus a perturbation resulting from the surrounding matrix [Eqn (15)], the potential V' has two distinct effects on the impurity, viz.

(i) The second derivatives of V' with respect to appropriate internal coordinates represent additional contributions to the effective force constants of the impurity, and

(ii) The first derivatives of V' constitute forces which modify the equilibrium geometry of the impurity. Although in the harmonic approximation, the latter effect would not alter the vibrational frequencies, when cubic and higher terms are included in the internal potential function, the changes in interatomic distances produced by the perturbation will cause further modifications of the effective quadratic force constants of the impurity.

If, following Bauer and Magat (**13**), it is assumed that impurity–lattice interactions arise solely from dipole–induced-dipole forces, the perturbing potential depends on the instantaneous dipole moment of the perturbed molecule and the dielectric properties of the surrounding solvent. If the solvent is represented by a continuous isotropic medium of dielectric constant ε, the perturbing potential is given by

$$V' = -\frac{\frac{1}{2}\gamma \mathbf{m}^2}{1 - \gamma\alpha}, \tag{20}$$

where $\gamma = \dfrac{2(\varepsilon - 1)}{(2\varepsilon + 1)a^3}$ and the solute molecule of average polarizability α and dipole moment $\mathbf{m}$ has been assumed to occupy the center of a spherical cavity of radius a in the dielectric medium. By carrying out the expansions

$$\mathbf{m} = \mathbf{m}_0 + \sum_s \left(\frac{\partial \mathbf{m}}{\partial S_s}\right)_0 S_s + \tfrac{1}{2}\sum_{s,s'} \left(\frac{\partial^2 \mathbf{m}}{\partial S_s \partial S_{s'}}\right)_0 S_s S_{s'} + \dots \tag{21}$$

and

$$\alpha = \alpha_0 + \sum_s \left(\frac{\partial \alpha}{\partial S_s}\right)_0 S_s + \tfrac{1}{2}\sum_{s,s'} \left(\frac{\partial^2 \alpha}{\partial S_s \partial S_{s'}}\right) S_s S_{s'} + \dots \tag{22}$$

and assuming that $\gamma\alpha \ll 1$, one finds, to second order,

$$V' = -\tfrac{1}{2}\gamma \Big((1 + \gamma\alpha_0)\mathbf{m}_0^2 + \sum_s [2(1 + \gamma\alpha_0)\mathbf{m}_0 \cdot \mathbf{m}_s' + \gamma \mathbf{m}_0^2 \alpha_s'] S_s$$
$$+ \sum_{s,s'} \{(1 + \gamma\alpha_0)[\mathbf{m}_s' \cdot \mathbf{m}_{s'}' + \mathbf{m}_0 \cdot \mathbf{m}_{ss'}''] + 2\gamma\alpha_s' \mathbf{m}_0 \cdot \mathbf{m}_{s'}'\} S_s S_{s'} + \dots \Big), \tag{23}$$

where $\mathbf{m}_s' \equiv \left(\dfrac{\partial \mathbf{m}}{\partial S_s}\right)_0$, $\mathbf{m}_{ss'}'' \equiv \left(\dfrac{\partial^2 \mathbf{m}}{\partial S_s \partial S_{s'}}\right)_0$, and similarly for α.

The derivatives needed in Eqns (17) and (19) are then given by

$$\left(\frac{\partial V'}{\partial \mathsf{S}_s}\right)_0 = -\gamma(1+\gamma\alpha_0)\mathbf{m}_0\cdot\mathbf{m}_s' - \frac{\gamma^2}{2}\mathbf{m}_0{}^2\alpha_s' + \ldots \tag{24}$$

and

$$\left(\frac{\partial^2 V'}{\partial \mathsf{S}_s \partial \mathsf{S}_{s'}}\right)_0 = -\gamma(1+\gamma\alpha_0)[\mathbf{m}_s'\cdot\mathbf{m}_{s'}' + \mathbf{m}_0\cdot\mathbf{m}_{ss'}''] - 2\gamma^2\alpha_s'\,\mathbf{m}_0\cdot\mathbf{m}_{s'}' + \ldots. \tag{25}$$

The theory outlined above has been applied with some success to the interpretation of solvent effects on the vibrational frequencies of small molecules in liquid solutions (**14**). It has also been modified to include variations in vibrational harmonicity with the nature of the solvent (**16**). However, in order to extend this approach to the case of a substitutional impurity in a crystal lattice it is necessary to take into consideration the geometry of the host crystal and the equilibrium orientation of the guest.

The interaction between a point dipole and spherical atom or molecule of polarizability α_s at a distance r is given by

$$V' = -\frac{\mathbf{m}^2\alpha_s}{2r^6}(3\cos^2\theta + 1), \tag{26}$$

where θ is the angle between the dipole and the line of centers between the two interacting particles. In a crystal, expressions of the form of Eqn (26) must be summed over all significant interactions. Since these forces are long-range in nature a considerable number of terms is necessary. General expressions for these sums have been evaluated by Lennard-Jones and Ingham (**17**). The cubic symmetry allows the factor $3\cos^2\theta + 1$ to be replaced by 2.

For the case of an impurity in an NaCl-type crystal of cell dimension a the anions are at face-centered positions, while the cations occupy all sites in a simple cubic lattice of dimension $\frac{1}{2}a$, less those positions occupied by the anions. The interaction potential is of the form

$$V' = -\frac{1}{a^6}[8C_6\alpha^- + (64A_6 - 8C_6)\alpha^+]\,\mathbf{m}^2, \tag{27}$$

where α^- and α^+ are the polarizabilities of the anion and cation, respectively, and $A_6 = 8{\cdot}4019$ and $C_6 = 14{\cdot}4539$ are the lattice sums calculated by Lennard-Jones and Ingham. It should be noted that the polarization of the impurity by the reaction of the lattice has been neglected in developing

Eqn (27). By expanding $\mathfrak{m}$ in Eqn (27), as before, one can easily obtain the derivatives

$$\left(\frac{\partial V'}{\partial \mathsf{S}_s}\right)_0 = -\frac{2}{a^6}[8C_6\alpha^- + (64A_6 - 8C_6)\alpha^+]\mathfrak{m}_0 \cdot \mathfrak{m}_s' \qquad (28)$$

and

$$\left(\frac{\partial^2 V'}{\partial \mathsf{S}_s \partial \mathsf{S}_{s'}}\right)_0 = -\frac{2}{a^6}[8C_6\alpha^- + (64A_6 - 8C_6)\alpha^+](\mathfrak{m}_s' \cdot \mathfrak{m}_{s'}' + \mathfrak{m}_0 \cdot \mathfrak{m}_{ss'}''), \qquad (29)$$

which are analogous to Eqns (24) and (25) obtained for the spherical cavity model.

The extension of the above model to include short-range repulsive forces is straightforward. In this case the forces fall off so rapidly that, ordinarily, only nearest-neighbor interactions need be considered.

III. Infrared Spectra of the Azide Ion in Alkali–Halide Lattices

The general theory outlined in the previous section was used to interpret the infrared spectra of the azide ion, N_3^-, in a number of alkali–halide matrices (**6**).

The azide ion, being linear and symmetric, belongs to the point group $\mathscr{D}_{\infty h}$. It has three fundamental vibrations, a symmetric stretching mode ν_1 ($A_{1g} \equiv \Sigma_g^+$), a doubly degenerate mode, ν_2 ($E_u \equiv \Pi_u$), and an asymmetric stretching mode, ν_3 ($A_{2u} \equiv \Sigma_u^+$). The symmetric stretching fundamental is of course inactive in the infrared spectrum.

The harmonic frequencies corresponding to the two infrared-active fundamentals of N_3^- are given by

$$\lambda_2{}^0 = 4\pi^2\omega_2{}^{02} = 6F_{22}/M \qquad (30)$$

and

$$\lambda_3{}^0 = 4\pi^2\omega_3{}^{02} = 3(F_{33} - F_{13})/M, \qquad (31)$$

where the $F_{tt'}$'s are the elements of the **F** matrix based on internal coordinates

When a polyatomic ion or molecule is introduced as a substitutiona· impurity in a lattice, the resulting symmetry depends on the symmetry of the available site and the symmetry and orientation of the impurity. The possible orientations of a linear and symmetric ion such as N_3^- substituted at an octahedral site are shown in Table 3, along with the resulting point groups, which are subgroups of $\mathscr{O}_h$. Table 3 indicates the symmetry species of both the internal and external vibrations of the azide impurity.

TABLE 3 [From reference (**6**)].

Subgroup→ (orientation)	$\mathscr{D}_{\infty h}$ ("free")	$\mathscr{D}_{4h}$ (on C_4)	$\mathscr{D}_{3d}$ (on C_3)	$\mathscr{D}_{2h} \equiv \mathscr{V}_h$ (on C_2)	$\mathscr{C}_i$ (on no C)	Activity
ν_1	Σ_g^+	A_{1g}	A_{1g}	A_g	A_g	Raman
ν_3	Σ_u^+	A_{2u}	A_{2u}	B_{1u}	A_u	ir
ν_2	Π_u	E_u	E_u	B_{2u}	A_u	ir
				B_{3u}	A_u	
(R_x, R_y)	Π_g	E_g	E_g	B_{2g}	A_g	Raman
				B_{3g}	A_g	
T_z	Σ_u^+	A_{2u}	A_{2u}	B_{1u}	A_u	ir
(T_x, T_y)	Π_u	E_u	E_u	B_{2u}	A_u	ir
				B_{3u}	A_u	

It can be seen from Table 3 that the perpendicular vibration of the azide ion should exhibit site splitting if the orientation is either along the twofold axis or a more general one. The apparent absence of splitting of the perpendicular vibrations of the cyanate ion, plus the observed temperature dependence of the frequencies of the vibrational fundamentals (increasing with decreasing

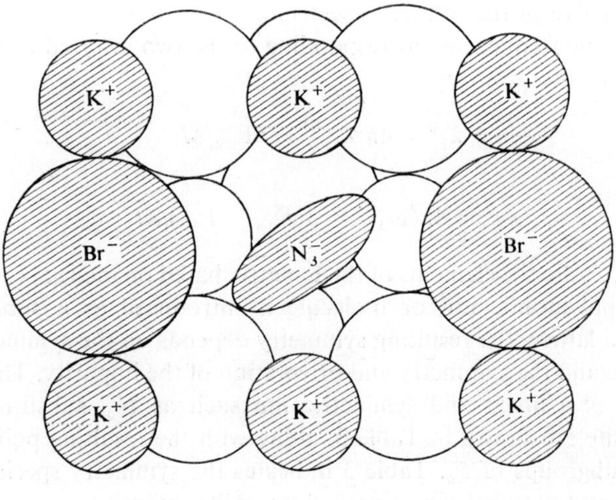

FIG. 2. Azide ion in a KBr lattice (C_3 orientation) [After Bryant and Turrell (**6**)].

temperature), strongly suggest that the ion is oriented along the threefold axis of the cubic cell. If the same orientation can be assumed for the azide ion in an alkali–halide matrix, the resulting subgroup is $\mathscr{D}_{3d}$ and the representations are given in Table 3. A diagram of the azide ion in such a matrix is shown in Fig. 2.

Spectra were obtained of the azide ion in KI, KBr, KCl, and NaCl host lattices. The frequency of the asymmetric stretching mode as a function of

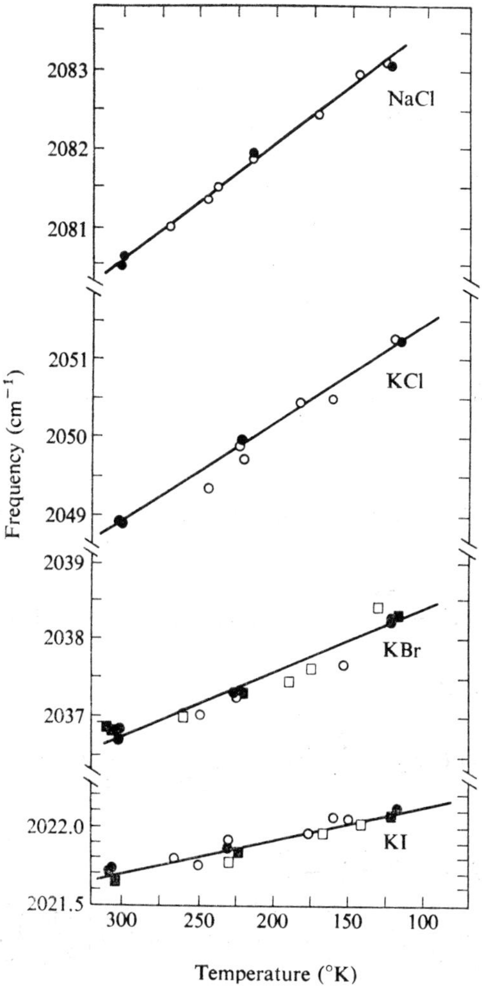

FIG. 3. Temperature dependence of the asymmetric stretching frequency ν_3 of the azide ion in various lattices. Note breaks in ordinate scale. Squares represent data taken on single crystals; solid points are at equilibrium temperatures [From reference (6)].

temperature and the nature of the lattice is shown in Fig. 3. These data were obtained using pressed-disk samples, as well as single crystals grown from aqueous solutions of the alkali halides containing approximately 1 % of the corresponding alkali azide. It was shown that solid solutions can be obtained by careful application of the pressed-disk technique.

The quantitative interpretation of the data of Fig. 3 requires a specific model of the impurity–lattice forces. Since each of the three vibrational modes of N_3^- is of a different symmetry species, the symmetry coordinates, when properly mass-weighted, become identical to the normal coordinates; that is,

$$Q_k = \sqrt{M}\, S_k \quad \text{and} \quad \left(\frac{\partial \mathfrak{m}}{\partial S_k}\right)_0 = \sqrt{M}\left(\frac{\partial \mathfrak{m}}{\partial Q_k}\right).$$

Furthermore, the azide ion has no permanent dipole moment. Equations (28) and (29) then become, respectively,

$$\left(\frac{\partial V'}{\partial S_k}\right)_0 = 0 \tag{32}$$

and

$$\left(\frac{\partial^2 V'}{\partial S_k^{\,2}}\right)_0 = -\frac{2M}{a^6}[8C_6\alpha^- + (64A_6 - 8C_6)\alpha^+]\left(\frac{\partial \mathfrak{m}}{\partial Q_k}\right)_0^2, \tag{33}$$

where $k = 2, 3$.

From Eqn (19) the effective force constants for the two infrared-active fundamentals become

$$F_k' = F_k - \frac{2M}{a^6}[8C_6\alpha^- + (64A_6 - 8C_6)\alpha^+]\left(\frac{\partial \mathfrak{m}}{\partial Q_k}\right)_0^2 \tag{34}$$

and, with $G_k = 1/M$

$$\lambda_k = G_k F_k' = G_k F_k - \frac{2}{a^6}[8C_6\alpha^- + (64A_6 - 8C_6)\alpha^+]\left(\frac{\partial \mathfrak{m}}{\partial Q_k}\right)_0^2 \tag{35}$$

and

$$\begin{aligned}\Delta\lambda_k &\equiv \lambda_k - \lambda_k^{\,0} = \lambda_k - G_k F_k \\ &= -\frac{2}{a^6}[8C_6\alpha^- + (64A_6 - 8C_6)\alpha^+]\left(\frac{\partial \mathfrak{m}}{\partial Q_k}\right)_0^2.\end{aligned} \tag{36}$$

As the frequency shifts are small, one can write

$$\Delta\lambda_k \approx 8\pi^2 \nu_k^{\,0} \Delta\nu_k, \tag{37}$$

or, since $\lambda_k^0 = 4\pi^2 \nu_k^{02}$,

$$\frac{\Delta\nu_k}{\nu_k^0} = -\frac{1}{a^6\lambda_k^0}[8C_6\alpha^- + (64A_6 - 8C_6)\alpha^+]\left(\frac{\partial\mathfrak{m}}{\partial Q_k}\right)_0^2. \tag{38}$$

The quantity $\left(\frac{\partial\mathfrak{m}}{\partial Q_k}\right)_0$ can be evaluated from the absolute intensity of the kth fundamental [See Eqns (68) and (75), Chap. 1]. Some results for the azide ion are shown in Table 4, where it is seen that the more intense fundamental, ν_3, is much more strongly perturbed than ν_2, and that the effect of inductive forces is to lower the vibrational frequencies from those of the free ion. The perturbed frequencies calculated from Eqn (38) decrease slightly with decreasing temperature, while the opposite effect is observed experimentally (See Fig. 3). Thus, inductive forces cannot account for the observed lattice perturbation.

TABLE 4 [From reference (6)].

Lattice	$(\partial\mathfrak{m}/\partial Q_3)_0^2$ (10^3 esu/g)[a]	$(\partial\mathfrak{m}/\partial Q_2)_0^2$ (10^3 esu/g)[a]	Temperature (°K)	$-\Delta\omega_3$ (cm^{-1})	$-\Delta\omega_2$ (cm^{-1})
KCl	30	1·9	120	5·4	1·1
			300	5·2	1·1
KBr	34	1·8	120	5·5	0·9
			300	5·3	0·9
KI	50	2·3	120	6·9	1·0
			300	6·6	1·0

[a] Estimated from observed intensity relative to cyanate and using the data of reference (**4**).

The short-range repulsive forces are usually represented by an exponential function of the distance between an atom of the impurity and a given ion of the lattice. The potential function in this case is of the form

$$V_s = \zeta\, e^{-R/\rho}, \tag{39}$$

where ζ and ρ are interpreted physically as a relative electronic charge density and an effective charge radius, respectively. These constants have been determined for a number of systems, including rare gases (**18**) and pure alkali–halide crystals (**19**).

Applying Eqn (39) to each ion of the lattice and summing, the total short-range potential becomes

$$V_s = \sum_{n_1,n_2,n_3} \zeta^{\ddagger} \exp[-R(n_1,n_2,n_3)/\rho^{\ddagger}], \tag{40}$$

where $\ddagger = +$ if $n_1 + n_2 + n_3$ is odd and $\ddagger = -$ if $n_1 + n_2 + n_3$ is even. The superscripts on ζ and ρ then indicate whether a given interaction is with a cation or an anion of the host lattice. In Eqn (40)

$$R(n_1, n_2, n_3) = \tfrac{1}{2}a[(n_1 - \eta\sqrt{3}\sin\theta\cos\phi)^2 + (n_2 - \eta\sqrt{3}\sin\theta\sin\phi)^2 + (n_3 - \eta\sqrt{3}\cos\theta)^2]^{\frac{1}{2}} \quad (41)$$

is a function of the crystallographic indices [See Eqn (65), Chap. 3] and $\eta = 2r/a\sqrt{3}$ for the C_3 orientation of the azide ion. The instantaneous distances between a nitrogen atom and the center of the site is given by r, while θ and ϕ have their usual significance in spherical polar coordinates with the Z axis colinear with a C_3 axis of the site.

Although the azide ion has no permanent dipole moment, it does have a quadrupole moment; thus, interactions of the type quadrupole–induced-dipole can occur. These forces can be treated in terms of a point quadrupole, although an extended quadrupole consisting of point charges offers some advantages in cases in which the charge distribution extends over dimensions comparable to the interionic distances in the crystal. If, as suggested by Bonnemay and Daudel (**20**), the azide ion can be represented by the charge distribution

$$\underset{-0{\cdot}83e}{\mathrm{N}} \text{———} \underset{0{\cdot}66e}{\mathrm{N}} \text{———} \underset{-0{\cdot}83e}{\mathrm{N}},$$

this contribution to the perturbing potential can be estimated by assuming that, for small displacements, the effective charges move with the nuclei. Then, for the interaction of an end nitrogen atom with a given atom of the lattice

$$V_Q = -0{\cdot}83e^2 \sum_{n_1,n_2,n_3}{}' (-1)^{n_1+n_2+n_3}/R(n_1, n_2, n_3), \quad (42)$$

where $R(n_1, n_2, n_3)$ is given by Eqn (41) and the sign alteration has been included to account for the charges on the anions and cations of the lattice. The prime on the summation sign indicates that the term $n_1 = n_2 = n_3 = 0$ (the impurity site) is omitted from the sum.

It was shown above that in the case of nonpolar impurities, polarization of the impurity by the reaction field does not result in frequency perturbations. However, when quadropule and higher moments are considered, the polarizability of the impurity can become important. This effect was treated by van der Elsken (**3**), who showed that if the nitrogen atoms are polarized by the reaction field, their interaction with the effective charge on the central nitrogen results in a contribution of $-\dfrac{0{\cdot}66(6\alpha_N)}{0{\cdot}83l^4}\left(\dfrac{\partial V_Q}{\partial r}\right)_0$ to the change in force constant. Here $\alpha_N \approx 1 \times 10^{-24}\ \mathrm{cm}^3$ is the polarizability of a nitrogen atom and l is the equilibrium N—N bond length. For the bending mode this effect vanishes in the limit of small vibrations.

The results of a numerical calculation of the frequency ν_3 of the azide ion using the method outlined above are compared with the experimental data in Fig. 4. As the force constant $k_3 \equiv F_{33} - F_{13}$ and the cubic constant β_{133} were not known, these parameters were adjusted to fit the observed frequencies of N_3^- at room temperature in KCl and KBr lattices. The temperature dependence was assumed to arise solely from variation of the lattice parameter, as determined from thermal expansion coefficients for the pure host crystals.

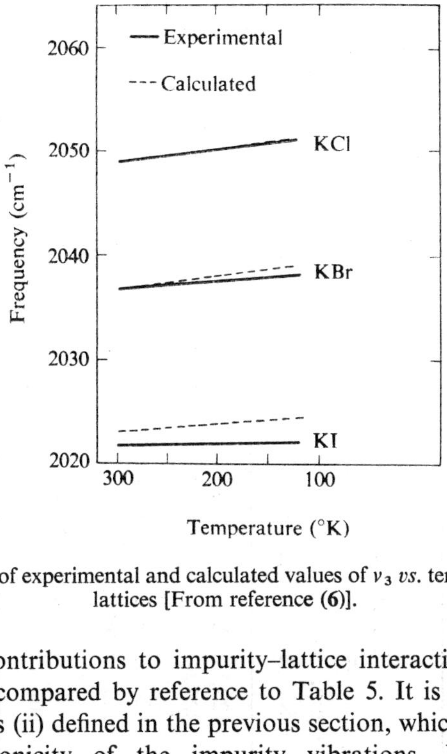

FIG. 4. Comparison of experimental and calculated values of ν_3 *vs.* temperature for various lattices [From reference (6)].

The various contributions to impurity–lattice interaction in the present example can be compared by reference to Table 5. It is apparent that the dominant effect is (ii) defined in the previous section, which depends directly on the anharmonicity of the impurity vibrations. Furthermore, the quadrupole–induced-dipole and the short-range terms provide approximately equal contributions to the perturbing forces. The direct effect (i) is seen to be much smaller than (ii) and to have a net contribution which tends to compensate effect (ii).

In this study of the azide ion in alkali–halide lattices the frequency of the asymmetric stretching mode was measured as a function of temperature. However, as the frequency shifts were assumed to depend solely on the

TABLE 5 [From reference (**6**)].

		Force (10^{-4} dyn)			Δk_3(i) (10^4 dyn/cm)			
Host lattice	Temp. (°K)	Coulombic	Short range	Δk_3 (ii)[a] (10^4 dyn/cm)	Coulombic	Short range	Charge-induced dipole	Δk_3[(i)&(ii)] (10^4 dyn/cm)
KCl	120	0.7805	0·9960	6·8567	0·1968	0·5888	−2·1291	5·5131
	300	0·7596	0·9477	6·5525	0·1919	0·5484	−2·0720	5·2208
KBr	120	0·6271	0·7330	5·2199	0·1606	0·4993	−1·7106	4·1602
	300	0·6087	0·6789	4·9419	0·1562	0·4444	−1·6604	3·8822
KI	120	0·4557	0·4504	3·4777	0·1191	0·3411	−1·2432	2·6947
	300	0·4407	0·4179	3·2955	0·1154	0·3102	−1·2022	2·5188

[a] Taking $\beta_{133} = -2{\cdot}04 \times 10^{10}$ dyn/cm^2.

changes in interionic distances, similar shifts would be expected as functions of pressure on the crystal. Such observations have recently been made by Cundill (**21**).

The frequency shifts of the fundamental of N_3^- in the same three potassium halides are plotted as functions of pressure in Fig. 5. The discontinuities in the region of 25 kbar result from the phase changes in the hosts, the structures of these crystals being of the CsCl type above the transition pressures.

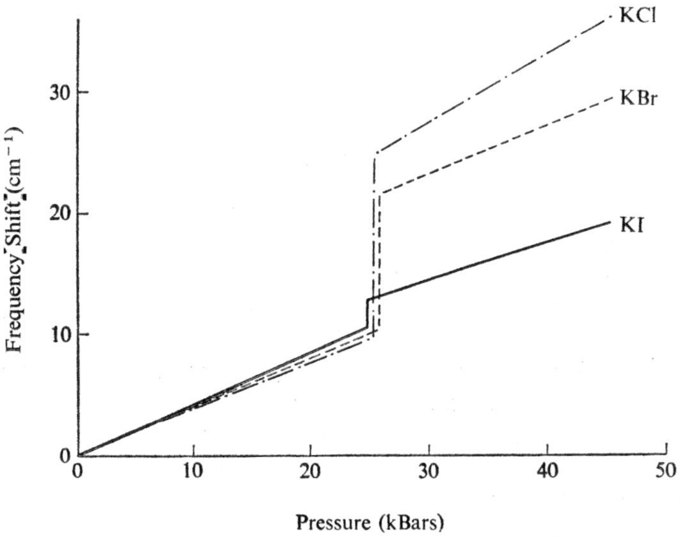

FIG. 5. Frequency shift of ν_3 of the azide ion with applied pressure [From Cundill (**21**)].

A model of the azide–lattice interaction which is similar to that described above was developed by Cundill to account for the pressure-induced frequency shifts. Although somewhat different repulsive–potential parameters were chosen and an effort was made to include the effect of lattice distortion in the neighborhood of the impurity, essentially the same values of the intra-ionic potential constants of N_3^- were obtained. This result may be largely fortuitous, as the effects of temperature and pressure on the lattice might not be expected to be entirely equivalent.

IV. DYNAMICS OF IMPERFECT LATTICES

The general treatment of the lattice dynamics of impure crystals is extremely difficult. However, as in the analysis of pure crystals presented in Chapter 3, a one-dimensional model serves to illustrate the principles of the method. In the case of the linear monatomic chain it was found that the frequency of the longitudinal mode is given as a function of the wave number, k, by solution of the equations of motion

$$M\ddot{\rho}_l = f_1(\rho_{l+1} - 2\rho_l + \rho_{l-1}), \tag{43}$$

where, for simplicity, only nearest-neighbor interactions ($m = 1$) have been included in Eqn (9), Chap. 3. Here M is the atomic mass, f_1 is the force constant between neighboring atoms, and ρ_l is the longitudinal displacement of the lth atom. Assuming solutions of the form

$$\rho_l = A\,\mathrm{e}^{2\pi i(\nu_0 t - kld)}, \tag{44}$$

where d is the equilibrium interatomic distance and ν_0 is the vibrational frequency of the unperturbed lattice, one obtains the secular equations

$$[2f_1(1 - \cos 2\pi s/\mathcal{N}) - \lambda_0 M]\,A\,\mathrm{e}^{2\pi isl/\mathcal{N}} = 0, \quad s = 0, 1, 2 \ldots \tag{45}$$

where multiplication by $\mathrm{e}^{-2\pi i\nu_0 t}$ has eliminated the time dependence, and the cyclic boundary condition has been imposed. This expression is a special case of the general equation given in Chapter 1,

$$[\mathbf{F}_\xi{}^0 - \mathbf{M}\mathbf{\Lambda}_0]\,\mathbf{L} = [\mathbf{F}_\xi{}^0 - \mathbf{G}_\xi{}^{0^{-1}}\,\mathbf{\Lambda}_0]\,\mathbf{L} = 0, \tag{46}$$

where $\mathbf{F}_\xi{}^0 = \mathbf{B}^\dagger\mathbf{F}_S{}^0\,\mathbf{B}$, as in Eqn (86), Chap. 3, and the subscripts ξ and S refer, respectively, to Cartesian and internal coordinates.

For the linear monatomic chain considered above, $\mathbf{M} = \mathbf{G}_\xi{}^{0^{-1}}$ is the diagonal matrix of atomic masses, while the transformation from Cartesian to internal coordinates is given by

$$\mathbf{B} = \begin{pmatrix} \ddots & & & & & & & \\ & -1 & 1 & 0 & 0 & 0 & 0 & 0 \\ & 0 & -1 & 1 & 0 & 0 & 0 & 0 \\ & 0 & 0 & -1 & 1 & 0 & 0 & 0 \\ & 0 & 0 & 0 & -1 & 1 & 0 & 0 \\ & 0 & 0 & 0 & 0 & -1 & 1 & 0 \\ & 0 & 0 & 0 & 0 & 0 & -1 & 1 \\ & 0 & 0 & 0 & 0 & 0 & 0 & -1 \\ & & & & & & & & \ddots \end{pmatrix}. \tag{47}$$

If the matrix in brackets in Eqn (46) is represented by $\mathbf{H}^0$, the secular determinant for the pure lattice is just

$$|\mathbf{H}^0| = 0. \tag{48}$$

When the lattice is perturbed by the introduction of impurities, the equations of motion take the form

$$\mathbf{HL} = \mathbf{IL}, \tag{49}$$

where **I** is a matrix which characterizes the impurities or other lattice defects. If the solid solution is very dilute, each guest is isolated, i.e., far from all others, and most of the elements of **I** will be zero.

It is convenient to introduce a matrix **g** defined by the relation

$$\mathbf{g}\,\mathbf{H} = \mathbf{E}. \tag{50}$$

The elements of **g** are known as Green's functions. Multiplication of Eqn (49) by **g** leads to the set of linear equations

$$\mathbf{L} = \mathbf{g}\,\mathbf{I}\,\mathbf{L} \tag{51}$$

or

$$(\mathbf{E} - \mathbf{g}\,\mathbf{I})\,\mathbf{L} = \mathbf{0}. \tag{52}$$

Here again, nontrivial solutions exist only if the determinant of the coefficients vanishes. Thus, the frequencies of the perturbed vibrational modes are determined by the relation

$$|\mathbf{E} - \mathbf{g}\,\mathbf{I}| = 0. \tag{53}$$

Clearly, only the submatrix of nonzero elements of **I** need be considered. Hence, the vast majority of lattice modes will be essentially unmodified by a low concentration of impurities.

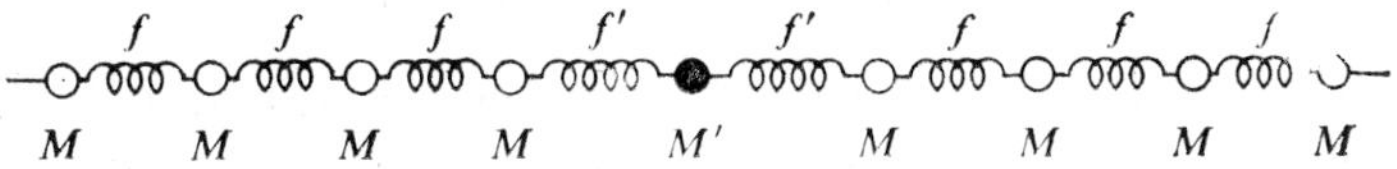

FIG. 6. Linear monatomic chain containing a single impurity.

Consider the simple case of a monatomic chain into which has been introduced a substitutional impurity atom of mass M', as shown in Fig. 6. The impurity is coupled to its nearest neighbors by a force constant f'. Hence, the two parameters

$$\eta \equiv \frac{M' - M}{M} \tag{54}$$

and

$$\zeta \equiv \frac{f' - f}{f} \tag{55}$$

characterize the impurity. In this case, then, the matrix **I** takes the form

$$\mathbf{I} = \begin{pmatrix} \ddots & & & & & \\ & 0 & 0 & 0 & 0 & 0 \\ & 0 & -f\zeta & f\zeta & 0 & 0 \\ & 0 & f\zeta & \eta M\lambda - 2f\zeta & f\zeta & 0 \\ & 0 & 0 & f\zeta & -f\zeta & 0 \\ & 0 & 0 & 0 & 0 & 0 \\ & & & & & \ddots \end{pmatrix}. \tag{56}$$

The application of Eqn (53) requires explicit expressions for the Green's function matrix, $\mathbf{g}$. In the one-dimensional case each element of $\mathbf{g}$ can be expanded in eigenvectors of Eqn (45); thus,

$$g_{ll'} = \frac{1}{\mathcal{N}} \sum_{s=-\frac{1}{2}\mathcal{N}+1}^{\mathcal{N}/2} \mathcal{A}_s \, e^{2\pi is(l-l')/\mathcal{N}} \tag{57}$$

or, more generally,

$$\mathbf{g} = \mathbf{L}^{-1} \mathcal{A} \, \mathbf{L}. \tag{58}$$

The coefficient $\mathcal{A}_s$ can be found by substituting Eqn (46) into Eqn (50). Then, using $\mathbf{H}^0 \mathbf{L} = \mathbf{0}$, one finds

$$\mathcal{A} = (\boldsymbol{\lambda}_0 - \boldsymbol{\lambda})^{-1} \mathbf{M}^{-1} \tag{59}$$

and

$$\mathbf{g} = \mathbf{L}^{-1} (\boldsymbol{\lambda}_0 - \boldsymbol{\lambda})^{-1} \mathbf{M}^{-1} \mathbf{L}. \tag{60}$$

It is seen from this result that the Green's function matrix, $\mathbf{g}$, contains all of the information concerning the pure lattice. For the monatomic chain Eqn (57) becomes

$$g_{ll'} = \frac{1}{\mathcal{N}} \sum_{s=-\frac{1}{2}\mathcal{N}+1}^{\mathcal{N}/2} \frac{e^{2\pi is(l-l')/\mathcal{N}}}{M\lambda - 2f_1 [1 - \cos(2\pi s/\mathcal{N})]}. \tag{61}$$

Here the eigenvectors in Eqn (45) have been normalized by writing $A = \mathcal{N}^{-1/2}$. Because the function $g_{ll'}$, depends only on the magnitude of the difference $\iota = l - l'$, it can be written as a function of a single index ι.

The Green's functions for this problem have been evaluated **(22)**, **(23)** and are given by

$$g_\iota = g_{-\iota} = \frac{1}{2f \sin\phi} [\cot \tfrac{1}{2}\mathcal{N}\phi \cos \iota\phi + \sin|\iota|\,\phi], \tag{62}$$

where ϕ is defined by the relation

$$\lambda = \frac{2f}{M}(1 - \cos\phi) = \frac{4f}{M}\sin^2 \tfrac{1}{2}\phi. \tag{63}$$

The physically significant solutions of the type of Eqn (63) are those for which λ is real and positive. The quantity ϕ is, in general, complex. Thus,

$$\phi = \phi' + i\phi'', \tag{64}$$

and, Eqn (63) becomes

$$\lambda = \frac{2f}{M}(1 - \cos\phi' \cosh\phi'' + i \sin\phi' \sinh\phi''). \tag{65}$$

Two cases can be distinguished,

$$\text{(i)}\ \phi'' = 0,\quad \lambda = \frac{2f}{M}(1 - \cos\phi') = \frac{4f}{M}\sin^2 \tfrac{1}{2}\phi', \tag{66}$$

where the corresponding frequencies given by

$$\lambda = 4\pi^2 \nu^2 \leqslant \frac{4f}{M} \tag{67}$$

are in the region of allowed frequencies of the perfect lattice, and

$$\text{(ii)}\quad \phi' = \pi n,\quad \lambda = \frac{4f}{M}(1 + \sinh^2 \tfrac{1}{2}\phi'') = \frac{4f}{M}\cosh^2 \tfrac{1}{2}\phi'', \tag{68}$$

where n must be an odd integer if the frequencies are to be real. Furthermore, one can put $n = 1$ in Eqn (68) because of the periodicity of the Green's functions. Thus, for case (ii) $\lambda \geqslant 4f/M$, and the frequencies are in the forbidden region above the band of frequencies of the perfect chain.

The frequencies of a monatomic chain perturbed by a single substitutional impurity can then be found from Eqn (53), which becomes

$$\begin{vmatrix} 1-f\zeta(g_1-g_0) & f\zeta(2g_1-g_2-g_0)-\eta M\lambda g_1 & f\zeta(g_2-g_1) \\ f\zeta(g_1-g_0) & 1-2f\zeta(g_1-g_0)-\eta M\lambda g_0 & f\zeta(g_1-g_0) \\ f\zeta(g_2-g_1) & f\zeta(2g_1-g_0-g_2)-\eta M\lambda g_1 & 1-f\zeta(g_1-g_0) \end{vmatrix} = 0, \tag{69}$$

with the Green's functions given by Eqn (62). Obviously the solutions are difficult to obtain, even for this simple example.

An even simpler case is that of an isotopic impurity in the monatomic chain. Here $\zeta = 0$ and Eqn (69) reduces to

$$\eta M \lambda g_0 = 1. \tag{70}$$

From Eqns (62) and (63)

$$g_0 = \frac{1}{2f \sin \phi} \cot \tfrac{1}{2}\mathcal{N}\phi, \tag{71}$$

and, substitution into Eqn (70) yields the transcendental equation

$$-\eta \tan \phi = \tan \tfrac{1}{2}\mathcal{N}\phi. \tag{72}$$

The numerical solution of Eqn (72) is illustrated in Fig. 7 for the case $\mathcal{N} = 8$. The intersections of the two sets of curves corresponding to the left- and right-hand sides of Eqn (72) give the desired solutions. In the absence of any defect, $\eta = 0$ and the allowed values of ϕ are given by the intersections of the tan $\frac{1}{2}\mathcal{N}\phi$ curves with the abscissa. When $\eta > 0$, corresponding to a heavy impurity, the ϕ's are lowered relative to their

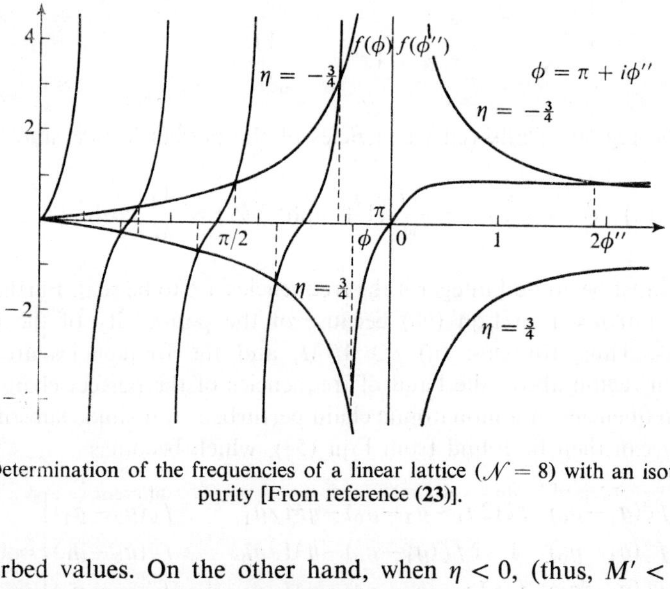

FIG. 7. Determination of the frequencies of a linear lattice ($\mathcal{N} = 8$) with an isotopic impurity [From reference (**23**)].

unperturbed values. On the other hand, when $\eta < 0$, (thus, $M' < M$) the allowed ϕ's are above their unperturbed values.

In the limit of large $\mathcal{N}$, Eqn (72) becomes

$$-\eta \tan \tfrac{1}{2}\phi = -i \tag{73}$$

and, substituting $\phi = \pi + i\,\phi''$, one finds

$$\phi'' = \ln\frac{1+\eta}{1-\eta}. \tag{74}$$

Then, by substitution of this expression into Eqn (68), the frequency above the band of unperturbed frequencies is found to be given by

$$\lambda = 4\pi^2 \nu^2 = \frac{4f}{M}\frac{1}{1-\eta^2}. \tag{75}$$

Since ϕ for this mode is now complex, the corresponding wave is damped, that is, the atomic displacement amplitudes decrease exponentially with increasing distance from the impurity site. If $M' \ll M$, this "localized" mode becomes a purely translational motion of the impurity in a rigid-lattice cage. This mode is clearly *ungerade* with respect to the impurity site, since *gerade* modes involve no displacement of the impurity, and, hence, are unaffected by isotopic substitution.

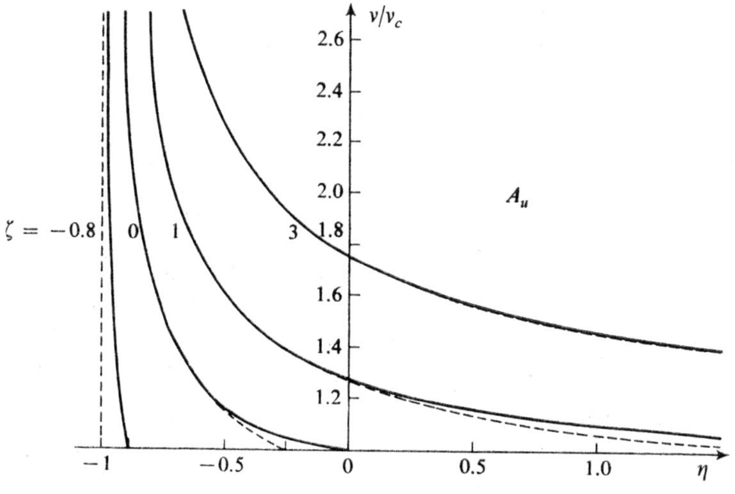

FIG. 8. Frequencies of localized A_u modes as functions of η for different values of ξ. [From reference (**23**)].

The results given by Eqn (75) for the case of isotopic substitution in a monatomic chain are shown in Fig. 8 by the curve $\zeta = 0$. The other curves, which are found by solution of Eqn (69), indicates the effect of varying the force constant f'. It should be noted that for positive values of ζ, *gerade* species localized modes become possible. In this case the frequency of the *ungerade* localized mode is given approximately by

$$\lambda = 4\pi^2 \nu^2 \approx \lambda_0 + \frac{f}{M}\left[1 + \frac{M'}{M}\left(\frac{\zeta - 1}{\zeta + 1}\right)\right], \tag{76}$$

where

$$\zeta \equiv \frac{f' - f}{f},$$

as before, and the condition for the localized mode to rise out of the band is found by requiring that the quantity in brackets in Eqn (76) remain positive; thus,

$$\zeta > \frac{\eta}{2 + \eta}, \qquad \text{where} \qquad \eta \equiv \frac{M' - M}{M}. \tag{77}$$

V. Infrared Spectra of Localized Impurity Modes

Several direct observations of localized modes in impure crystals have been made in recent years. The simplest examples are those involving monatomic ionic impurities such as H^- or D^-. These impurities, when found in alkali halides are often referred to as *U centers*.

Table 6.

Impurity	Transition	Temperature[a] 300°K	77°K	20°K
H^-	Fundamental	958·0 ± 0·6	965·1 ± 0·3	956·6 ± 0·5
	First overtone	1903 ± 1	1919·0 ± 0·3	1919·8 ± 0·3
	Second overtone	—	2910·8 ± 0·6	2912·2 ± 0·5
		—	2825·4 ± 0·8	2825·6 ± 0·8
D^-	Fundamental	—	694 ± 0·5	—
	First overtone	1374 ± 3	1383·9 ± 0·5	1384·5 ± 0·4
	Second overtone	—	—	2093 ± 1
		—	—	2048 ± 2

[a] Data are from reference (**24**).

Infrared spectra of the impurities H^- and D^- in calcium fluoride have been observed by Hayes, *et al.* [(**24**), p. 475]. Their data, which include the frequencies of the first and second overtones of the fundamental localized mode, are shown in Table 6. The interpretation of these results is comparatively simple because the mass of the hydrogen impurity is small compared to those of the lattice ions. Thus, the localized mode is quite accurately represented by the translational motion of H^- in a rigid cage.

The impurity is assumed to occupy a fluoride site in the CaF_2 host lattice. The space group of CaF_2 is $Fm3m \equiv \mathcal{O}_h^5$ and the fluoride sites have the symmetry of point group $\mathcal{T}_d$. Using the character table for this point group the three translational degrees of freedom of the impurity are represented by the structures $\boldsymbol{\Gamma} = F_2$. Thus, a single triply degenerate fundamental is predicted in the infrared spectrum.

The selection rules for the overtones of the localized mode can also be found using the character table for $\mathcal{T}_d$. As the fundamental is triply degenerate, Eqn (75), Chap. 2, must be employed, leading to the results

$$\boldsymbol{\Gamma}_2 = A_1 + E + F_2 \tag{78}$$

and

$$\boldsymbol{\Gamma}_3 = A_1 + F_1 + 2F_2, \tag{79}$$

for the first and second overtones, respectively. Hence, the first overtones should appear as a single absorption feature in the infrared spectrum, while the second overtone should be split into a doublet. The data of Table 6 are in agreement with these predictions, lending support to the assumption of a $\mathcal{T}_d$ impurity site.

The data of Table 6 can be used to determine the anharmonic potential function which governs the translational motion of the impurity ion. Because of the symmetry of the site, many of the potential constants vanish. For example, since the components of the polarizability tensor, α_{xx}, α_{xy}, etc. transform in the same way as xx, xy, etc. respectively, inspection of the character table for $\mathcal{T}_d$ indicates that only x^2, y^2, and z^2 belong to the totally symmetric representation, A_1. Hence, the terms xy, yz and zx cannot appear in the harmonic part of the potential function. Similarly, only the cubic terms in xyz and the two quartic terms of types x^4, y^4, z^4 and x^2y^2, y^2z^2, and z^2x^2 are nonvanishing in the anharmonic contributions.

On the basis of the above symmetry considerations the general anharmonic potential function for this problem becomes

$$V = V_0 + \tfrac{1}{2}F(x^2 + y^2 + z^2) + \beta xyz + \gamma_1(x^4 + y^4 + z^4) + \gamma_2(x^2y^2 + y^2z^2 + z^2x^2), \tag{80}$$

to fourth degree in the displacements. The force constants appearing in Eqn (80) can be related to a harmonic frequency and a set of anharmonic constants using standard perturbation theory (**25**), the term in β requiring a second-order treatment. The data of Table 6 for H^- lead to the following values for the force constants:

$$F = 1{\cdot}11 \text{ mdyn/Å} \qquad \gamma_1 = 8{\cdot}7 \times 10^{-4} \text{ mdyn/Å}^3$$
$$\beta = 0{\cdot}0471 \text{ mdyn/Å}^2 \quad \gamma_2 - 9{\cdot}9 \times 10^{-6} \text{ mdyn/Å}^3.$$

These force constants represent effective values for the translational motion of H^- in the tetrahedral cage, rather than interactions with specific lattice ions. A more chemically significant set of force constants would require further experimental data and the evaluation of lattice sums involving both H^-–Ca^{++} and H^-–F^- interactions.

It should be noted that the set of force constants determined above for H^- does not provide a good fit to the experimental data for D^- in CaF_2. In the case of the heavier isotope, the rigid-cage model is a poorer approximation, that is, the localized mode is said to extend into the lattice to some extent.

VI. Vibrational Selection Rules for Impure Crystals

As shown earlier (Chapter 4), a perfect crystal lattice is invariant under a group of translation operations. This translational invariance is, of course, destroyed by the introduction of any type of imperfection. In the case of a single substitutional impurity in an otherwise-perfect lattice, the symmetry of the system is described by a point group which is determined by the symmetry of the (unoccupied) impurity site, the point symmetry of the impurity, and the orientation of the impurity in the site. This situation was illustrated earlier in this chapter in the discussion of the infrared spectra of the azide ion in alkali–halide lattices.

In general, the guest is oriented in such a way that its point group and the site group have elements in common, which form a group $\mathfrak{L}$. Thus, the operations of $\mathfrak{L}$ are those which leave the ion invariant and $\mathfrak{L}$ is a subgroup of both the molecular point group and the site group. It is this local symmetry, $\mathfrak{L}$, which forms the basis for the determination of vibrational selection rules in impurity systems.

The vibrational degrees of freedom of a dilute impurity system can be classified as (1) internal vibrations of the impurity, (2) localized lattice modes, and (3) nonlocalized modes of the lattice. The internal modes are essentially the $3N - 6$ vibrations ($3N - 5$ for a linear impurity) of the hypothetical free impurity, which must now, however, be classified under symmetry group $\mathfrak{L}$. The selection rules for these internal vibrational fundamentals and their combinations and overtones can be found following thc method of Chapter 2. As usual, infrared or Raman activity of a given transition can be determined by noting the symmetry species of the components of the dipole moment and the polarizability, respectively, in the character table for $\mathfrak{L}$.

From the symmetry point of view the localized modes correspond to the six (or five) external degrees of freedom of the impurity, although, as pointed out above, these phonons may extend into the lattice somewhat. The vibrational selection rules for these localized modes again follow by inspection of the character table for $\mathfrak{L}$. Thus, the localized translatory mode of an impurity is always infrared-active, since the translational coordinates and the dipole moment components always appear together in the same species of a character table. Whether or not a given localized mode can be observed, however, depends on its frequency relative to the band of lattice frequencies. If the mass of the impurity and the force constants coupling it to the lattice are not such that the localized mode rises out of the band, it will ordinarily be masked by host-lattice absorption.

In order to determine selection rules for nonlocalized phonons it is necessary to establish the correlations between the irreducible representations of the space group and those of the local symmetry group, $\mathfrak{L}$. Once this correlation has been determined, selection rules for combinations of internal or external vibrations of the impurity with nonlocalized lattice modes can be found in the usual way (See Chapter 2).

It is convenient to establish first the correlation between the space group and the point group of the impurity site. The method will be illustrated for

TABLE 7.

Point in Brillouin zone	Group of **k**	No. of points in star
Γ	$\mathcal{O}_h$	1
X	$\mathcal{D}_{4h}$	3
L	$\mathcal{D}_{3d}$	4
W	$\mathcal{D}_{2d}$	6
Δ	$\mathcal{C}_{4v}$	6
Λ	$\mathcal{C}_{3v}$	8
Z	$\mathcal{C}_{2v}$	12
Σ	$\mathcal{C}_{2v}$	12
Q	$\mathcal{C}_2$	24
S	$\mathcal{C}_{2v}$	12
U	$\mathcal{C}_{2v}$	12
K	$\mathcal{C}_{2v}$	12
V	$\mathcal{C}_s$	24
k (in arbitrary direction)	$\mathcal{C}_1$	48

the case of an $\mathcal{O}_h$ site in an NaCl-type lattice (space group $Fm3m \equiv \mathcal{O}_h^5$) using the symmetry properties of the wave vector described briefly in Chapter 4. There it was shown that the irreducible representations of the symmorphic space groups are specified by the groups of the possible wave vectors and their stars. As specific examples, consider the wave vector represented by the point $\mathbf{\Delta}$ in the first Brillouin zone of a face-centered cubic lattice (Fig. 9). The group of this vector is $\mathscr{C}_{4v}$ and the star is six-pointed, as indicated in Table 7.

TABLE 8.

$\mathscr{C}_{4v}$	E	$2C_4$	C_2	$2\sigma_v$	$2\sigma_d$
χ_j'	1	-1	1	-1	1
χ_j	6	-2	2	-4	2

Now consider a particular representation of $\mathscr{C}_{4v}$, say B_2. The characters χ_j', obtained from the character table for this group (Appendix C) are given in Table 8. Since $\mathscr{C}_{4v}$ is not an invariant subgroup of $\mathcal{O}_h$, not all operations of a given class of $\mathcal{O}_h$ are found in the group $\mathscr{C}_{4v}$. Hence, if p is the number of points in the star, the characters of the reducible representations in $\mathcal{O}_h$ will be given by

$$\chi_j = \frac{p\, g_j'\, \chi_j'}{g_j}, \tag{81}$$

where the g_j''s are the number of operations in class j and the primes refer to the group of the wave vector, $\mathfrak{W}$. Equation (81) yields the characters χ_j given in the last line of Table 8. Using these results with the magic formula, the mapping $B_2(\mathbf{\Delta} = \mathscr{C}_{4v}) \Rightarrow F_{2g} + A_{2u} + E_u$ is easily established. The correlations of all points in the Brillouin zone of this lattice with the $\mathcal{O}_h$ site group is given in Table 9. Similar correlations involving other space groups and sites do not appear to have been developed in the published literature.

Reconsider now the example of an azide ion in a KBr lattice. It was shown that if the C_3 orientation is assumed, the local symmetry group is $\mathscr{D}_{3d}$. As the azide ion occupies an $\mathcal{O}_h$ site, the correlation found above between $\mathcal{O}_h^5$ and $\mathcal{O}_h$ can be used in conjunction with the correlation between the site group and its subgroup $\mathfrak{L} = \mathscr{D}_{3d}$. The latter corrclation, which is easily found from the corresponding character tables in Appendix C, is shown in Table 10.

TABLE 9.

$A_{1g}(\Gamma) = A_{1g}$	$A_{1g}(X) = A_{1g} + E_g$
$A_{2g}(\Gamma) = A_{2g}$	$A_{2g}(X) = F_{1g}$
$E_g\ (\Gamma) = E_g$	$B_{1g}(X) = A_{2g} + E_g$
$F_{1g}(\Gamma) = F_{1g}$	$B_{2g}(X) = F_{2g}$
$F_{2g}(\Gamma) = F_{2g}$	$E_g\ (X) = F_{1g} + F_{2g}$
$A_{1u}(\Gamma) = A_{1u}$	$A_{1u}(X) = A_{1u} + E_u$
$A_{2u}(\Gamma) = A_{2u}$	$A_{2u}(X) = F_{1u}$
$E_u\ (\Gamma) = E_u$	$B_{1u}(X) = A_{2u} + E_u$
$F_{1u}(\Gamma) = F_{1u}$	$B_{2u}(X) = F_{2u}$
$F_{2u}(\Gamma) = F_{2u}$	$E_u\ (X) = F_{1u} + F_{2u}$
$A_1(W) = A_{1g} + E_g + F_{2u}$	$A_1(\Delta) = A_{1g} + E_g + F_{1u}$
$A_2(W) = A_{2g} + E_g + F_{1u}$	$A_2(\Delta) = A_{1u} + E_u + F_{1g}$
$B_1(W) = A_{1u} + E_u + F_{2g}$	$B_1(\Delta) = A_{2g} + E_g + F_{2u}$
$B_2(W) = A_{2u} + E_u + F_{1g}$	$B_2(\Delta) + A_{2u} = E_u + F_{2g}$
$E(W) = F_{1g} + F_{2g} + F_{1u} + F_{2u}$	$E\ (\Delta) = F_{1g} + F_{2g} + F_{1u} + F_{2u}$
$A_1(\Lambda) = A_{1g} + A_{2u} + F_{2g} + F_{1u}$	$A_1(\Sigma) = A_{1g} + E_g + E_{2g} + F_{1u} + F_{2u}$
$A_2(\Lambda) = A_{1u} + A_{2g} + F_{1g} + F_{2u}$	$A_2(\Sigma) = A_{1u} + E_u + F_{2u} + F_{1g} + F_{2g}$
$E\ (\Lambda) = E_g + F_{1g} + F_{2g} + E_u + F_{1u} + F_{2u}$	$B_1(\Sigma) = A_{2g} + E_g + F_{1g} + F_{1u} + F_{2u}$
	$B_2(\Sigma) = A_{2u} + E_u + F_{1u} + F_{1g} + F_{2g}$
$A_1(Z) = A_{1g} + A_{2g} + F_{1u} + F_{2u}$	$A_{1g}(L) = A_{1g} + F_{2g}$
$A_2(Z) = F_{1g} + F_{2g} + A_{1u} + A_{2u} + 2E_u$	$A_{2g}(L) = A_{2g} + F_{1g}$
$B_1(Z) = F_{1g} + F_{2g} + F_{1u} + F_{2u}$	$E_g\ (L) = E_g + F_{1g} + F_{2g}$
$B_2(Z) = F_{1g} + F_{2g} + F_{1u} + F_{2u}$	$A_{1u}(L) = A_{1u} + F_{2u}$
	$A_{2u}(L) = A_{2u} + F_{1u}$
	$E_u\ (L) = E_u + F_{2u}$

$A(Q) = A_{1g} + A_{1u} + E_g + E_u + F_{1g} + F_{1u} + 2F_{2g} + 2F_{2u}$
$B(Q) = A_{2g} + A_{2u} + E_g + E_u + 2F_{1g} + 2F_{1u} + F_{2g} + F_{2u}$

$A'\ (V) = A_{1g} + A_{2g} + 2E_g + F_{1g} + F_{2g} + 2F_{1u} + 2F_{2u}$
$A''(V) = A_{1u} + A_{2u} + 2E_u + F_{1u} + F_{2u} + 2F_{1g} + 2F_{2g}$

$\mathbf{k} = A_{1g} + A_{2g} + A_{1u} + A_{2u} + 2E_g + 2E_u + 3F_{1g} + 3F_{2g} + 3F_{1u} + 3F_{2u}$

For the points **K**, **U** and **S** the reduction is the same as for the point **Σ**.

TABLE 10

$\mathcal{O}_h$	$\mathcal{D}_{3d}$
A_{1g}	A_{1g}
A_{2g}	A_{2g}
E_g	E_g
F_{1g}	A_{2g}, E_g
F_{2g}	A_{1g}, E_g
A_{1u}	A_{1u}
A_{2u}	A_{2u}
E_u	E_u
F_{1u}	A_{2u}, E_u
F_{2u}	A_{1u}, E_u

Infrared activity is limited to modes of species A_{2u} and E_u in $\mathcal{D}_{3d}$. Thus, lattice vibrations whose structures in $\mathcal{O}_h$ contain the reduced representations A_{2u}, E_u, F_{1u}, or F_{2u} will be infrared-active. Referring to the classification of the lattice modes of $\mathcal{O}_h^5$ given in Table 11, it is seen that only the LO and TO vibrations corresponding to the wave vector **L** in Fig. 9 are forbidden in the infrared spectrum. In many cases impurity systems have even less symmetry, allowing all lattice vibrations to be active (e.g. NCO^- in KBr). Thus, in general, the far-infrared spectra of impure crystals is

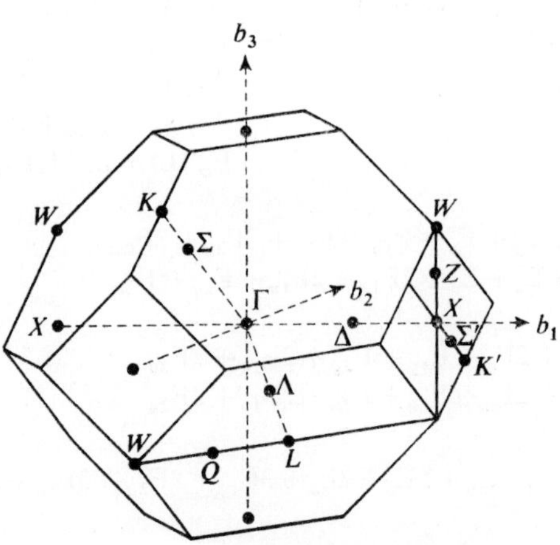

FIG. 9. First Brillouin zone for a face-centered cubic lattice showing special points in reciprocal space.

extremely rich, the actual form of the absorption bands being determined by the transition dipole moments and the density of states, as in the case of pure crystals.

TABLE 11.

Point in Brillouin zone	Group of k ($\mathfrak{W}$)	Optical branches	Acoustical branches
L	$\mathscr{D}_{3d}$	$A_{1g}(LO) + E_g(TO)$	$A_{2u}(LA) + E_u(TA)$
W	$\mathscr{D}_{2d}$	$A_1 + E$	$B_2 + E$
Z	$\mathscr{C}_{2v}$	$A_1 + B_1 + B_2$	$A_1 + B_1 + B_2$
Δ	$\mathscr{C}_{4v}$	$A_1(LO) + E(TO)$	$A_1(LA) + E(TA)$
Λ	$\mathscr{C}_{3v}$	$A_1(LO) + E(TO)$	$A_1(LA) + E(TA)$
Σ	$\mathscr{C}_{2v}$	$A_1 + B_1 + B_2$	$A_1 + B_1 + B_2$
S	$\mathscr{C}_{2v}$	$A_1 + B_1 + B_2$	$A_1 + B_1 + B_2$
X	$\mathscr{D}_{4h}$	$A_{2u}(LO) + E_u(TO)$	$A_{2u}(LA) + E_u(TA)$
Q	$\mathscr{C}_2$	$2A + B$	$A + 2B$
Γ	$\mathscr{O}_h$	F_{1u}	F_{1u}

VII. COMBINATION SPECTRA: CYANATE ION IN KBr

As the lattice vibrations of impure crystals, localized as well as nonlocalized, give rise to vibrational fundamentals (first-order transitions) in the far-infrared spectral region, they are not always easy to observe. However, most, if not all, lattice modes can combine with internal vibrations of an impurity to produce "satellite" absorption bands which are more-or-less symmetrically located with respect to a given internal fundamental. Again, the spectra of the cyanate ion in alkali–halide lattices provide excellent examples of these combination bands (**12**), (**26**).

In the region of the ν_3 stretching fundamental of NCO^- in KBr numerous absorption features are observed which can be assigned to the various isotopic species of NCO^- and/or "hot bands". In addition, several sharp satellites, as well as a broad absorption background, are observed on each side of the fundamental absorption. A spectrum of this system at 90° K which is due to Cundill and Sherman is shown in Fig. 10.

The frequencies of the sharp satellites are found to be very sensitive to pressure, thus providing conclusive evidence that they arise from combinations of localized modes with the ν_3 fundamental of NCO^-. Furthermore, an investigation of the spectra of five different isotopic species of this ion showed that the localized modes involved are due to librations of the NCO^- guest about axes which are not principal axes of the free ion. These modes appear to include displacements of lattice atoms beyond the nearest neighbors and, hence, are not accurately described as librations of the guest in a rigid cage. In fact, a quantitative treatment of the isotopic frequency shifts allowed an estimate to be made of the delocalization of these modes. It was found, for example, that the high-frequency satellite ($\nu_3 \pm 183{\cdot}6\ \text{cm}^{-1}$) arises from a libration involving large nitrogen displacement. This mode is fairly well-localized, approximately 3/4 of its energy being associated with motion of the impurity. For the ($\nu_3 \pm 97{\cdot}4\ \text{cm}^{-1}$) band, on the other hand, only 1/7 of the mode energy is localized in the librational motion of the impurity.

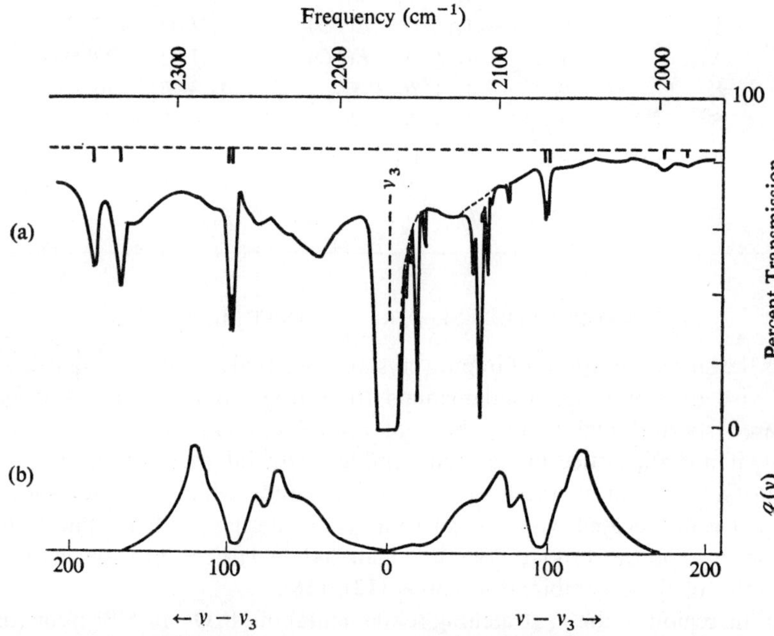

FIG. 10.(a) Infrared absorption spectrum of NCO^- in KBr at 90°K. (b) Density of states as a function of $\nu - \nu_3$ [From reference (**26**)].

It is interesting to note that although spectra have been recorded of N_3^-, BO_2^-, NO_3^-, and CN^- impurities, in addition to cyanate, in nearly a dozen different host lattices, sharp satellites have never been detected in cases in which the impurity is heavier than the ion which it replaces. This result is in agreement with a condition for the occurrence of localized modes, such as that given by Eqn (77) for the case of the impure linear lattice.

The spectrum of Fig. 10 shows, in addition to the sharp satellites, a broad absorption feature on each side of the ν_3 fundamental. These broad absorptions are undoubtedly due to combination of ν_3 with the non-localized lattice modes of the system. The integrated density of states for pure KBr, as determined by neutron diffraction (**27**) is shown in Fig. 10 for reference. It is understood, however, that the actual density of states of the impure lattice would be expected to be somewhat modified by the presence of the impurity.

VIII. Spectra of Rare-Gas Matrix-Isolated Species

Although the matrix-isolation technique using rare-gas matrices was originally applied by Pimental to the study of free radicals, it has been extended to the investigation of other chemical species. The method has been employed in the analysis of dimer and multimer formation in hydrogen-bonded systems, the isolation of high-temperature species, and in the study of diffusion and chemical reaction in solid matrices. Nevertheless, use in the spectroscopic determination of the structures of free radicals is still the most important chemical application of the matrix-isolation technique.

Free radicals are ordinarily, and now almost routinely, produced by the photolysis of a molecular species trapped in an inert matrix. The rare gases and nitrogen are the preferred host materials due to their inertness and infrared transparence, although a number of other matrices have been employed.

The first free radical to be observed in matrices using infrared spectroscopy was HCO. It was produced by photolizing HI in a CO matrix at 20°K, presumably by the direct reaction of hydrogen atoms with CO. In this example the matrix was not inert, as it played an essential role in the synthesis of the free radical. In subsequent work inert matrices have been more often used, although the radicals have still been prepared *in situ* by photolysis of either a well-isolated molecule such as CH_3NO_2 in Ar to give HNO or of an isolated species which subsequently reacts with another molecule in an adjacent site (e.g. NCN to give C which reacts with HCl to yield HCCl). Some representative matrix-isolation studies of free radicals are indicated in Table 12. A complete list of free radicals investigated by the matrix-isolation technique through 1967 has been compiled by Hallam [(**28**), p. 329].

Another early application of the matrix-isolation technique was to the investigation of molecular association. In Pimentel's classic study of matrix-isolated methanol, bands were identified of the monomer, dimer, trimer, tetramer and higher polymers (**1**). Each oligomer, when isolated in a low-temperature matrix, exhibits relatively sharp bands which are free of rotational structure. By starting with a well-isolated monomer, controlled diffusion in the matrix can be used to produce the desired polymeric species. In the limit at which a high degree of polymerization is achieved, the spectrum becomes similar to that of the pure solid.

TABLE 12.

Free radical	Method of production	Vibrational frequencies	Structure	References
HCO	HI + CO in Ar; HI, HBr, or H_2S in CO	$\nu_1 = 2482$ $\nu_2 = 1091$ $\nu_3 = 1862$	bent ($\mathscr{C}_s$)	**(29)**, **(30)**, **(31)**
CH_3	CH_4 in Ar, N_2; CH_3X + Li, Na, or K in Ar	$\nu_2 = 619$ $\nu_3 =$ — $\nu_4 = 1383$	planar ($\mathscr{D}_{3h}$)	**(32)**
SiF_3	$SiHF_3$ in Ar, N_2, CO_2	$\nu_1 = 832$ $\nu_2 = 406$ $\nu_3 = 954$ $\nu_4 = 290$	pyramidal ($\mathscr{C}_{3v}$)	**(33)**
NCO	HNCO in Ne, Ar, N_2, CO	$\nu_1 = 1275$ $\nu_2 = 487$ $\nu_3 = 1933$	linear ($\mathscr{C}_{\infty v}$)	**(34)**
CF	CH_3F in Ar, N_2	$\nu = 1279$		**(35)**
CCl	CH_3Cl in Ar, N_2	$\nu(^{12}C^{35}Cl) = 876$		**(36)**
CCO	N_3CN + CO in Ar; C_3O_2 in Ar	$\nu_1 = 1074$ $\nu_2 = 381$ $\nu_3 = 1978$	linear ($\mathscr{C}_{\infty v}$)	**(37)**, **(38)**
NH_2	NH_3 in Ar, N_2	$\nu_1 =$ — $\nu_2 = 1499$ $\nu_3 = 3220$	bent ($\mathscr{C}_{2v}$)	**(39)**
OH	H_2O in Ar or Kr	$\nu = 3452$		**(40)**

In a recent, very extensive series of experiments Barnes, *et al.* **(41)** have studied the infrared spectra of matrix-isolated multimers of the hydrogen halides at 20°K. By careful control of the composition and decomposition rate, as well as numerous diffusion experiments, these authors were able to make detailed assignments of the very complex spectra of these systems. Dimers, open and cyclic trimers, open and cyclic tetramers, and polymers were identified, in addition to the isolated HX monomers. This work was extended to include matrices of O_2, CO, CH_4, SF_6, C_2H_4, and CF_4, besides the more familiar matrix materials, Ar and N_2.

Another application of the matrix-isolation technique which has developed rapidly in recent years is to the analysis of the vapor phase in equilibrium with materials such as metal oxides and halides at high temperatures. Conventional spectroscopic studies of vapors at high temperatures (often several thousand degrees) are difficult due to their chemical reactivity with cells and windows and the complexity introduced in the spectra by "hot bands". Furthermore, it is often difficult to obtain sufficient concentrations of the absorbing species. These difficulties are readily overcome using the matrix-isolation technique, usually in conjunction with a Knudsen cell, which allows controlled effusion of the vapor, followed by condensation with a jet of inert gas on a cold window. For details of this application of the matrix-isolation technique the reader is referred to the review article by Barnes and Hallam (**42**) and to the series of papers by Linevsky and coworkers (**43**).

The problems outlined above are representative of chemical applications of the matrix-isolation method in molecular spectroscopy. However, the most studied systems are those in which a diatomic molecule, usually a hydrogen halide, is trapped in a low-temperature, inert matrix. Such

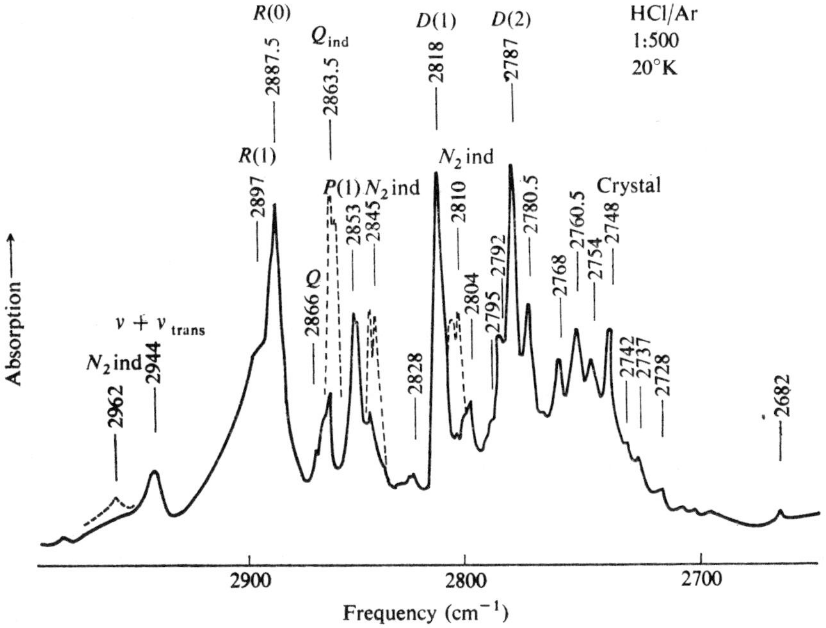

FIG. 11. Infrared absorption spectrum of HCl in Ar at 20°K [After Barnes and Hallam (**42**)].

systems, which are perhaps the simplest of spectroscopic significance, can be treated using a dynamical model. They are, therefore, of great interest in the study of intermolecular forces in solids.

When a molecule such as HCl is isolated in, say, an argon matrix, its fundamental and overtone vibrational frequencies are perturbed as a result of interactions of the type discussed earlier in this chapter. Furthermore, each vibrational transition is accompanied by fine structure which is not unlike the rotational structure observed in the spectra of gaseous molecules. However, as a consequence of the low sample temperatures employed, only a very few rotational levels are populated, leading to a very simple fine-structure pattern.

The matrix spectrum of HCl in Ar is very complicated, as evidenced by Fig. 11. However, many of the components arise from HCl dimers or higher multimers, as well as from perturbation by N_2 impurities which result in the formation of HCl–N_2 pairs, with the two molecules occupying adjacent sites. The remaining components can be qualitatively accounted for using a model of a diatomic rotor perturbed by an anisotropic hindering potential.

From a quantitative point of view, however, the hindered-rotor model does not account for the frequency shifts produced by isotopic substitution. A more sophisticated model which includes all of the external degrees of freedom of the trapped molecule must be considered in order to obtain a quantitative description of the system. This treatment, sometimes referred to as the rotation–translation coupling model, will now be discussed in some detail, as it provides an excellent example of the application of molecular dynamics to a simple spectroscopic problem.

IX. External Motions of Trapped Hydrogen Halides

If a guest molecule occupies a site in a host lattice, its classical kinetic energy obtained from Eqn (4), Chap. 1, becomes

$$2T = \dot{\mathbf{R}}^2 \sum_\alpha m_\alpha + \sum_\alpha m_\alpha(\boldsymbol{\omega} \times \mathbf{r}_\alpha)\cdot(\boldsymbol{\omega} \times \mathbf{r}_\alpha) + 2\,\dot{\mathbf{R}}\cdot\boldsymbol{\omega} \times \sum_\alpha m_\alpha \mathbf{r}_\alpha, \qquad (82)$$

where terms involving the high-frequency molecular vibrations have been dropped in the rigid-rotor approximation. In this case the third term on the right-hand side of Eqn (82), which represents coupling between rotational and translational motions, cannot be easily eliminated. The presence of an external potential function precludes application of the Eckart conditions in the usual form, because the molecule is no longer required to rotate about its center of mass.

The external potential function, which governs the rotational and translational motions of the guest, is determined by the various guest–host interactions. If the host lattice is assumed to be rigid, the external potential will have the symmetry of the site in question. In the more general case in which the localized modes extend into the lattice, this potential will be a function of the instantaneous positions of the atoms of the host lattice.

The new center of rotation does not, in general, coincide with the center of mass. Its position is determined by the center of gravity of the molecule and the "center of interaction" of the external potential (**44**), (**45**). The center of interaction, which is essentially the center of electrical symmetry of the guest molecule, must lie on all of its symmetry elements.

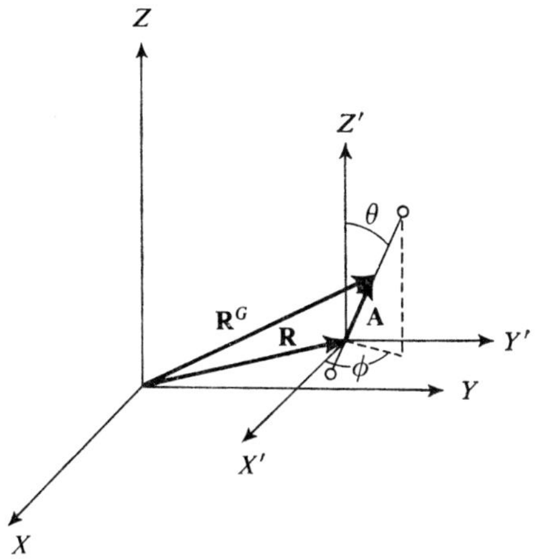

FIG. 12. Coordinates used to describe the motions of a trapped diatomic molecule.

In the case of a diatomic (or, in general, linear) impurity the center of interaction lies on the molecular axis. Thus, using the nonrotating coordinate system X', Y', Z' whose origin is at the center of interaction (Fig. 12), the position of the center of gravity is given by the vector **A**. The orientation of the molecule is specified by the angles θ and ϕ and the angular velocity is then given by

$$\boldsymbol{\omega} = \dot{\boldsymbol{\theta}} + \dot{\boldsymbol{\phi}} = (-\dot{\theta} \sin \phi)\, \mathbf{i} + (\dot{\theta} \cos \phi)\, \mathbf{j} + \dot{\phi}\, \mathbf{k}, \tag{83}$$

and the position of the αth atom is

$$\mathbf{r}_\alpha = (z_\alpha \sin \theta \cos \phi)\, \mathbf{i} + (z_\alpha \sin \theta \sin \phi)\, \mathbf{j} + (z_\alpha \cos \theta)\, \mathbf{k}, \tag{84}$$

where the z axis is a rotating axis colinear with the axis of the molecule.

Substitution of Eqns (83) and (84) into Eqn (82) yields the kinetic-energy expression

$$2T = \dot{R}^2 \sum_\alpha m_\alpha + (\dot{\theta}^2 + \dot{\phi}^2 \sin^2\theta)\left(\sum_\alpha m_\alpha z_\alpha{}^2\right)$$
$$+ 2[\dot{X}\,(\dot{\theta}\cos\theta\cos\phi - \dot{\phi}\sin\theta\sin\phi)$$
$$+ \dot{Y}\,(\dot{\theta}\cos\theta\sin\phi + \dot{\phi}\sin\theta\cos\phi) - \dot{Z}\dot{\theta}\sin\theta]\left(\sum_\alpha m_\alpha z_\alpha\right). \quad (85)$$

Since the center-of-interaction coordinates and the center-of-mass coordinates are related by

$$z_i = z_i{}^G + A, \quad (86)$$

$$\sum_\alpha m_\alpha z_\alpha = \sum_\alpha m_\alpha(z_\alpha{}^G + A) = MA \quad (87)$$

and

$$\sum_\alpha m_\alpha z_\alpha{}^2 = \sum_\alpha m_\alpha[(z_\alpha{}^G)^2 + 2\,z_\alpha{}^G A + A^2] = I + MA^2, \quad (88)$$

where

$$M \equiv \sum_\alpha m_\alpha, \quad I \equiv \sum_\alpha m_\alpha(z_\alpha{}^G)^2, \quad \text{and} \quad \sum_\alpha m_\alpha z_\alpha{}^G = 0.$$

Equation (85) then becomes

$$2\,T = M\dot{R}^2 + (\dot{\theta}^2 + \dot{\phi}^2\sin^2\theta)\,(I + MA^2)$$
$$+ 2MA[\dot{X}\,(\dot{\theta}\cos\theta\cos\phi - \dot{\phi}\sin\theta\sin\phi)$$
$$+ \dot{Y}\,(\dot{\theta}\cos\theta\sin\phi + \dot{\phi}\sin\theta\cos\theta) - \dot{Z}\,\dot{\theta}\sin\theta]. \quad (89)$$

The momenta conjugate to the various coordinates appearing in Eqn (89) are found by differentiating T with respect to the corresponding velocities. The kinetic energy in terms of these momenta becomes

$$2T = \frac{1}{M}(P_X{}^2 + P_Y{}^2 + P_Z{}^2) + \frac{1}{I}(P_\theta{}^2 + \csc^2\theta\,P_\phi{}^2)$$
$$+ \frac{2A}{I}[\sin\theta\,P_Z P_\theta - \cos\theta\cos\phi\,P_X P_\theta + \csc\theta\sin\theta\,P_X P_\phi$$
$$- \cos\theta\sin\phi\,P_Y P_\theta - \csc\theta\cos\phi\,P_Y P_\phi]$$
$$+ \frac{A^2}{I}[(1 - \sin^2\theta\cos^2\phi)\,P_X{}^2 + (1 - \sin^2\theta\sin^2\phi)\,P_Y{}^2 + \sin^2\phi\,P_Z{}^2$$
$$- 2\sin^2\theta\sin\phi\cos\phi\,P_X P_Y - 2\sin\theta\cos\theta\cos\phi\,P_X P_Z$$
$$- 2\sin\theta\cos\theta\sin\phi\,P_Y P_Z] + \ldots, \quad (90)$$

where terms cubic and higher in A have been neglected. It is seen that the first term is just the energy of translation of the molecule, while the second is the rotational kinetic energy. The remaining terms, which couple translational and rotational motions, vanish when the center of interaction coincides with the center of mass ($A = 0$).

The passage from Eqn (90) to the corresponding quantum-mechanical operator for the kinetic energy is complicated by the fact that the classical expression contains momenta which are not conjugate to Cartesian co-ordinates. The reader who is not familiar with this problem is referred to the text by Kemble [(**46**), p. 237], or to the paper of Keyser and Robinson (**47**). The resulting Hamiltonian is given by

$$\begin{aligned}
\mathscr{H} = &-\frac{h^2}{8\pi^2 M}\left[\frac{\partial^2}{\partial X^2}+\frac{\partial^2}{\partial Y^2}+\frac{\partial^2}{\partial Z^2}\right] \\
&-\frac{h^2}{8\pi^2 I}\left[\frac{1}{\sin\theta}\frac{\partial}{\partial\theta}\left(\sin\theta\frac{\partial}{\partial\theta}\right)+\frac{1}{\sin\theta^2}\frac{\partial^2}{\partial\phi^2}\right] \\
&+V(X,Y,Z,\theta,\phi)-\frac{h^2 A}{8\pi^2 I}\left[\sin\theta\frac{\partial}{\partial Z}\frac{\partial}{\partial\theta}+\csc\theta\frac{\partial}{\partial Z}\frac{\partial}{\partial\theta}(\sin^2\theta)\right. \\
&-\cos\theta\cos\phi\frac{\partial}{\partial X}\frac{\partial}{\partial\theta}-\csc\theta\cos\phi\frac{\partial}{\partial X}\frac{\partial}{\partial\theta}(\sin\theta\cos\theta)+\csc\theta\sin\phi\frac{\partial}{\partial X}\frac{\partial}{\partial\phi} \\
&+\csc\theta\frac{\partial}{\partial X}\frac{\partial}{\partial\phi}(\sin\phi)-\cos\theta\sin\phi\frac{\partial}{\partial Y}\frac{\partial}{\partial\theta}-\csc\theta\sin\phi\frac{\partial}{\partial Y}\frac{\partial}{\partial\theta}(\sin\theta\cos\theta) \\
&\left.-\csc\theta\cos\phi\frac{\partial}{\partial Y}\frac{\partial}{\partial\phi}-\csc\theta\frac{\partial}{\partial Y}\frac{\partial}{\partial\phi}(\cos\phi)\right] \\
&-\frac{h^2 A^2}{8\pi^2 I}\left[(1-\sin^2\theta\cos^2\phi)\frac{\partial^2}{\partial X^2}+(1-\sin^2\theta\sin^2\phi)\frac{\partial^2}{\partial Y^2}+\sin^2\phi\frac{\partial^2}{\partial Z^2}\right. \\
&-2\sin^2\theta\sin\phi\cos\phi\frac{\partial}{\partial X}\frac{\partial}{\partial Y}-2\sin\theta\cos\theta\cos\phi\frac{\partial}{\partial X}\frac{\partial}{\partial Z} \\
&\left.-2\sin\theta\cos\theta\sin\phi\frac{\partial}{\partial Y}\frac{\partial}{\partial Z}\right],
\end{aligned} \tag{91}$$

where, in the rigid-cage approximation the external potential function depends only on the position of the center of interaction and the molecular orientation.

As pointed out above, observed spectral data on hydrogen halides in rare-gas matrices have been interpreted either on the basis of the rotational barrier model or the rotation–translation coupling model. Although these two models are very different, each can be considered to be a special case of the general problem represented by the Hamiltonian of Eqn (91).

If the center of gravity and the center of interaction are assumed to be coincident ($A = 0$) and fixed to the center of the site, only the second and third terms remain in Eqn (90). The second term is just the kinetic energy of a linear rotor, while $V(\theta, \phi)$ is the angularly dependent part of the external potential. For the free rotor $V(\theta, \phi) = 0$ and the rotational energy is given by

$$E_{\text{rot}} = \frac{h^2}{8\pi^2 I} J(J+1) \qquad J = 0, 1, 2, \ldots, \tag{92}$$

where the rotational levels are defined by the quantum number J. The degeneracy of the Jth level is given by $2J + 1$.

If $V(\theta, \phi) \neq 0$, the rotational levels will be perturbed and some of the degeneracy will be lifted. The energy levels of a linear rotor in a potential field having octahedral symmetry were calculated many years ago by Devonshire (**48**). These results are shown in Fig. 13, where the perturbed levels have been classified under the symmetry of the octahedral point group, $\mathcal{O}_h$. Similar calculations have been made by Bowers and Flygare (**49**) for the case of the linear rotor in a $\mathcal{D}_{3h}$ site.

In Fig. 13 the reduced rotational energy is given as a function of a parameter K which can be identified with the "barrier height". This quantity can, in principle, be determined by assigning the initial and final states of an observed transition and finding the value of K which gives the desired separation between the corresponding levels in Fig. 13. In practice, since the rotational transitions are superimposed on the fundamental vibrational transition $v = 0 \rightarrow v = 1$, whose exact frequency is not usually known, two observed rotation–vibration lines must be assigned in order to determine the barrier. From Fig. 11 it is seen that the two lines $R(0)$ and $P(1)$ are the most prominent. Therefore, in most studies of matrix-isolated diatomic molecules, the separation between these two lines has been used to estimate a barrier height.

If, as suggested above, all of the perturbation effects are attributed to the presence of an anisotropic external potential $V(\theta, \phi)$ the barriers calculated from experimental spectra are seriously overestimated. It has been shown by Kimel and coworkers (**50**) that in the case of noncentrosymmetric molecules the rotation–translation coupling terms are much more important than the impurity–lattice interaction potential. In fact, in Friedmann and Kimel's treatment of the problem, the angularly dependent part of $V(X, Y, Z, \theta, \phi)$ was neglected and all of the perturbation of rotational levels was attributed to the terms in A in Eqn (91).

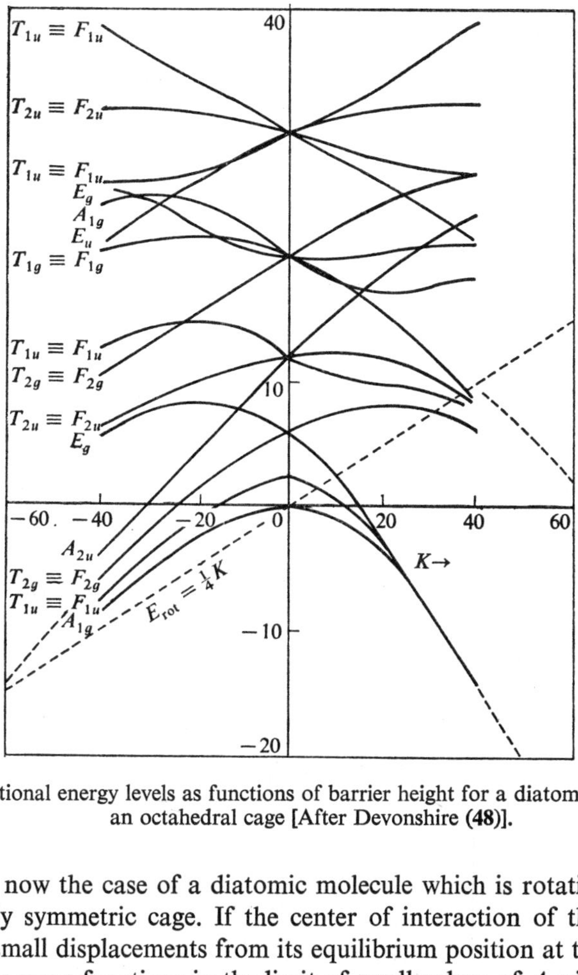

FIG. 13. Rotational energy levels as functions of barrier height for a diatomic molecule in an octahedral cage [After Devonshire (48)].

Consider now the case of a diatomic molecule which is rotating freely in a spherically symmetric cage. If the center of interaction of the molecule undergoes small displacements from its equilibrium position at the center of the cage, the wave functions in the limit of small values of A will be simply the product of rigid-rotor wave functions and three-dimensional harmonic-oscillator wave functions. In this, zero-order approximation, the rotation–translation energy is then given by

$$E^{(0)}{}_{v_X, v_Y, v_Z, J} = h\nu_T(v_X + v_Y + v_Z + 3/2) + \frac{h^2}{8\pi^2 I} J(J+1)$$
$$= h\nu_T(T + 3/2) + hBJ(J+1), \tag{93}$$

where

$$\nu_T = \frac{1}{2\pi}\sqrt{\frac{k}{M}}$$

is the frequency of the translational mode, $T \equiv v_X + v_Y + v_Z$ is the translational quantum number, k is the corresponding force constant, and

$$B \equiv \frac{h}{8\pi^2 I}$$

is the rotational constant. These energy levels are diagrammed in Fig. 14 for a typical case.

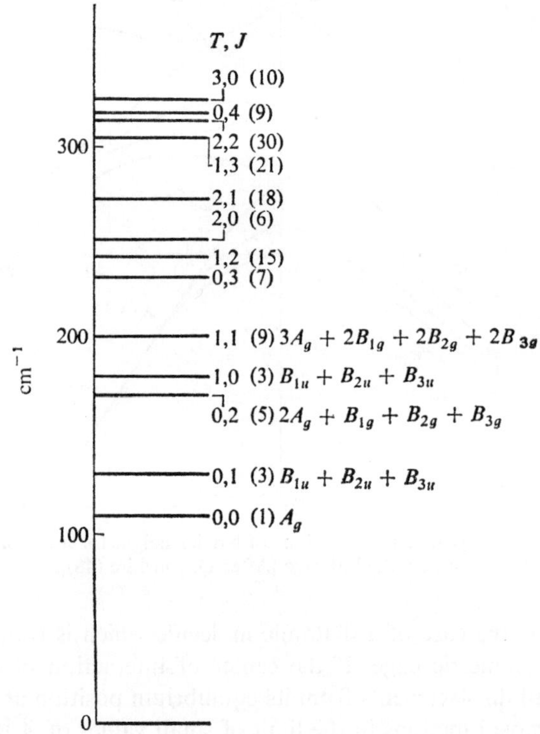

FIG. 14. Rotational-translational energy levels (neglecting interaction) for $B = 10{\cdot}5$ cm^{-1} and $\nu_T = 72$ cm^{-1}. The degeneracy of each level is shown in parentheses.

It has been shown by Keyser and Robinson that the Hamiltonian of Eqn (91) is invariant under the operations of point group $\mathcal{D}_{2h}$. Hence, the levels can be classified by using the representations of this group and the selection rules can be determined by the usual method.

The effect of the translation–rotation coupling terms in Eqn (91) can be estimated using second-order perturbation theory. In the original development by Friedmann and Kimel the expression

$$\frac{\Delta E}{hcB} = \frac{MA^2\xi}{2I}\left[1 - \frac{2\xi(J^2 + J + 1) - 4J(J + 1)}{(\xi - 2J)(\xi + 2J + 2)}\right], \tag{94}$$

where $\xi \equiv \nu_T/B$, was proposed to account for the shifts of the rotation–vibration lines of diatomic molecules in rare-gas matrices.

The matrix shift of a rotational line corresponding to the transition $J \rightarrow J'$ is then given by

$$\begin{aligned}\frac{\Delta E_{J\rightarrow J'}}{h} &= \frac{E_{0,J'}{}^{(0)} + E_{0,J'}{}^{(2)} - (E_{0,J}{}^{(0)} + E_{0,J}{}^{(2)})}{h} \\ &= B[J'(J' + 1) - J(J + 1)] \\ &\quad \times \left\{1 - \frac{MA^2\xi^4}{I[\xi + 2(J + 1)][\xi - 2J'][\xi - 2J][\xi - 2(J + 1)]}\right\}. \end{aligned} \tag{95}$$

For $\xi \gg J'$,

$$\frac{\Delta E_{J\rightarrow J'}}{h} \approx B[J'(J' + 1) - J(J + 1)](1 - MA^2/I), \tag{96}$$

which is simply the zero-order rotational energy change for molecules having an effective rotational constant given by

$$B_{\text{eff}} = B(1 - MA^2/I). \tag{97}$$

It should be noted, however, that the perturbation treatment breaks down for $\xi \approx 2J$ or $\xi \approx 2J'$ due to the "resonance denominator" in Eqn (95).

A more general solution to the problem of rotation–translation coupling was obtained by Keyser and Robinson (**47**). These authors factored the secular determinant using symmetry coordinates based on the $\mathscr{D}_{2h}$ symmetry of the Hamiltonian of Eqn (91) and carried out numerical diagonalization of each block to find the frequency shifts as functions of A for various values of ξ. A few of their results for HCl are shown in Fig. 15.

In order to compare the experimental results for HCl with those for DCl, it should first be noted that

$$A_{\text{DCl}} = A_{\text{HCl}} \pm 0{\cdot}034\ \text{Å}, \tag{98}$$

where 0·034 Å is the change in the position of the center of mass in going from HCl to DCl. The plus sign is applicable if the center of interaction lies on the same side of the center of mass as the chlorine atom, while the minus sign corresponds to a center of interaction lying on the hydrogen side of the center of mass. Although attempts have been made to estimate A values from theory (**51**), the results depend on the type of interaction being considered. Therefore, at the present time the most reasonable approach appears to be to determine from experimental spectra the values of A_{HCl} and A_{DCl} subject to the conditions of Eqn (98). With $A_{HCl} = 0{\cdot}081$ Å and $A_{DCl} = 0{\cdot}047$ Å, fair agreement is obtained between observed and calculated rotational line shifts for HCl and DCl in various rare-gas matrices. In particular, Eqn (95) accounts for the observed fact that the shifts for DCl are greater than those for HCl. The opposite result would be expected on the basis of a hindered-rotation model.

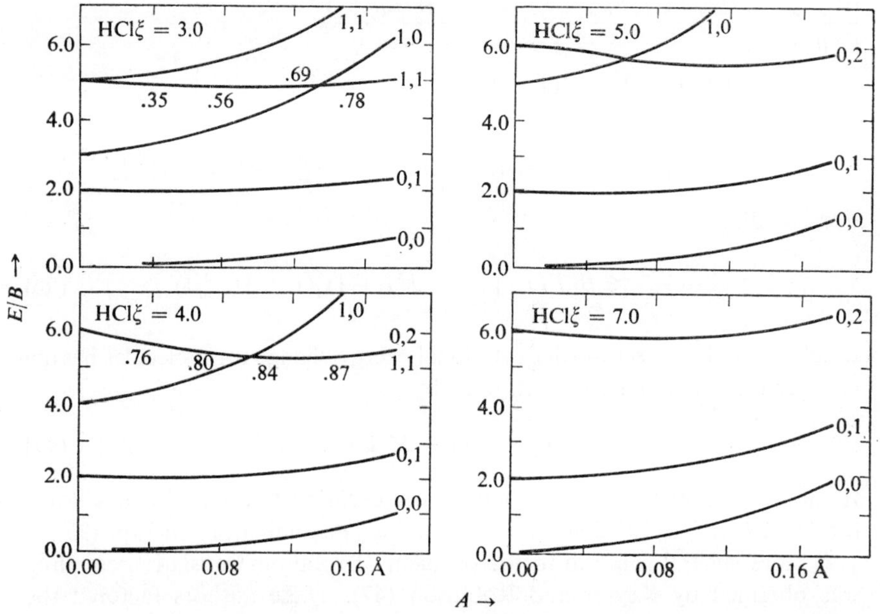

FIG. 15. Rotational-translational energy levels as functions of A [From Keyser and Robinson (**47**)].

From the above discussions of the hindered-rotation model and the rotation–translation coupling model it is clear that a general treatment of the problem would require solutions to the Schrödinger equation which include both effects in the Hamiltonian. Furthermore, as pointed out by Pandey (**52**), the possibility of interaction of the external degrees of freedom of the guest with the lattice vibrations of the host lattice should be considered.

In summary, some recent results by Pandey shown in Table 13 indicate that all three effects mentioned above are significant. The direct observation

TABLE 13.

System	Exptl shift	Calculated shift[d] Crystalline field part	Lattice vibrational part	RTC part	Total shift	A (in Å)
HCl–Ar	5·6[b]	0·7	0·8	4·1	5·6	0·081
HCl–Kr	5·4[b]	0·7	0·6	4·1	5·4	
HCl–Xe	...	0·7	0·7	5·1	6·5	
DCl–Ar	1·3,[b] 2·1[c]	0·7	0·8	0·4	1·9	0·047
DCl–Kr	1·9,[b] 1·3[c]	0·7	0·6	0·4	1·7	
DCl–Xe	2·5[c]	0·7	0·7	0·3	1·8	
HBr–Ar	12·8[c]	0·4	0·7	12·1	13·2	0·110
HBr–K4	14·1,[c] 13·3[b]	0·4	0·6	11·7	12·7	
HBr–Xe	...	0·3	0·6	11·6	12·5	

[a] All shifts are in cm^{-1} units. [b] Reference (**54**). [c] Reference (**49**). [d] Reference (**52**).

of the translational frequencies of HCl at 72, 61, and 46 cm^{-1} in Ar, Kr, and Xe lattices, respectively, provides strong support for the validity of the rotation–translation coupling model (**53**). In addition, however, the assignment of the two components of the $R(0)$ doublet (**49**), whose presence can only be accounted for by a rotational barrier, indicates that angular anisotropy of the external potential cannot be neglected.

REFERENCES

1. van Thiel, M., Becker, E. D. and Pimentel, G. C. *J. Chem. Phys.* **27,** 95 (1957).
2. Maslakowez, A. *Z. Physik* **51,** 696 (1928).
3. Ketelaar, J. A. A., Haas, C. and van der Elsken, J. *J. Chem. Phys.* **24,** 624 (1956); vander Elsken, J. Thesis, Amsterdam (1959).
4. Maki, A. and Decius, J. C. *J. Chem. Phys.* **31,** 772 (1959).
5. Price, W. C., Sherman, W. F. and Wilkinson, G. R. *Proc. Roy. Soc.* (*London*) **255,** 5 (1960).
6. Bryant, J. I. and Turrell, G. C. *J. Chem. Phys.* **37,** 1069 (1962).
7. Kyropoulos, S. *Zeit, Anorg, Allgem. Chem.* **154,** 308 (1926); *Zeit. Physik* **63,** 849 (1930).
8. Herzberg, G. "Molecular Spectra and Molecular Structure II. Infrared and Raman Spectra of Polyatomic Molecules", Van Nostrand, Princeton (1945).
9. Taylor, J. H., Benedict, W. S. and Strong, J. *J. Chem. Phys.* **20,** 1884 (1952).
10. Amat, G. and Grenier–Bresson, M.-L. *J. Phys. Radium* **15,** 591 (1954).
11. Decius, J. C. *J. Chem. Phys.* **20,** 511 (1952).
12. Decius, J. C., Jacobson, J. L., Sherman, W. F. and Wilkinson, G. R. *J. Chem. Phys.* **43,** 2180 (1965).
13. Bauer, E. and Magat, M. *J. Phys. Radium* **9,** 319 (1938); *Physica* **5,** 718 (1938).
14. Pullin, A. D. E. *Spectrochim. Acta.* **13,** 125 (1958).
15. Wilson, E. B., Jr., Decius, J. C. and Cross, P. C. "Molecular Vibrations", McGraw-Hill, New York (1955).
16. Huong, P. V., Perrot, M. and Tyrrell, G. *J. Mol. Spectry.* **28,** 341 (1968).
17. Lennard-Jones, J. E. and Ingham, A. E. *Proc. Roy. Soc.* (*London*) **A107,** 636 (1925).

18. Hirschfelder, J. O., Curtiss, C. F. and Bird, R. B. "Molecular Theory of Gases and Liquids", Wiley, New York (1954).

19. Coulson, C. A. "Valence", Oxford (1960).

20. Bonnemay, A. and Daudel. R, *Compt. Rend.* **230,** 2300 (1950).

21. Cundill, M. A. Thesis, London (1968).

22. Maradudin, A. A. Phonons and lattice imperfections. *In* "Phonons and Phonon Interactions" (T. A. Bak, ed.), Benjamin, New York (1964).

23. Ludwig, W. Dynamics of a crystal lattice with defects. *In* "Theory of Crystal Defects", Academia, Prague (1966).

24. Hayes, W., Jones, G. D., Elliott, R. J. and Sennett, C. T. *In* "Lattice Dynamics" (R. F. Wallis, ed.), Pergamon, London (1965).

25. Pauling, L. and Wilson, E. B., Jr., "Introduction to Quantum Mechanics", McGraw-Hill, New York (1935).

26. Cundill, M. A. and Sherman, W. F. *Phys. Rev. Letters* **16,** 570 (1966); *Phys. Rev,* **168,** 1007 (1968).

27. Woods, A. D. B., Brockhouse, B. N., Cowley, R. A. and Cochran, R. *Phys. Rev.* **131,** 1025 (1963).

28. Hallam, H. E. *In* "Molecular Spectroscopy" (P. Hepple, ed.), Institute of Petroleum, London (1968).

29. Ewing, G. E., Thompson, W. E. and Pimentel, G. C. *J. Chem. Phys.* **32,** 927 (1960).

30. Milligan, D. E. and Jacox, M. E. *J. Chem. Phys.* **41,** 3032 (1964).

31. Ogilvie, J. F. *Spectrochim. Acta.* **23A,** 737 (1967).

32. Andrews, W. L. S. and Pimental, G. C. *J. Chem. Phys.* **44,** 2527 (1966); **47,** 3637 (1967).

33. Milligan, D. E., Jacox, M. E. and Guillory, W. A. *J. Chem. Phys.* **49,** 5330 (1968).

34. Milligan, D. E. and Jacox, M. E. *J. Chem. Phys.* **47,** 5157 (1967).

35. Jacox, M. E. and Milligan, D. E. *J. Chem. Phys.* **50,** 3252 (1969).

36. Jacox, M. E. and Milligan, D. E. *J. Chem. Phys.* **53,** 2588 (1970).

37. Jacox, M. E., Milligan, D. E., Moll, N, G. and Thompson, W. E. *J. Chem. Phys.* **43,** 3734 (1965).

38. Moll, N. G. and Thompson, W. E. *J. Chem. Phys.* **44,** 2684 (1966).

39. Milligan, D. E. and Jacox, M. E. *J. Chem. Phys.* **43,** 4487 (1965).

40. Ogilvie, J. F. *Nature, Lond.* **204,** 572 (1964).

41. Barnes, A. J., Hallam, H. E. and Scrimshaw, G. F. *Trans. Faraday Soc.* **65,** 3150 (1969).

42. Barnes, A. J. and Hallam, H. E. *Quart. Revs.* (*London*) **23,** 392 (1969).

43. Linevsky, M. L. *J. Chem. Phys.* **34,** 587 (1961).

44. Babloyantz, A. *Mol. Phys.* **2,** 39 (1959).

45. Friedmann, H. *Advan. Chem. Phys.* **4,** 225 (1962).

46. Kemble, E. C. "The Fundamental Principles of Quantum Mechanics", Dover, New York (1968).

47. Keyser, L. F. and Robinson, G. W. *J. Chem. Phys.* **44,** 3225 (1966).

48. Devonshire, A. F. *Proc. Roy. Soc.* (*London*) **A153,** 601 (1936).

49. Bowers, M. T. and Flygare, W. H. *J. Chem. Phys.* **44,** 1389 (1966).

50. Friedmann, H. and Kimel, S. *J. Chem. Phys.* **42,** 2552 (1965); **43,** 3925 (1965).

51. Herman, R. *J. Chem. Phys.* **44,** 1346 (1966).

52. Pandey, G. K. *J. Chem. Phys.* **49,** 1555 (1965).

53. Verstegen, J. M. P. J., Goldring, H., Kimel, S. and Katz, B. *J. Chem. Phys.* **44,** 3216 (1966).

54. Mann, D. E., Acquista, N. and White, D. *J. Chem. Phys.* **44,** 3453 (1966).

Appendix A

G-Matrix Elements for Stretching and Bending Coordinates

The inverse kinetic-energy matrix was introduced in Chapter 1 [See Eqn (16)], where a general method was developed for its construction. This method was used by Decius to derive several very convenient tables of **G**-matrix elements for the various types of internal coordinates in common use, viz., bond stretching, valence-angle bending [(**1**), p. 303], out-of-plane bending, and torsion (**2**).

General expressions for the **G** elements based on bond-stretching (r) and valence angle-bending (ϕ) coordinates are included in Table I. The specific combinations of coordinates which are involved are defined in Fig. 1. Since each element of **G** depends on two coordinates, a double subscript is necessary to specify the general types, G_{rr}, $G_{r\phi}$, and $G_{\phi\phi}$. Furthermore, a superscript is added to indicate the number of atoms common to the two coordinates.

The atoms are numbered following a standard convention as shown in Fig. 1. Those atoms which are common to the two coordinates being considered are indicated by a double circle and placed along a horizontal line. Finally, for elements such as $G_{r\phi}{}^{1}$ it is necessary to distinguish the cases in which the common atom is an end atom or a central atom of ϕ. Following Decius' convention these cases are specified by the addition of a pair of indices in parentheses giving the numbers of atoms to the left of the common atoms, above and below the horizontal line, respectively, as shown in Fig. 1.

In the entries in Table 1, μ_α is the reciprocal mass of the αth atom and $\rho_{\alpha\delta}$ the reciprocal of the equilibrium length of the bond between atoms α and δ. The equilibrium angle between bonds $\alpha\delta$ and $\beta\delta$ is denoted $\phi_{\alpha\delta\beta}$, while $\psi_{\alpha\beta\gamma}$ is the equilibrium value of the dihedral angle defined by

$$\cos\psi_{\alpha\beta\gamma} = \frac{\cos\phi_{\alpha\delta\gamma} - \cos\phi_{\alpha\delta\beta}\cos\phi_{\beta\delta\gamma}}{\sin\phi_{\alpha\delta\beta}\sin\phi_{\beta\delta\gamma}}, \qquad (1)$$

where the atoms are designated as shown in Fig. 2.

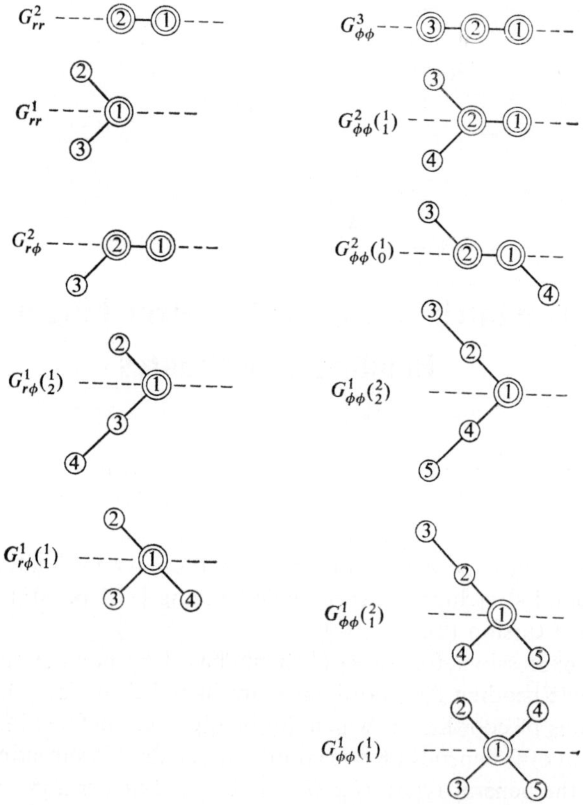

FIG. 1. Schematic representation of kinetic-energy matrix elements [From reference (**1**)].

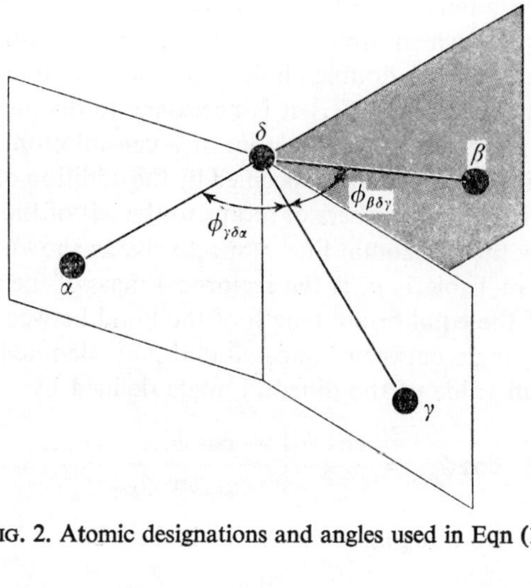

FIG. 2. Atomic designations and angles used in Eqn (1).

TABLE 1 [From reference (1)]

$(c\phi = \cos\phi,\ s\phi = \sin\phi)$

$G_{rr}{}^{2}$	$\mu_1 + \mu_2$
$G_{rr}{}^{1}$	$\mu_1 c\phi$
$G_{r\phi}{}^{2}$	$-\rho_{23}\mu_2 s\phi$
$G_{r\phi}{}^{1}\binom{1}{2}$	$\rho_{13}\mu_1 s\phi_1 c\tau$
$\binom{1}{1}$	$-(\rho_{13} s\phi_{213} c\psi_{234} + \rho_{14} s\phi_{214} c\psi_{243})\mu_1$
$G_{\phi\phi}{}^{3}$	$\rho_{12}{}^{2}\mu_1 + \rho_{23}{}^{2}\mu_3 + (\rho_{12}{}^{2} + \rho_{23}{}^{2} - 2\rho_{12}\rho_{23} c\phi)\mu_2$
$G_{\phi\phi}{}^{2}\binom{1}{1}$	$(\rho_{12}{}^{2} c\psi_{314})\mu_1 + [(\rho_{12} - \rho_{23} c\phi_{123} - \rho_{24} c\phi_{124})\rho_{12} c\psi_{314} + (s\phi_{123} s\phi_{124} s^2\psi_{314} + c\phi_{324} c\psi_{314})\rho_{23}\rho_{24}]\mu_2$
$\binom{1}{0}$	$-\rho_{12} c\tau[(\rho_{12} - \rho_{14} c\phi_1)\mu_1 + (\rho_{12} - \rho_{23} c\phi_2)\mu_2]$
$G_{\phi\phi}{}^{1}\binom{2}{2}$	$-(s\tau_{25} s\tau_{34} + c\tau_{25} c\tau_{34} c\phi_1)\rho_{12}\rho_{14}\mu_1$
$\binom{2}{1}$	$[(s\phi_{214} c\phi_{415} c\tau_{34} - s\phi_{215} c\tau_{35})\rho_{14} + (s\phi_{215} c\phi_{415} c\tau_{35} - s\phi_{214} c\tau_{34})\rho_{15}]\dfrac{\rho_{12}\mu_1}{s\phi_{415}}$
$\binom{1}{1}$	$[(c\phi_{415} - c\phi_{314} c\phi_{315} - c\phi_{214} c\phi_{215} + c\phi_{213} c\phi_{214} c\phi_{315})\rho_{12}\rho_{13} + (c\phi_{413} - c\phi_{514} c\phi_{513} - c\phi_{214} c\phi_{213} + c\phi_{215} c\phi_{214} c\phi_{513})\rho_{12}\rho_{15} + (c\phi_{215} - c\phi_{312} c\phi_{315} - c\phi_{412} c\phi_{415} + c\phi_{413} c\phi_{412} c\phi_{315})\rho_{14}\rho_{13} + (c\phi_{213} - c\phi_{512} c\phi_{513} - c\phi_{412} c\phi_{413} + c\phi_{415} c\phi_{412} c\phi_{513})\rho_{14}\rho_{15}]\dfrac{\mu_1}{s\phi_{214} s\phi_{315}}$

Certain of the elements of the type $G_{\phi\phi}$ also involve a torsion angle, $\tau_{\alpha\delta}$, which is defined by

$$\cos\tau_{\alpha\delta} = \frac{(\mathbf{e}_{\alpha\beta} \times \mathbf{e}_{\beta\gamma}) \cdot (\mathbf{e}_{\beta\gamma} \times \mathbf{e}_{\gamma\delta})}{\sin\phi_{\alpha\beta\gamma}\sin\phi_{\beta\gamma\delta}}, \tag{2}$$

where $\mathbf{e}_{\alpha\beta}$ is a unit vector directed from atom α toward atom β, etc. In this case the atoms are identified in Fig. 3. The reader is warned that some care must be taken in determining the sign of $\tau_{\alpha\delta}$, and for further details is referred to Wilson, Decius, and Cross [See reference (**1**), p. 59].

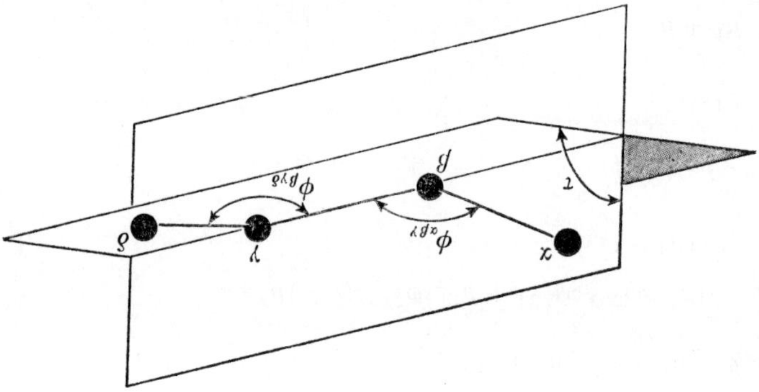

FIG. 3. Atomic designations and angles used in Eqn (2).

Appendix B

Calculation of Force Constants

The secular equations for the harmonic vibrational frequencies of a polyatomic molecule were developed in Chapter 1 using the Wilson **F** – **G** method. For example, in **GF** form, these equations are given by Eqn (33), Chap. 1,

$$\mathbf{L}^{-1}\,\mathbf{GFL} = \mathbf{\Lambda}, \tag{1}$$

and the problem is one of finding the eigenvalues of the nonsymmetric matrix, **GF**. In the early days of normal-coordinate calculations the diagonalization of **GF** was carried out numerically using a desk calculator, a long and tedious procedure, even for very small molecules. With the application of digital computers to this problem a little over a decade ago, the determination of a molecular force field from observed vibrational frequencies has become almost routine. Several somewhat different computer programs have been written and discussed in the literature (**3**)–(**10**). The program developed by J. H. Schachtschneider (**11**) will be outlined here, as it is in use in various modified forms by numerous workers in the field.

As pointed out in Section V, Chap. 1, in the special case in which **G** is diagonal (for example, when Cartesian basis coordinates are used) the equivalent problem represented by Eqn (46), Chap. 1, can be solved. However, in order to use this method in the general case it is first necessary to diagonalize **G**. Following the suggestion by Miyazawa (**11**), **G** is diagonalized by an orthogonal matrix **D**; thus,

$$\mathbf{D}^{-1}\,\mathbf{GD} = \mathbf{D}^{\dagger}\mathbf{GD} = \mathbf{\Gamma}, \tag{2}$$

where $\mathbf{\Gamma}$ is the diagonal matrix of the eigenvalues of **G**. Now define a matrix

$$\mathbf{W} \equiv \mathbf{D}\mathbf{\Gamma}^{1/2} \tag{3}$$

and note that from the orthogonality of **D**,

$$\mathbf{W}\mathbf{W}^{\dagger} = \mathbf{G}. \tag{4}$$

Finally, define a symmetric matrix $\mathbf{H}$ by the relation

$$\mathbf{H} \equiv \mathbf{W}^{\dagger}\mathbf{F}\,\mathbf{W}. \tag{5}$$

This matrix can be diagonalized by a matrix $\mathbf{C}$ to yield a set of eigenvalues $\mathbf{\Lambda}'$, viz.,

$$\mathbf{C}^{-1}\,\mathbf{HC} = \mathbf{\Lambda}' \tag{6}$$

or, using Eqn (5),

$$\mathbf{W}^{\dagger}\,\mathbf{FWC} = \mathbf{C\Lambda}'. \tag{7}$$

Multiplying on the left by $\mathbf{W}$, one finds

$$\mathbf{WW}^{\dagger}\,\mathbf{FWC} = \mathbf{GFWC} = \mathbf{WC\Lambda}' \tag{8}$$

or

$$(\mathbf{WC})^{-1}\,\mathbf{GF}(\mathbf{WC}) = \mathbf{\Lambda}' \tag{9}$$

By comparison of Eqns (9) and (1) it is evident that

$$\mathbf{WC} = \mathbf{L} \tag{10}$$

and

$$\mathbf{\Lambda}' = \mathbf{\Lambda}. \tag{11}$$

The procedure outlined above allows the eigenvectors and eigenvalues of the product $\mathbf{GF}$ to be determined by first finding the eigenvalues $\mathbf{\Gamma}$ of $\mathbf{G}$ and then constructing the matrix $\mathbf{W}$ via Eqn (3). This calculation can be carried out initially for each isotopic species and used throughout the iterative computation of $\mathbf{F}$ from a set of observed vibrational frequencies. The symmetric matrices $\mathbf{G}$ and $\mathbf{H}$ can each be diagonalized using available programs for the calculation of the eigenvalues of symmetric matrices, e.g., using Jacobi's method. An additional advantage of this procedure is that the inverse eigenvector matrix can be easily obtained from the relation

$$\mathbf{L}^{-1} = \mathbf{C}^{\dagger}\mathbf{\Gamma}^{-1}\,\mathbf{W}^{\dagger}, \tag{12}$$

which follows from Eqns (3) and (10).

The determination of a suitable set of force constants for a molecule from its observed vibrational frequencies invariably starts with an estimate of the force field, $\mathbf{F}^0$. With the molecular geometry (and, hence, $\mathbf{G}$) known† the eigenvalues of $\mathbf{GF}^0$ correspond to the calculated frequency parameters $\lambda_k\,(\text{cal}) = 4\pi^2[\nu_k\,(\text{cal})]^2$. Ordinarily the $\nu_k\,(\text{cal})$'s do not agree with the

† Most computer programs for the molecular vibration problem contain a subroutine for the construction of the $\mathbf{G}$ matrix from the molecular geometry.

observed frequencies, and the force constants must be refined by iteration until the quantities $\Delta\lambda_k \equiv \lambda_k\,(\mathrm{cal}) - \lambda_k\,(\mathrm{obs})$ vanish. If $\mathbf{F}^0$ is well chosen, the $\Delta\lambda_k$'s will be small and first-order perturbation theory will provide the needed corrections to the **F** matrix. Thus,

$$\mathbf{G}(\mathbf{F}^0 + \Delta\mathbf{F})\,\mathbf{L}^0 \approx \mathbf{L}^0\,\mathbf{\Lambda}_{\mathrm{obs}} \tag{13}$$

or

$$\mathbf{G}(\Delta\mathbf{F})\,\mathbf{L}^0 \approx \mathbf{L}^0(\mathbf{\Lambda}_{\mathrm{cal}} - \mathbf{\Lambda}_{\mathrm{obs}}) = \mathbf{L}^0\,(\Delta\mathbf{\Lambda}). \tag{14}$$

Since $\mathbf{L}^0\mathbf{L}^{0\dagger} = \mathbf{G}$ [See Eqn (32), Chap. 1], Eqn (14) can be written in the form

$$\mathbf{L}^0(\Delta\mathbf{F})\,\mathbf{L}^0 \approx \Delta\mathbf{\Lambda}. \tag{15}$$

If this equation is written out, it is immediately seen that the coefficients of the elements of $\Delta\mathbf{F}$ are simply the elements of the Jacobian of $\mathbf{\Lambda}$ with respect to **F**. Hence, Eqn. (15) can be rewritten as

$$\mathbf{J}(\Delta\mathbf{F}) \approx \Delta\mathbf{\Lambda}. \tag{16}$$

Equation (16) forms the basis of the iteration method. However, as the elements of **F** are not always independent, it is more convenient to work with a matrix $\mathbf{\Phi}$ of independent force constants defined by

$$\mathbf{F} = \mathbf{Z}\,\mathbf{\Phi}. \tag{17}$$

The force field used may be of the valence, the Urey–Bradley, or any other form, and, if desired, symmetry coordinates can be introduced (Section XII, Chap. 2). It should be noted, however, that in computer calculations the factorization of the secular determinant using molecular symmetry is of little value. However, symmetry coordinates are often still useful if the observed frequencies have been classified by symmetry species using for example polarization data, since this classification can be retained throughout the calculation.

Substituting Eqn (17) into Eqn (16), one finds

$$\mathbf{JZ}\,(\Delta\mathbf{\Phi}) = \Delta\mathbf{\Lambda}. \tag{18}$$

The dimensions of the matrix **JZ** are $(3N - 6)$ by s, where s is the number of independent force constants. If $s > 3N - 6$, Eqn (16) cannot be solved without introducing constraints to reduce the number of columns in **JZ** or adding the frequencies of other isotopic species in order to increase the number of rows.

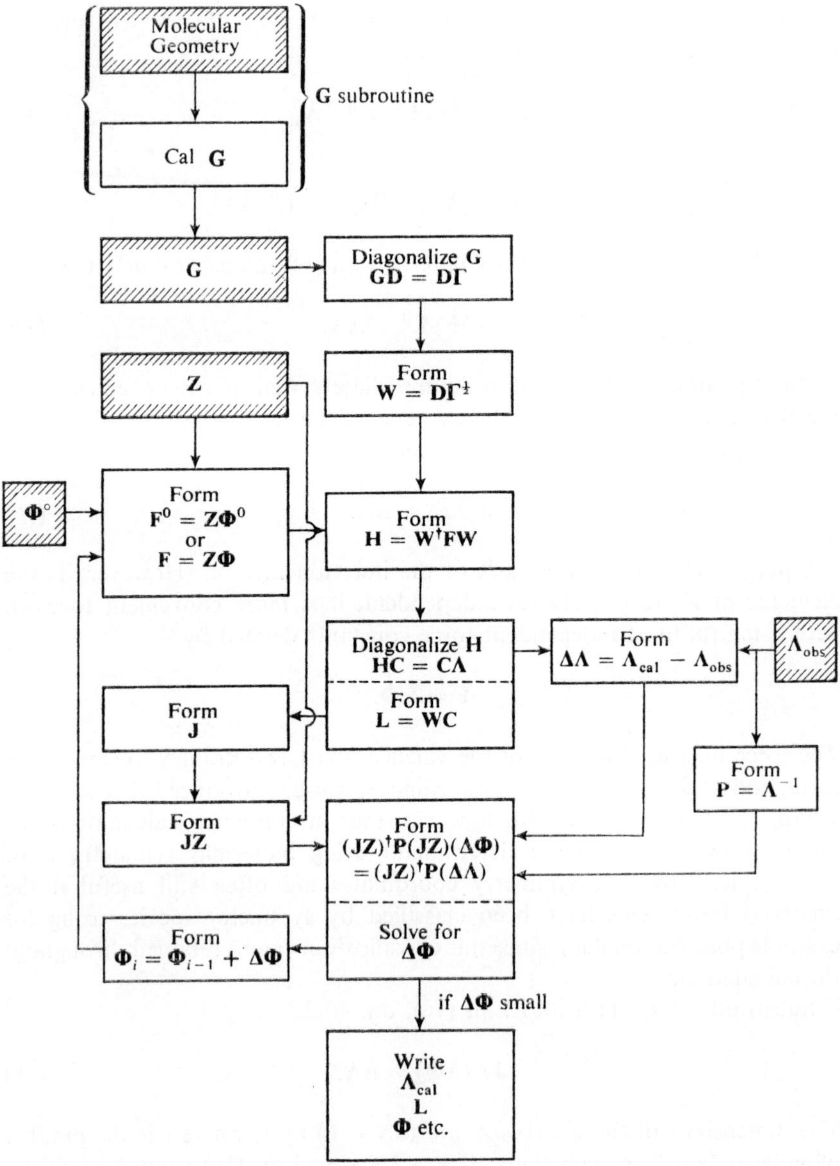

FIG. 1. Flow diagram of normal-coordinate calculation.

If $s < 3N - 6$, the problem is over-determined and it is usual to introduce a least-squares criterion on the frequency fit. Defining a diagonal weight matrix, $\mathbf{P}$, and multiplying Eqn (18) by $(\mathbf{JZ})^\dagger \mathbf{P}$, one obtains the expression

$$(\mathbf{JZ})^\dagger \mathbf{P}(\mathbf{JZ})(\mathbf{\Delta\Phi}) = (\mathbf{JZ})^\dagger \mathbf{P}(\mathbf{\Delta\Lambda}), \tag{19}$$

whose solution, $\mathbf{\Delta\Phi}$, is the set of least-square corrections to the force constants. The weight matrix is usually taken to be $\mathbf{P} = \mathbf{\Lambda}_{\text{obs}}{}^{-1}$.

As Eqn (19) is only approximate, the perturbation must be repeated using revised normal coordinates to set up a new $\mathbf{J}$ matrix. This cycle is repeated until the calculated λ's converge to the observed frequency parameters, or in the cases in which the potential function is over-determined, until further refinement produces no change in the resulting force constants. This iteration process is summarized in the flow diagram of Fig. 1.

In the above outline of the force–constant calculation the question of uniqueness of the solutions is not considered. However, for a secular determinant of order $3N - 6$ there exists as many as $3N - 6$ different matrices $\mathbf{F}$ which can, in principle, reproduce the observed frequencies. Fortunately, the vast majority of these solutions corresponds to physically unacceptable force fields. Furthermore, a judicious choice of the initial $\mathbf{F}$ matrix will usually direct the computer toward a chemically satisfactory, although not necessarily unique, solution. This problem is one which undoubtedly deserves further attention.

Appendix C

Character Tables

Character tables are given in this appendix for the following point groups: (1) the groups which correspond to the 32 crystal classes, (2) groups containing five-fold axes which may be needed to describe the symmetries of certain molecules or complex ions, and (3) the infinite groups $\mathscr{C}_{\infty v}$ and $\mathscr{D}_{\infty h}$ which are appropriate to linear structures.

Some comment on the notation for the irreducible representations or symmetry species is appropriate, as certain variations will be found in the literature. Molecular spectroscopists usually designate one-dimensional symmetry species by A or B, two-dimensional species by E, and three-dimensional species by F. However, solid-state physicists often use the symbol T in place of F for the triply degenerate species, while some authors prefer lower-case letters to capitals. The symmetry or antisymmetry of a species with respect to the generating rotation operation (See Table 2, Chap. 2) is specified by A or B, respectively, in groups such as $\mathscr{C}_n$ and $\mathscr{D}_n$. Subscripts 1 and 2 are used on A and B to distinguish symmetry with respect to rotation about a C_2 axis perpendicular to the generating axis. Symmetry with respect to inversion is indicated by a subscript g or u (German: *gerade*, *ungerade*), while a prime or double prime represents species which are, respectively, symmetric or antisymmetric with respect to a horizontal plane. Finally, numerical subscripts are used to distinguish various doubly and triply degenerate species. The reader is referred to Wilson *et al.* (**1**), p. 322 for the significance of these symbols.

In the case of linear molecules (point groups $\mathscr{C}_{\infty v}$ and $\mathscr{D}_{\infty h}$) these symbols are usually replaced by a system derived from the term sysmbol for the electronic states of diatomic and other linear molecules. A capital Greek letter Σ, Π, Δ, Φ, ..., is used, corresponding to $l = 0, 1, 2, 3, \ldots$, where l is the quantum number for rotation about the molecular axis. These symbols are equivalent to A_1, E_1, E_2, E_3, ..., respectively, in the Mulliken notation described above. For Σ species a superscript $+$ or $-$ is added to indicate the symmetry with respect to a plane which contains the molecular axis. Species of point group $\mathscr{D}_{\infty h}$ are further distinguished by the subscripts g or u which, as usual, specify the symmetry with respect to inversion.

The components of the translation and rotation vectors are given as T_x, T_y, T_z, and R_x, R_y, R_z, respectively. As indicated in Chapter 2, the components of **T** correspond directly to the components of the dipole moment vector and are used in the determination of infrared selection rules. The components of the polarizability tensor appear as linear combinations such as $\alpha_{xx} + \alpha_{yy}$, etc., which have the symmetry of the indicated irreducible representation. These components are used to determine Raman activity of molecular and crystal vibrations, as explained in Chapters 2 and 4, respectively.

TABLE 1. Character tables for the cyclic groups, $\mathscr{C}_n (n = 2, 3, 4, 5, 6)$.

$\mathscr{C}_2$	E	C_2		
A	1	1	T_z, R_z	$\alpha_{xx}, \alpha_{yy}, \alpha_{zz}, \alpha_{xy}$
B	1	−1	T_x, T_y, R_x, R_y	α_{yz}, α_{zx}

$\mathscr{C}_3$	E	C_3	C_3^2		$\varepsilon = \exp(2\pi i/3)$
A	1	1	1	T_z, R_z	$\alpha_{xx} + \alpha_{yy}, \alpha_{zz}$
E	$\begin{cases}1 \\ 1\end{cases}$	ε ε^*	ε^* ε	$(T_x, T_y), (R_x, R_y)$	$(\alpha_{xx} - \alpha_{yy}, \alpha_{xy}), (\alpha_{yz}, \alpha_{zx})$

$\mathscr{C}_4$	E	C_4	C_2	C_4^3		
A	1	1	1	1	T_z, R_z	$\alpha_{xx} + \alpha_{yy}, \alpha_{zz}$
B	1	−1	1	−1		$\alpha_{xx} - \alpha_{yy}, \alpha_{xy}$
E	$\begin{cases}1 \\ 1\end{cases}$	i $-i$	−1 −1	$-i$ i	$(T_x, T_y), (R_x, R_y)$	$(\alpha_{yz}, \alpha_{zx})$

$\mathscr{C}_5$	E	C_5	C_5^2	C_5^3	C_5^4		$\varepsilon = \exp(2\pi i/5)$
A	1	1	1	1	1	T_z, R_z	$\alpha_{xx} + \alpha_{yy}, \alpha_{zz}$
E_1	$\begin{cases}1 \\ 1\end{cases}$	ε ε^*	ε^2 ε^{2*}	ε^{2*} ε^2	ε^* ε	$(T_x, T_y), (R_x, R_y)$	$(\alpha_{yz}, \alpha_{zx})$
E_2	$\begin{cases}1 \\ 1\end{cases}$	ε^2 ε^{2*}	ε^* ε	ε ε^*	ε^{2*} ε^2		$(\alpha_{xx} - \alpha_{yy}, \alpha_{xy})$

TABLE 1 (*Continued*)

$\mathscr{C}_6$	E	C_6	C_3	C_2	C_3^2	C_6^5		$\varepsilon = \exp(2\pi i/6)$
A	1	1	1	1	1	1	T_z, R_z	$\alpha_{xx} + \alpha_{yy}, \alpha_{zz}$
B	1	−1	1	−1	1	−1		
E_1	1	ε	$-\varepsilon^*$	−1	$-\varepsilon$	ε^*	$(T_x, T_y), (R_x, R_y)$	$(\alpha_{yz}, \alpha_{zx})$
	1	ε^*	$-\varepsilon$	−1	$-\varepsilon^*$	ε		
E_2	1	$-\varepsilon^*$	$-\varepsilon$	1	$-\varepsilon^*$	$-\varepsilon$		$(\alpha_{xx} - \alpha_{yy}, \alpha_{xy})$
	1	$-\varepsilon$	$-\varepsilon^*$	1	$-\varepsilon$	$-\varepsilon^*$		

TABLE 2. Character tables for the dihedral groups, $\mathscr{D}_n (n = 2, 3, 4, 5, 6)$.

$\mathscr{D}_2 = \mathscr{V}$	E	$C_2(z)$	$C_2(y)$	$C_2(x)$		
A	1	1	1	1		$\alpha_{xx}, \alpha_{yy}, \alpha_{zz}$
B_1	1	1	−1	−1	T_z, R_z	α_{xy}
B_2	1	−1	1	−1	T_y, R_y	α_{zx}
B_3	1	−1	−1	1	T_x, R_x	α_{yz}

$\mathscr{D}_3$	E	$2C_3$	$3C_2'$		
A_1	1	1	1		$\alpha_{xx} + \alpha_{yy}, \alpha_{zz}$
A_2	1	1	−1	T_z, R_z	
E	2	−1	0	$(T_x, T_y), (R_x, R_y)$	$(\alpha_{xx} - \alpha_{yy}, \alpha_{xy}), (\alpha_{yz}, \alpha_{zx})$

$\mathscr{D}_4$	E	$2C_4$	C_2	$2C_2'$	$2C_2''$		
A_1	1	1	1	1	1		$\alpha_{xx} + \alpha_{yy}, \alpha_{zz}$
A_2	1	1	1	−1	−1	T_z, R_z	
B_1	1	−1	1	1	−1		$\alpha_{xx} - \alpha_{yy}$
B_2	1	−1	1	−1	1		α_{xy}
E	2	0	−2	0	0	$(T_x, T_y), (R_x, R_y)$	$(\alpha_{yz}, \alpha_{zx})$

TABLE 2 (*Continued*)

$\mathscr{D}_5$	E	$2C_5$	$2C_5^2$	$5C_2'$		$\alpha = 72°$
A_1	1	1	1	1		$\alpha_{xx}+\alpha_{yy}, \alpha_{zz}$
A_2	1	1	1	−1	T_z, R_z	
E_1	2	$2\cos\alpha$	$2\cos 2\alpha$	0	$(T_x, T_y), (R_x, R_y)$	$(\alpha_{yz}, \alpha_{zx})$
E_2	2	$2\cos 2\alpha$	$2\cos\alpha$	0		$(\alpha_{xx}-\alpha_{yy}, \alpha_{xy})$

$\mathscr{D}_6$	E	$2C_6$	$2C_3$	C_2	$3C_2'$	$3C_2''$		
A_1	1	1	1	1	1	1		$\alpha_{xx}+\alpha_{yy}, \alpha_{zz}$
A_2	1	1	1	1	−1	−1	T_z, R_z	
B_1	1	−1	1	−1	1	−1		
B_2	1	−1	1	−1	−1	1		
E_1	2	1	−1	−2	0	0	$(T_x, T_y), (R_x, R_y)$	$(\alpha_{yz}, \alpha_{zx})$
E_2	2	−1	−1	2	0	0		$(\alpha_{xx}-\alpha_{yy}, \alpha_{xy})$

TABLE 3. Character tables for the groups $\mathscr{D}_{nh} (n = 2, 3, 4, 5, 6)$

$\mathscr{D}_{2h} = \mathscr{V}_h$	E	$C_2(z)$	$C_2(y)$	$C_2(x)$	i	$\sigma(xy)$	$\sigma(xz)$	$\sigma(yz)$		
A_g	1	1	1	1	1	1	1	1		$\alpha_{xx}, \alpha_{yy}, \alpha_{zz}$
B_{1g}	1	1	−1	−1	1	1	−1	−1	R_z	α_{xy}
B_{2g}	1	−1	1	−1	1	−1	1	−1	R_y	α_{zx}
B_{3g}	1	−1	−1	1	1	−1	−1	1	R_x	α_{yz}
A_u	1	1	1	1	−1	−1	−1	−1		
B_{1u}	1	1	−1	−1	−1	−1	1	1	T_z	
B_{2u}	1	−1	1	−1	−1	1	−1	1	T_y	
B_{3u}	1	−1	−1	1	−1	1	1	−1	T_x	

$\mathscr{D}_{3h}$	E	$2C_3$	$3C_2'$	σ_h	$2S_3$	$3\sigma_v$		
A_1'	1	1	1	1	1	1		$\alpha_{xx}+\alpha_{yy}, \alpha_{zz}$
A_2'	1	1	−1	1	1	−1	R_z	
E'	2	−1	0	2	−1	0	(T_x, T_y)	$(\alpha_{xx}-\alpha_{yy}, \alpha_{xy})$
A_1''	1	1	1	−1	−1	−1		
A_2''	1	1	−1	−1	−1	1	T_z	
E''	2	−1	0	−2	1	0	(R_x, R_y)	$(\alpha_{yz}, \alpha_{zx})$

TABLE 3 (*Continued*)

$\mathscr{D}_{4h}$	E	$2C_4$	C_2	$2C_2'$	$2C_2''$	i	$2S_4$	σ_h	$2\sigma_v$	$2\sigma_d$		
A_{1g}	1	1	1	1	1	1	1	1	1	1		$\alpha_{xx}+\alpha_{yy}, \alpha_{zz}$
A_{2g}	1	1	1	−1	−1	1	1	1	−1	−1	R_z	
B_{1g}	1	−1	1	1	−1	1	−1	1	1	−1		$\alpha_{xx}-\alpha_{yy}$
B_{2g}	1	−1	1	−1	1	1	−1	1	−1	1		α_{xy}
E_g	2	0	−2	0	0	2	0	−2	0	0	(R_x, R_y)	$(\alpha_{yz}, \alpha_{zx})$
A_{1u}	1	1	1	1	1	−1	−1	−1	−1	−1		
A_{2u}	1	1	1	−1	−1	−1	−1	−1	1	1	T_z	
B_{1u}	1	−1	1	1	−1	−1	1	−1	−1	1		
B_{2u}	1	−1	1	−1	1	−1	1	−1	1	−1		
E_u	2	0	−2	0	0	−2	0	2	0	0	(T_x, T_y)	

$\mathscr{D}_{5h}$	E	$2C_5$	$2C_5^2$	$5C_2'$	σ_h	$2S_5$	$2S_5^3$	$5\sigma_v$		$\alpha = 72°$
A_1'	1	1	1	1	1	1	1	1		$\alpha_{xx}+\alpha_{yy}$, α_{zz}
A_2'	1	1	1	−1	1	1	1	−1	R_z	
E_1'	2	$2\cos\alpha$	$2\cos 2\alpha$	0	2	$2\cos\alpha$	$2\cos 2\alpha$	0	(T_x, T_y)	
E_2'	2	$2\cos 2\alpha$	$2\cos\alpha$	0	2	$2\cos 2\alpha$	$2\cos\alpha$	0		$(\alpha_{xx}-\alpha_{yy}$, $\alpha_{xy})$
A_1''	1	1	1	1	−1	−1	−1	−1		
A_2''	1	1	1	−1	−1	−1	−1	1	T_z	
E_1''	2	$2\cos\alpha$	$2\cos 2\alpha$	0	−2	$-2\cos\alpha$	$-2\cos 2\alpha$	0	(R_x, R_y)	$(\alpha_{yz}, \alpha_{zx})$
E_2''	2	$2\cos 2\alpha$	$2\cos\alpha$	0	−2	$-2\cos 2\alpha$	$-2\cos\alpha$	0		

$\mathscr{D}_{6h}$	E	$2C_6$	$2C_3$	C_2	$3C_2'$	$3C_2''$	i	$2S_3$	$2S_6$	σ_h	$3\sigma_d$	$3\sigma_v$		
A_{1g}	1	1	1	1	1	1	1	1	1	1	1	1		$\alpha_{xx}+\alpha_{yy}, \alpha_{zz}$
A_{2g}	1	1	1	1	−1	−1	1	1	1	1	−1	−1	R_z	
B_{1g}	1	−1	1	−1	1	−1	1	−1	1	−1	1	−1		
B_{2g}	1	−1	1	−1	−1	1	1	−1	1	−1	−1	1		
E_{1g}	2	1	−1	−2	0	0	2	1	−1	−2	0	0	(R_x, R_y)	$(\alpha_{yz}, \alpha_{zx})$
E_{2g}	2	−1	−1	2	0	0	2	−1	−1	2	0	0		$(\alpha_{xx}-\alpha_{yy}$, $\alpha_{xy})$
A_{1u}	1	1	1	1	1	1	−1	−1	−1	−1	−1	−1		
A_{2u}	1	1	1	1	−1	−1	−1	−1	−1	−1	1	1	T_z	
B_{1u}	1	−1	1	−1	1	−1	−1	1	−1	1	−1	1		
B_{2u}	1	−1	1	−1	−1	1	−1	1	−1	1	1	−1		
E_{1u}	2	1	−1	−2	0	0	−2	−1	1	2	0	0	(T_x, T_y)	
E_{2u}	2	−1	−1	2	0	0	−2	1	1	−2	0	0		

TABLE 4. Character tables for the groups $\mathscr{S}_n (n = 2, 4, 6, 8)$.

$\mathscr{S}_2 \equiv \mathscr{C}_1$	E	i		
A_g	1	1	R_x, R_y, R_z	$\alpha_{xx}, \alpha_{yy}, \alpha_{zz}, \alpha_{xy}$, α_{yz}, α_{zx}
A_u	1	-1	T_x, T_y, T_z	

$\mathscr{S}_4$	E	S_4	C_2	S_4^3		
A	1	1	1	1	R_z	$\alpha_{xx} + \alpha_{yy}, \alpha_{zz}$
B	1	-1	1	-1	T_z	$\alpha_{xx} - \alpha_{yy}, \alpha_{xy}$
E	1 1	i $-i$	-1 -1	$-i$ i	$(T_x, T_y), (R_x, R_y)$	$(\alpha_{yz}, \alpha_{zx})$

$\mathscr{S}_6$	E	C_3	C_3^2	i	S_6^5	S_6		$\varepsilon = \exp(2\pi i/3)$
A_g	1	1	1	1	1	1	R_z	$\alpha_{xx} + \alpha_{yy}, \alpha_{zz}$
E_g	1 1	ε ε^*	ε^* ε	1 1	ε ε^*	ε^* ε	(R_x, R_y)	$(\alpha_{xx} - \alpha_{yy}, \alpha_{xy}), (\alpha_{yz}, \alpha_{zx})$
A_u	1	1	1	-1	-1	-1	T_z	
E_u	1 1	ε ε^*	ε^* ε	-1 -1	$-\varepsilon$ $-\varepsilon^*$	$-\varepsilon^*$ $-\varepsilon$	(T_x, T_y)	

$\mathscr{S}_8$	E	S_8	C_4	S_8^3	C_2	S_8^5	C_4^3	S_8^7	$\varepsilon = \exp(2\pi i/8)$	
A	1	1	1	1	1	1	1	1	R_z	$\alpha_{xx} + \alpha_{yy}, \alpha_{zz}$
B	1	-1	1	-1	1	-1	1	-1	T_z	
E_1	1 1	ε ε^*	i $-i$	$-\varepsilon^*$ $-\varepsilon$	-1 -1	$-\varepsilon$ $-\varepsilon^*$	$-i$ i	ε^* ε	(T_x, T_y)	
E_2	1 1	i $-i$	-1 -1	$-i$ i	1 1	i $-i$	-1 -1	$-i$ i		$(\alpha_{xx} - \alpha_{yy}, \alpha_{xy})$
E_3	1 1	$-\varepsilon^*$ $-\varepsilon$	$-i$ i	ε ε^*	-1 -1	ε^* ε	i $-i$	$-\varepsilon$ $-\varepsilon^*$	(R_x, R_y)	$(\alpha_{yz}, \alpha_{zx})$

TABLE 5. Character tables for the groups $\mathscr{C}_{nh}(n = 1, 2, 3, 4, 5, 6)$.

$\mathscr{C}_{1h} \equiv \mathscr{C}_s$	E	σ_h		
A'	1	1	T_x, T_y, R_z	$\alpha_{xx}, \alpha_{yy}, \alpha_{zz}, \alpha_{xy}$
A''	1	-1	T_z, R_x, R_y	α_{yz}, α_{zz}

$\mathscr{C}_{2h}$	E	C_2	i	σ_h		
A_g	1	1	1	1	R_z	$\alpha_{xx}, \alpha_{yy}, \alpha_{zz}, \alpha_{xy}$
B_g	1	-1	1	-1	R_x, R_y	α_{yz}, α_{zz}
A_u	1	1	-1	-1	T_z	
B_u	1	-1	-1	1	T_x, T_y	

$\mathscr{C}_{3h}$	E	C_3	C_3^2	σ_h	S_3	S_3^5		$\varepsilon = \exp(2\pi i/3)$
A'	1	1	1	1	1	1	R_z	$\alpha_{xx} + \alpha_{yy}, \alpha_{zz}$
E'	$\begin{Bmatrix}1 \\ 1\end{Bmatrix}$	$\begin{matrix}\varepsilon \\ \varepsilon^*\end{matrix}$	$\begin{matrix}\varepsilon^* \\ \varepsilon\end{matrix}$	$\begin{matrix}1 \\ 1\end{matrix}$	$\begin{matrix}\varepsilon \\ \varepsilon^*\end{matrix}$	$\begin{matrix}\varepsilon^* \\ \varepsilon\end{matrix}$	(T_x, T_y)	$(\alpha_{xx} - \alpha_{yy}, \alpha_{xy})$
A''	1	1	1	-1	-1	-1	T_z	
E''	$\begin{Bmatrix}1 \\ 1\end{Bmatrix}$	$\begin{matrix}\varepsilon \\ \varepsilon^*\end{matrix}$	$\begin{matrix}\varepsilon^* \\ \varepsilon\end{matrix}$	$\begin{matrix}-1 \\ -1\end{matrix}$	$\begin{matrix}-\varepsilon \\ -\varepsilon^*\end{matrix}$	$\begin{matrix}-\varepsilon^* \\ -\varepsilon\end{matrix}$	(R_x, R_y)	$(\alpha_{yz}, \alpha_{zx})$

$\mathscr{C}_{4h}$	E	C_4	C_2	C_4^3	i	S_4^3	σ_h	S_4		
A_g	1	1	1	1	1	1	1	1	R_z	$\alpha_{xx} + \alpha_{yy}, \alpha_{zz}$
B_g	1	-1	1	-1	1	-1	1	-1		$\alpha_{xx} - \alpha_{yy}, \alpha_{xy}$
E_g	$\begin{Bmatrix}1 \\ 1\end{Bmatrix}$	$\begin{matrix}i \\ -i\end{matrix}$	$\begin{matrix}-1 \\ -1\end{matrix}$	$\begin{matrix}-i \\ i\end{matrix}$	$\begin{matrix}1 \\ 1\end{matrix}$	$\begin{matrix}i \\ -i\end{matrix}$	$\begin{matrix}-1 \\ -1\end{matrix}$	$\begin{matrix}-i \\ i\end{matrix}$	(R_x, R_y)	$(\alpha_{yz}, \alpha_{zx})$
A_u	1	1	1	1	-1	-1	-1	-1	T_z	
B_u	1	-1	1	-1	-1	1	-1	1		
E_u	$\begin{Bmatrix}1 \\ 1\end{Bmatrix}$	$\begin{matrix}i \\ -i\end{matrix}$	$\begin{matrix}-1 \\ -1\end{matrix}$	$\begin{matrix}-i \\ i\end{matrix}$	$\begin{matrix}-1 \\ -1\end{matrix}$	$\begin{matrix}-i \\ i\end{matrix}$	$\begin{matrix}1 \\ 1\end{matrix}$	$\begin{matrix}i \\ -i\end{matrix}$	(T_x, T_y)	

TABLE 5 (*Continued*)

h	E	C_5	C_5^2	C_5^3	C_5^4	σ_h	S_5	S_5^7	S_5^3	S_5^9		$\varepsilon = \exp(2\pi i/5)$
	1	1	1	1	1	1	1	1	1	1	R_z	$\alpha_{xx} + \alpha_{yy}, \alpha_{zz}$
′	1	ε	ε^2	ε^{2*}	ε^*	1	ε	ε^2	ε^{2*}	ε^*	(T_z, T_y)	
	1	ε^*	ε^{2*}	ε^2	ε	1	ε^*	ε^{2*}	ε^2	ε		
′	1	ε^2	ε^*	ε	ε^{2*}	1	ε^2	ε^*	ε	ε^{2*}		$(\alpha_{xx} - \alpha_{yy}, \alpha_{xy})$
	1	ε^{2*}	ε	ε^*	ε^2	1	ε^{2*}	ε	ε^*	ε^2		
′	1	1	1	1	1	−1	−1	−1	−1	−1	T_z	
″	1	ε	ε^2	ε^{2*}	ε^*	−1	$-\varepsilon$	$-\varepsilon^2$	$-\varepsilon^{2*}$	$-\varepsilon^*$	(R_x, R_y)	$(\alpha_{yz}, \alpha_{zx})$
	1	ε^*	ε^{2*}	ε^2	ε	−1	$-\varepsilon^*$	$-\varepsilon^{2*}$	$-\varepsilon^2$	$-\varepsilon$		
″	1	ε^2	ε^*	ε	ε^{2*}	−1	$-\varepsilon^2$	$-\varepsilon^*$	$-\varepsilon$	$-\varepsilon^{2*}$		
	1	ε^{2*}	ε	ε^*	ε^2	−1	$-\varepsilon^{2*}$	$-\varepsilon$	$-\varepsilon^*$	$-\varepsilon^2$		

h	E	C_6	C_3	C_2	C_3^2	C_6^5	i	S_3^5	S_6^5	σ_h	S_6	S_3		$\varepsilon = \exp(2\pi i/6)$
	1	1	1	1	1	1	1	1	1	1	1	1	R_z	$\alpha_{xx} + \alpha_{yy}, \alpha_{zz}$
	1	−1	1	−1	1	−1	1	−1	1	−1	1	−1		
g	1	ε	$-\varepsilon^*$	−1	$-\varepsilon$	ε^*	1	ε	$-\varepsilon^*$	−1	$-\varepsilon$	ε^*	(R_x, R_y)	$(\alpha_{yz}, \alpha_{zx})$
	1	ε^*	$-\varepsilon$	−1	$-\varepsilon^*$	ε	1	ε^*	$-\varepsilon$	−1	$-\varepsilon^*$	ε		
g	1	$-\varepsilon^*$	$-\varepsilon$	1	$-\varepsilon^*$	$-\varepsilon$	1	$-\varepsilon^*$	$-\varepsilon$	1	$-\varepsilon^*$	$-\varepsilon$		$(\alpha_{xx} - \alpha_{yy}, \alpha_{xy})$
	1	$-\varepsilon$	$-\varepsilon^*$	1	$-\varepsilon$	$-\varepsilon^*$	1	$-\varepsilon$	$-\varepsilon^*$	1	$-\varepsilon$	$-\varepsilon^*$		
	1	1	1	1	1	1	−1	−1	−1	−1	−1	−1	T_z	
	1	−1	1	−1	1	−1	−1	1	−1	1	−1	1		
u	1	ε	$-\varepsilon^*$	−1	$-\varepsilon$	ε^*	−1	$-\varepsilon$	ε^*	1	ε	$-\varepsilon^*$	(T_x, T_y)	
	1	ε^*	$-\varepsilon$	−1	$-\varepsilon^*$	ε	−1	$-\varepsilon^*$	ε	1	ε^*	$-\varepsilon$		
u	1	$-\varepsilon^*$	$-\varepsilon$	1	$-\varepsilon^*$	$-\varepsilon$	−1	ε^*	ε	−1	ε^*	ε		
	1	$-\varepsilon$	$-\varepsilon^*$	1	$-\varepsilon$	$-\varepsilon^*$	−1	ε	ε^*	−1	ε	ε^*		

TABLE 6. Character tables for the groups $\mathscr{C}_{nv}(n = 2, 3, 4, 5, 6)$

v	E	C_2	$\sigma_v(xz)$	$\sigma_v(yz)$		
	1	1	1	1	T_z	$\alpha_{xx}, \alpha_{yy}, \alpha_{zz}$
	1	1	−1	−1	R_z	α_{xy}
	1	−1	1	−1	T_x, R_y	α_{zx}
	1	−1	−1	1	T_y, R_x	α_{yz}

TABLE 6 (*Continued*)

$\mathscr{C}_{3v}$	E	$2C_3$	$3\sigma_v$		
A_1	1	1	1	T_z	$\alpha_{xx}+\alpha_{yy}, \alpha_{zz}$
A_2	1	1	−1	R_z	
E	2	−1	0	$(T_x, T_y), (R_x, R_y)$	$(\alpha_{xx}-\alpha_{yy}, \alpha_{xy}), (\alpha_{yz}, \alpha_{zx})$

$\mathscr{C}_{4v}$	E	$2C_4$	C_2	$2\sigma_v$	$2\sigma_d$		
A_1	1	1	1	1	1	T_z	$\alpha_{xx}+\alpha_{yy}, \alpha_{zz}$
A_2	1	1	1	−1	−1	R_z	
B_1	1	−1	1	1	−1		$\alpha_{xx}-\alpha_{yy}$
B_2	1	−1	1	−1	1		α_{xy}
E	2	0	−2	0	0	$(T_x, T_y), (R_x, R_y)$	$(\alpha_{yz}, \alpha_{zx})$

$\mathscr{C}_{5v}$	E	$2C_5$	$2C_5^2$	$5\sigma_v$		$\alpha = 72°$
A_1	1	1	1	1	T_z	$\alpha_{xx}+\alpha_{yy}, \alpha_{zz}$
A_2	1	1	1	−1	R_z	
E_1	2	$2\cos\alpha$	$2\cos 2\alpha$	0	$(T_x, T_y), (R_x, R_y)$	$(\alpha_{yz}, \alpha_{zx})$
E_2	2	$2\cos 2\alpha$	$2\cos\alpha$	0		$(\alpha_{xx}-\alpha_{yy}, \alpha_{xy})$

$\mathscr{C}_{6v}$	E	$2C_6$	$2C_3$	C_2	$3\sigma_v$	$3\sigma_d$		
A_1	1	1	1	1	1	1	T_z	$\alpha_{xx}+\alpha_{yy}, \alpha_{xy}$
A_2	1	1	1	1	−1	−1	R_z	
B_1	1	−1	1	−1	1	−1		
B_2	1	−1	1	−1	−1	1		
E_1	2	1	−1	−2	0	0	$(T_x, T_y), (R_x, R_y)$	$(\alpha_{yz}, \alpha_{zx})$
E_2	2	−1	−1	2	0	0		$(\alpha_{xx}-\alpha_{yy}, \alpha_{xy})$

TABLE 7. Character tables for the groups $\mathscr{D}_{nd}(n = 2, 3, 4, 5, 6)$.

$\mathscr{D}_{2d} = \mathscr{V}_d$	E	$2S_4$	C_2	$2C_2'$	$2\sigma_d$		
A_1	1	1	1	1	1		$\alpha_{xx}+\alpha_{yy}, \alpha_{zz}$
A_2	1	1	1	−1	−1	R_z	
B_1	1	−1	1	1	−1		$\alpha_{xx}-\alpha_{yy}$
B_2	1	−1	1	−1	1	T_z	α_{xy}
E	2	0	−2	0	0	$(T_x, T_y), (R_x, R_y)$	$(\alpha_{yz}, \alpha_{zx})$

TABLE 7 (*Continued*)

$\mathscr{D}_{3d}$	E	$2C_3$	$3C_2$	i	$2S_6$	$3\sigma_d$		
A_{1g}	1	1	1	1	1	1		$\alpha_{xx}+\alpha_{yy}, \alpha_{zz}$
A_{2g}	1	1	-1	1	1	-1	R_z	
E_g	2	-1	0	2	-1	0	(R_x, R_y)	$(\alpha_{xx}-\alpha_{yy}, \alpha_{xy}), (\alpha_{yz}, \alpha_{zx})$
A_{1u}	1	1	1	-1	-1	-1		
A_{2u}	1	1	-1	-1	-1	1	T_z	
E_u	2	-1	0	-2	1	0	(T_x, T_y)	

$\mathscr{D}_{4d}$	E	$2S_8$	$2C_4$	$2S_8{}^3$	C_2	$4C_2'$	$4\sigma_d$		
A_1	1	1	1	1	1	1	1		$\alpha_{xx}+\alpha_{yy}, \alpha_{zz}$
A_2	1	1	1	1	1	-1	-1	R_z	
B_1	1	-1	1	-1	1	1	-1		
B_2	1	-1	1	-1	1	-1	1	T_z	
E_1	2	$\sqrt{2}$	0	$-\sqrt{2}$	-2	0	0	(T_x, T_y)	
E_2	2	0	-2	0	2	0	0		$(\alpha_{xx}-\alpha_{yy}, \alpha_{xy})$
E_3	2	$-\sqrt{2}$	0	$\sqrt{2}$	-2	0	0	(R_x, R_y)	$(\alpha_{yz}, \alpha_{zx})$

$\mathscr{D}_{5d}$	E	$2C_5$	$2C_5{}^2$	$5C_2$	i	$2S_{10}{}^3$	$2S_{10}$	$5\sigma_d$	$\alpha = 72°$	
A_{1g}	1	1	1	1	1	1	1	1		$\alpha_{xx}+\alpha_{yy}, \alpha_{zz}$
A_{2g}	1	1	1	-1	1	1	1	-1	R_z	
E_{1g}	2	$2\cos\alpha$	$2\cos 2\alpha$	0	2	$2\cos\alpha$	$2\cos 2\alpha$	0	(R_x, R_y)	$(\alpha_{yz}, \alpha_{zx})$
E_{2g}	2	$2\cos 2\alpha$	$2\cos\alpha$	0	2	$2\cos 2\alpha$	$2\cos\alpha$	0		$(\alpha_{xx}-\alpha_{yy}, \alpha_{xy})$
A_{1u}	1	1	1	1	-1	-1	-1	-1		
A_{2u}	1	1	1	-1	-1	-1	-1	1	T_z	
E_{1u}	2	$2\cos\alpha$	$2\cos 2\alpha$	0	-2	$-2\cos\alpha$	$-2\cos 2\alpha$	0	(T_x, T_y)	
E_{2u}	2	$2\cos 2\alpha$	$2\cos\alpha$	0	-2	$-2\cos 2\alpha$	$-2\cos\alpha$	0		

$\mathscr{D}_{6d}$	E	$2S_{12}$	$2C_6$	$2S_4$	$2C_3$	$2S_{12}{}^5$	C_2	$6C_2'$	$6\sigma_d$		
A_1	1	1	1	1	1	1	1	1	1		$\alpha_{xx}+\alpha_{yy}, \alpha_{zz}$
A_2	1	1	1	1	1	1	1	-1	-1	R_z	
B_1	1	-1	1	-1	1	-1	1	1	-1		
B_2	1	-1	1	-1	1	-1	1	-1	1	T_z	
E_1	2	$\sqrt{3}$	1	0	-1	$-\sqrt{3}$	-2	0	0	(T_x, T_y)	
E_2	2	1	-1	-2	-1	1	2	0	0		$(\alpha_{xx}-\alpha_{yy}, \alpha_{xy})$
E_3	2	0	-2	0	2	0	-2	0	0		
E_4	2	-1	-1	2	-1	-1	2	0	0		
E_5	2	$-\sqrt{3}$	1	0	-1	$\sqrt{3}$	-2	0	0	(R_x, R_y)	$(\alpha_{yz}, \alpha_{zx})$

TABLE 8. Character tables for the cubic groups.

$\mathcal{T}$	E	$4C_3$	$4C_3^2$	$3C_2$		$\varepsilon = \exp(2\pi i/3)$
A	1	1	1	1		$\alpha_{xx} + \alpha_{yy} + \alpha_{zz}$
E	1	ε	ε^*	1		$(\alpha_{xx} + \alpha_{yy} - 2\alpha_{zz}, \alpha_{zz} - \alpha_{yy})$
	1	ε^*	ε	1		
$T \equiv F$	3	0	0	-1	**T, R**	$(\alpha_{xy}, \alpha_{yz}, \alpha_{zx})$

$\mathcal{T}_h$	E	$4C_3$	$4C_3^2$	$3C_2$	i	$4S_6$	$4S_6^5$	3σ		$\varepsilon = \exp(2\pi i/3)$
A_g	1	1	1	1	1	1	1	1		$\alpha_{xx} + \alpha_{yy} + \alpha_{zz}$
E_g	1	ε	ε^*	1	1	ε^*	ε	1		$(\alpha_{xx} + \alpha_{yy} - 2\alpha_{zz},$
	1	ε^*	ε	1	1	ε	ε^*	1		$\alpha_{xx} - \alpha_{yy})$
$T_g \equiv F_g$	3	0	0	-1	3	0	0	-1	**R**	$(\alpha_{xy}, \alpha_{yz}, \alpha_{zx})$
A_u	1	1	1	1	-1	-1	-1	-1		
E_u	1	ε	ε^*	1	-1	$-\varepsilon^*$	$-\varepsilon$	-1		
	1	ε^*	ε	1	-1	$-\varepsilon$	$-\varepsilon^*$	-1		
$T_u \equiv F_u$	3	0	0	-1	-3	0	0	1	**T**	

$\mathcal{T}_d$	E	$8C_3$	$3C_2$	$6S_4$	$6\sigma_d$		
A_1	1	1	1	1	1		$\alpha_{xx} + \alpha_{yy} + \alpha_{zz}$
A_2	1	1	1	-1	-1		
E	2	-1	2	0	0		$(\alpha_{xx} + \alpha_{yy} - 2\alpha_{zz}, \alpha_{xx} - \alpha_{yy})$
$T_1 \equiv F_1$	3	0	-1	1	-1	**R**	
$T_2 \equiv F_2$	3	0	-1	-1	1	**T**	$(\alpha_{xy}, \alpha_{yz}, \alpha_{zx})$

$\mathcal{O}$	E	$8C_3$	$3C_2$	$6C_4$	$6C_2'$		
A_1	1	1	1	1	1		$\alpha_{xx} + \alpha_{yy} + \alpha_{zz}$
A_2	1	1	1	-1	-1		
E	2	-1	2	0	0		$(\alpha_{xx} + \alpha_{yy} - 2\alpha_{zz}, \alpha_{xx} - \alpha_{yy})$
$T_1 \equiv F_1$	3	0	-1	1	-1	**T, R**	
$T_2 \equiv F_2$	3	0	-1	-1	1		$(\alpha_{xy}, \alpha_{yz}, \alpha_{zx})$

TABLE 8 (*Continued*)

$\mathcal{O}_h$	E	$8C_3$	$3C_2$	$6C_4$	$6C_2'$	i	$8S_6$	$3\sigma_h$	$6S_4$	$6\sigma_d$		
A_{1g}	1	1	1	1	1	1	1	1	1	1		$\alpha_{xx} + \alpha_{yy} + \alpha_{zz}$
A_{2g}	1	1	1	−1	−1	1	1	1	−1	−1		
E_g	2	−1	2	0	0	2	−1	2	0	0		$(\alpha_{xx} + \alpha_{yy} - 2\alpha_{zz}, \alpha_{xx} - \alpha_{yy})$
$\equiv F_{1g}$	3	0	−1	1	−1	3	0	−1	1	−1	**R**	
$\equiv F_{2g}$	3	0	−1	−1	1	3	0	−1	−1	1		$(\alpha_{xy}, \alpha_{yz}, \alpha_{zx})$
A_{1u}	1	1	1	1	1	−1	−1	−1	−1	−1		
A_{2u}	1	1	1	−1	−1	−1	−1	−1	1	1		
E_u	2	−1	2	0	0	−2	1	−2	0	0		
$\equiv F_{1u}$	3	0	−1	1	−1	−3	0	1	−1	1	**T**	
$\equiv F_{2u}$	3	0	−1	−1	1	−3	0	1	1	−1		

TABLE 9. Character tables for the icosahedral groups.

$\mathscr{I}$	E	$12C_5$	$12C_5^2$	$20C_3$	$15C_2$		
A	1	1	1	1	1		$\alpha_{zz} + \alpha_{yy} + \alpha_{zz}$
$\equiv F_1$	3	$\frac{1+\sqrt{5}}{2}$	$\frac{1-\sqrt{5}}{2}$	0	−1	**T, R**	
$\equiv F_2$	3	$\frac{1-\sqrt{5}}{2}$	$\frac{1+\sqrt{5}}{2}$	0	−1		
G	4	−1	−1	1	0		
H	5	0	0	−1	1		$(\alpha_{xx} + \alpha_{yy} - 2\alpha_{zz}, \alpha_{xx} - \alpha_{yy}, \alpha_{xy}, \alpha_{yz}, \alpha_{zx})$

TABLE 9 (*Continued*)

$\mathscr{I}_h$	E	$12C_5$	$12C_5^2$	$20C_3$	$15C_2$	i	$12S_{10}$	$12S_{10}^3$	$20S_6$	15σ		
A_g	1	1	1	1	1	1	1	1	1	1		$\alpha_{xx} + \alpha_{yy} + \alpha_{zz}$
$T_{1g} \equiv F_{1g}$	3	$\frac{1}{2}(1 + \sqrt{5})$	$\frac{1}{2}(1 - \sqrt{5})$	0	-1	3	$\frac{1}{2}(1 - \sqrt{5})$	$\frac{1}{2}(1 + \sqrt{5})$	0	-1	**R**	
$T_{2g} \equiv F_{2g}$	3	$\frac{1}{2}(1 - \sqrt{5})$	$\frac{1}{2}(1 + \sqrt{5})$	0	-1	3	$\frac{1}{2}(1 + \sqrt{5})$	$\frac{1}{2}(1 - \sqrt{5})$	0	-1		
G_g	4	-1	-1	1	0	4	-1	-1	1	0		
H_g	5	0	0	-1	1	5	0	0	-1	1		$(\alpha_{xx} + \alpha_{yy} - 2\alpha_{zz}, \alpha_{xx} - \alpha_{yy}, \alpha_{xy}, \alpha_{yz}, \alpha_{zx})$
A_u	1	1	1	1	1	-1	-1	-1	-1	-1		
$T_{1u} \equiv F_{1u}$	3	$\frac{1}{2}(1 + \sqrt{5})$	$\frac{1}{2}(1 - \sqrt{5})$	0	-1	-3	$-\frac{1}{2}(1 - \sqrt{5})$	$-\frac{1}{2}(1 + \sqrt{5})$	0	1	**T**	
$T_{2u} \equiv F_{2u}$	3	$\frac{1}{2}(1 - \sqrt{5})$	$\frac{1}{2}(1 + \sqrt{5})$	0	-1	-3	$-\frac{1}{2}(1 + \sqrt{5})$	$-\frac{1}{2}(1 - \sqrt{5})$	0	1		
G_u	4	-1	-1	1	0	-4	1	1	-1	0		
H_u	5	0	0	-1	1	-5	0	0	1	-1		

TABLE 10. Character tables for the infinite groups of linear molecules.

$\mathscr{C}_{\infty v}$	E	$2C_{\infty}^{\phi}$	$\ldots$	$\infty\sigma_v$		
$A_1 \equiv \Sigma^+$	1	1	$\ldots$	1	T_z	$\alpha_{zz} + \alpha_{yy}, \alpha_{zz}$
$A_2 \equiv \Sigma^-$	1	1	$\ldots$	-1	R_z	
$E_1 \equiv \Pi$	2	$2\cos\phi$	$\ldots$	0	$(T_x, T_y), (R_x, R_y)$	$(\alpha_{yz}, \alpha_{zx})$
$E_2 \equiv \Delta$	2	$2\cos 2\phi$	$\ldots$	0		$(\alpha_{xx} - \alpha_{yy}, \alpha_{xy})$
$E_3 \equiv \Phi$	2	$2\cos 3\phi$	$\ldots$	0		
$\ldots$	$\ldots$	$\ldots$	$\ldots$	$\ldots$		

$\mathscr{D}_{\infty h}$	E	$2C_{\infty}^{\phi}$	$\ldots$	$\infty\sigma_v$	i	$2S_{\infty}^{\phi}$	$\ldots$	∞C_2		
Σ_g^+	1	1	$\ldots$	1	1	1	$\ldots$	1		$\alpha_{xx} + \alpha_{yy}, \alpha_{zz}$
Σ_g^-	1	1	$\ldots$	-1	1	1	$\ldots$	-1	R_z	
Π_g	2	$2\cos\phi$	$\ldots$	0	2	$-2\cos\phi$	$\ldots$	0	(R_z, R_y)	$(\alpha_{yz}, \alpha_{zx})$
Δ_g	2	$2\cos 2\phi$	$\ldots$	0	2	$2\cos 2\phi$	$\ldots$	0		$(\alpha_{xx} - \alpha_{yy}, \chi_{xy})$
$\ldots$	$\ldots$	$\ldots$	$\ldots$	$\ldots$	$\ldots$	$\ldots$	$\ldots$	$\ldots$		
Σ_u^+	1	1	$\ldots$	1	-1	-1	$\ldots$	-1	T_z	
Σ_u^-	1	1	$\ldots$	-1	-1	-1	$\ldots$	1		
Π_u	2	$2\cos\phi$	$\ldots$	0	-2	$2\cos\phi$	$\ldots$	0	(T_x, T_y)	
Δ_u	2	$2\cos 2\phi$	$\ldots$	0	-2	$-2\cos 2\phi$	$\ldots$	0		
$\ldots$	$\ldots$	$\ldots$	$\ldots$	$\ldots$	$\ldots$	$\ldots$	$\ldots$	$\ldots$		

Appendix D

Subgroups of the Crystallographic Point Groups†

The thirty-two point groups which constitute the crystal classes are represented in Fig. 1 as functions of group order. Each group is connected to its subgroups by one or more lines. A dotted line indicates that the subgroup is not invariant. Solid lines connect a group to its invariant subgroups, the number of lines indicating the number of different ways of establishing the correlation.

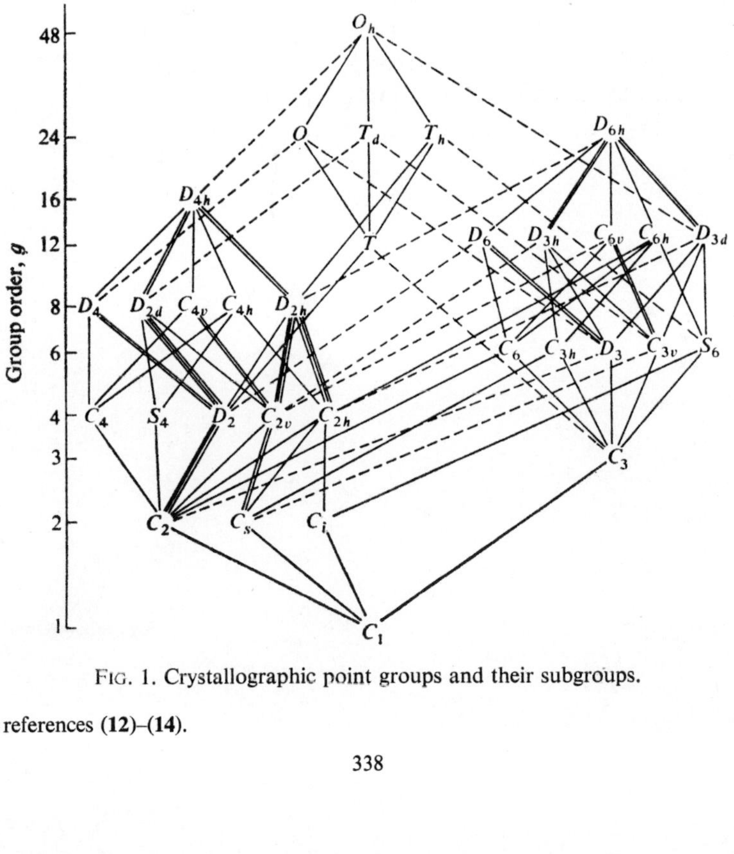

FIG. 1. Crystallographic point groups and their subgroups.

† See references (12)–(14).

Figure 1 can be used to find the possible site groups for a given factor group or molecular symmetry group, as explained in Chapter 4. Furthermore, this figure is applicable to the reduction of symmetry of π crystal due to application of an external field. In the latter case the symmetry group of the perturbed crystal will depend on the direction of the external field with respect to the crystallographic axes.

Appendix E

Space Groups and Crystallographic Sites

In the following table (**14**)–(**16**) the 230 crystallographic space groups have been arranged by crystal system following the convention adopted in the International Tables. Both the International (Hermann–Mauguin) and Schönflies notations are given. The numbers in the column headed "Lattice" refer to the diagram of the Bravais lattices given in Appendix F.

Following the symbols for each space group will be found the Schönflies symbols for the symmetries of the various sites. Each site-group symbol is preceded by an integer indicating the number of distinct sets of each symmetry and followed by the multiplicity in parentheses. Thus, an an entry such as C_1 (4), $4C_2$ (2) following the space-group symbol $P222_1$ indicates that this space group contains one set having symmetry C_1 and consisting of four sites, and four distinct sets of sites of symmetry C_2, each set being composed of two sites.

A site whose position is restricted to a point is referred to in crystallography as a *special position*. For these sites the point-group symbol is shown in boldface type in Table 1. The reader should consult the "International Tables" for the locations of sites in the unit cells.

TABLE 1

System	Lattice	Space Group		Sites
Triclinic	1	$P1$	$\mathscr{C}_1^1$	$\mathbf{C_1}$
	1	$P\bar{1}$	$\mathscr{C}_i^1$	$C_1(2)$, $8\ \mathbf{C_i}$
Monoclinic	2	$P2$	$\mathscr{C}_2^1$	$C_1(2)$, $4\ C_2$
	2	$P2_1$	$\mathscr{C}_2^2$	$C_1(2)$
	3	$B2(C2)$	$\mathscr{C}_2^3$	$C_1(4)$, $2\ C_2(2)$
	2	Pm	$\mathscr{C}_s^1$	$C_1(2)$, $2\ C_s$

TABLE 1. (*Continued*)

Monoclinic (*Cont.*)	2	*Pb(Pc)*	$\mathscr{C}_s^2$	$C_1(2)$
	3	*Bm(Cm)*	$\mathscr{C}_s^3$	$C_1(4)$, $C_s(2)$
	3	*Bb(Cc)*	$\mathscr{C}_s^4$	$C_1(4)$
	2	*P2/m*	$\mathscr{C}_{2h}^1$	$C_1(4)$, $2\,C_s(2)$, $4\,C_2(2)$, $\mathbf{8\,C_{2h}}$
	2	$P2_1/m$	$\mathscr{C}_{2h}^2$	$C_1(4)$, $C_s(2)$, $\mathbf{4\,C_i(2)}$
	3	*B2/m (C2/m)*	$\mathscr{C}_{2h}^3$	$C_1(8)$, $C_s(4)$, $2\,C_2(4)$, $\mathbf{2\,C_i(4)}$, $\mathbf{4\,C_{2h}(2)}$
	2	*P2/b (P2/c)*	$\mathscr{C}_{2h}^4$	$C_1(4)$, $2C_2(2)$, $\mathbf{4C_i(2)}$
	2	$P2_1/b\ (P2_1/c)$	$\mathscr{C}_{2h}^5$	$C_1(4)$ $\mathbf{4\,C_i(2)}$
	3	*B2/b (C2/c)*	$\mathscr{C}_{2h}^6$	$C_1(8)$, $C_2(4)$, $\mathbf{4\,C_i(4)}$
Orthorhombic	4	*P*222	$\mathscr{D}_2^1 \equiv V^1$	$C_1(4)$, $12\,C_2(2)$, $\mathbf{8\,D_2}$
	4	$P222_1$	$\mathscr{D}_2^2$	$C_1(4)$, $4\,C_2(2)$
	4	$P2_1 2_1 2$	$\mathscr{D}_2^3$	$C_1(4)$, $2\,C_2(2)$
	4	$P2_1 2_1 2_1$	$\mathscr{D}_2^4$	$C_1(4)$
	5	$C222_1$	$\mathscr{D}_2^5$	$C_1(8)$, $2\,C_2(4)$
	5	*C*222	$\mathscr{D}_2^6$	$C_1(8)$, $7\,C_2(4)$, $4\,D_2(2)$
	7	*F*222	$\mathscr{D}_2^7$	$C_1(16)$, $6\,C_2(8)$, $\mathbf{4\,D_2(4)}$
	6	*I*222	$\mathscr{D}_2^8$	$C_1(8)$, $6\,C_2(4)$, $\mathbf{4\,D_2(2)}$
	6	$I2_1 2_1 2_1$	$\mathscr{D}_2^9$	$C_1(8)$, $3\,C_2(4)$
	4	*Pmm*2	$\mathscr{C}_{2v}^1$	$C_1(4)$, $4\,C_s(2)$, $4\,C_{2v}$
	4	$Pmc2_1$	$\mathscr{C}_{2v}^2$	$C_1(4)$, $2\,C_s(2)$
	4	*Pcc*2	$\mathscr{C}_{2v}^3$	$C_1(4)$, $4\,C_2(2)$
	4	*Pma*2	$\mathscr{C}_{2v}^4$	$C_1(4)$, $C_s(2)$, $2\,C_2(2)$
	4	$Pca2_1$	$\mathscr{C}_{2v}^5$	$C_1(4)$
	4	*Pnc*2	$\mathscr{C}_{2v}^6$	$C_1(4)$, $2\,C_2(2)$
	4	$Pmn2_1$	$\mathscr{C}_{2v}^7$	$C_1(4)$, $C_s(2)$
	4	*Pba*2	$\mathscr{C}_{2v}^8$	$C_1(4)$, $2\,C_2(2)$
	4	$Pna2_1$	$\mathscr{C}_{2v}^9$	$C_1(4)$
	4	*Pnn*2	$\mathscr{C}_{2v}^{10}$	$C_1(4)$, $2\,C_2(2)$
	5	*Cmm*2	$\mathscr{C}_{2v}^{11}$	$C_1(8)$, $2\,C_s(4)$, $C_2(4)$, $2\,C_{2v}(2)$
	5	$Cmc2_1$	$\mathscr{C}_{2v}^{12}$	$C_1(8)$, $C_s(4)$
	5	*Ccc*2	$\mathscr{C}_{2v}^{13}$	$C_1(8)$, $3\,C_2(4)$
	5	*Amm*2	$\mathscr{C}_{2v}^{14}$	$C_1(8)$, $3\,C_s(4)$, $2\,C_{2v}(2)$
	5	*Abm*2	$\mathscr{C}_{2v}^{15}$	$C_1(8)$, $C_s(4)$, $2\,C_2(4)$
	5	*Ama*2	$\mathscr{C}_{2v}^{16}$	$C_1(8)$, $C_s(4)$, $C_2(4)$
	5	*Aba*2	$\mathscr{C}_{2v}^{17}$	$C_1(8)$, $C_2(4)$
	7	*Fmm*2	$\mathscr{C}_{2v}^{18}$	$C_1(16)$, $2\,C_s(8)$, $C_2(8)$, $C_{2v}(4)$
	7	*Fdd*2	$\mathscr{C}_{2v}^{19}$	$C_1(16)$, $C_2(8)$
	6	*Imm*2	$\mathscr{C}_{2v}^{20}$	$C_1(8)$, $2\,C_s(4)$, $2\,C_{2v}(2)$
	6	*Iba*2	$\mathscr{C}_{2v}^{21}$	$C_1(8)$, $2\,C_2(4)$

TABLE 1. (*Continued*)

Orthorhombic (*Cont.*)	6	*Ima*2	$\mathscr{C}_{2v}^{22}$	$C_1(8)$, $C_s(4)$, $C_2(4)$
	4	*Pmmm*	$\mathscr{D}_{2h}^{1} \equiv \mathscr{V}_h^{1}$	$C_1(8)$, 6 $C_s(4)$, 12 $C_{2v}(2)$, **8 $\mathbf{D}_{2h}$**
	4	*Pnnn*	$\mathscr{D}_{2h}^{2}$	$C_1(8)$, 6 $C_2(4)$, **2 $\mathbf{C}_i(4)$**, **4 $\mathbf{D}_2(2)$**
	4	*Pccm*	$\mathscr{D}_{2h}^{3}$	$C_1(8)$, $C_s(4)$, 8 $C_2(4)$, **4 $\mathbf{D}_2(2)$**, **4 $\mathbf{C}_{2h}(2)$**
	4	*Pban*	$\mathscr{D}_{2h}^{4}$	$C_1(8)$, 6 $C_2(4)$, **2 $\mathbf{C}_i(4)$**, **4 $\mathbf{D}_2(2)$**
	4	*Pmma*	$\mathscr{D}_{2h}^{5}$	$C_1(8)$, 3 $C_s(4)$, 2 $C_2(4)$, 2 $C_{2v}(2)$, **4 $\mathbf{C}_{2h}(2)$**
	4	*Pnna*	$\mathscr{D}_{2h}^{6}$	$C_1(8)$, 2 $C_2(4)$, **2 $\mathbf{C}_i(4)$**
	4	*Pmna*	$\mathscr{D}_{2h}^{7}$	$C_1(8)$, $C_s(4)$, 3 $C_2(4)$, **4 $\mathbf{C}_{2h}(2)$**
	4	*Pcca*	$\mathscr{D}_{2h}^{8}$	$C_1(8)$, 3 $C_2(4)$, **2 $\mathbf{C}_i(4)$**
	4	*Pbam*	$\mathscr{D}_{2h}^{9}$	$C_1(8)$, 2 $C_s(4)$, 2 $C_2(4)$, **4 $\mathbf{C}_{2h}(2)$**
	4	*Pccn*	$\mathscr{D}_{2h}^{10}$	$C_1(8)$, 2 $C_2(4)$, **2$\mathbf{C}_i(4)$**
	4	*Pbcm*	$\mathscr{D}_{2h}^{11}$	$C_1(8)$, $C_s(4)$, $C_2(4)$, **2 $\mathbf{C}_i(4)$**
	4	*Pnnm*	$\mathscr{D}_{2h}^{12}$	$C_1(8)$, $C_s(4)$, 2 $C_2(4)$, **4 $\mathbf{C}_{2h}(2)$**
	4	*Pmmn*	$\mathscr{D}_{2h}^{13}$	$C_1(8)$, 2 $C_s(4)$, **2 $\mathbf{C}_i(4)$**, 2 $C_{2v}(2)$
	4	*Pbcn*	$\mathscr{D}_{2h}^{14}$	$C_1(8)$, $C_2(4)$, **2 $\mathbf{C}_i(4)$**
	4	*Pbca*	$\mathscr{D}_{2h}^{15}$	$C_1(8)$, **2 $\mathbf{C}_i(4)$**
	4	*Pnma*	$\mathscr{D}_{2h}^{16}$	$C_1(8)$, $C_s(4)$, **2 $\mathbf{C}_i(4)$**
	5	*Cmcm*	$\mathscr{D}_{2h}^{17}$	$C_1(16)$, 2 $C_s(8)$, $C_2(8)$, **$\mathbf{C}_i(8)$**, $C_{2v}(4)$, **2 $\mathbf{C}_{2h}(4)$**
	5	*Cmca*	$\mathscr{D}_{2h}^{18}$	$C_1(16)$, $C_s(8)$, 2 $C_2(8)$, **$\mathbf{C}_i(8)$**, **2 $\mathbf{C}_{2h}(4)$**
	5	*Cmmm*	$\mathscr{D}_{2h}^{19}$	$C_1(16)$, 4 $C_s(8)$, $C_2(8)$, 6 $C_{2v}(4)$, **2 $\mathbf{C}_{2h}(4)$**, **4 $\mathbf{D}_{2h}(2)$**
	5	*Cccm*	$\mathscr{D}_{2h}^{20}$	$C_1(16)$, $C_s(8)$, 5 $C_2(8)$, **4 $\mathbf{C}_{2h}(4)$**, **2 $\mathbf{D}_2(4)$**
	5	*Cmma*	$\mathscr{D}_{2h}^{21}$	$C_1(16)$, 2 $C_s(8)$, 5 $C_2(8)$, $C_{2v}(4)$, **4 $\mathbf{C}_{2h}(4)$**, **2 $\mathbf{D}_2(4)$**
	5	*Ccca*	$\mathscr{D}_{2h}^{22}$	$C_1(16)$, 4 $C_2(8)$, **2 $\mathbf{C}_i(8)$**, **2 $\mathbf{D}_2(4)$**
	7	*Fmmm*	$\mathscr{D}_{2h}^{23}$	$C_1(32)$, 3 $C_s(16)$, 3 $C_2(16)$, 3 $C_{2v}(8)$, **$\mathbf{D}_2(8)$**, **3 $\mathbf{C}_{2h}(8)$**, **2 $\mathbf{D}_{2h}(4)$**
	7	*Fddd*	$\mathscr{D}_{2h}^{24}$	$C_1(32)$, 3 $C_2(16)$, **2 $\mathbf{C}_i(16)$**, **2 $\mathbf{D}_2(8)$**

TABLE 1. (*Continued*)

Orthorhombic (*Cont.*)	6	*I mmm*	$\mathscr{D}_{2h}^{25}$	$C_1(16)$, $3\,C_s(8)$, $\mathbf{C_i(8)}$, $\mathbf{6\,C_{2v}(4)}$, $\mathbf{4\,D_{2h}(2)}$
	6	*I bam*	$\mathscr{D}_{2h}^{26}$	$C_1(16)$, $C_s(8)$, $4\,C_2(8)$, $\mathbf{C_i(8)}$, $\mathbf{2\,C_{2h}(4)}$, $\mathbf{2\,D_2(4)}$
	6	*I bca*	$\mathscr{D}_{2h}^{27}$	$C_1(16)$, $3\,C_2(8)$, $\mathbf{2\,C_i(8)}$
	6	*I mma*	$\mathscr{D}_{2h}^{28}$	$C_1(16)$, $2\,C_s(8)$, $2\,C_2(8)$, $C_{2v}(4)$ $\mathbf{4\,C_{2h}(4)}$
Tetragonal	8	$P\,4$	$\mathscr{C}_4^{1}$	$C_1(4)$, $C_2(2)$, $2\,C_4$
	8	$P\,4_1$	$\mathscr{C}_4^{2}$	$C_1(4)$
	8	$P\,4_2$	$\mathscr{C}_4^{3}$	$C_1(4)$, $3\,C_2(2)$
	8	$P\,4_3$	$\mathscr{C}_4^{4}$	$C_1(4)$
	9	$I\,4$	$\mathscr{C}_4^{5}$	$C_1(8)$, $C_2(4)$, $C_4(2)$
	9	$I\,4_1$	$\mathscr{C}_4^{6}$	$C_1(8)$, $C_2(4)$
	8	$P\,\bar{4}$	$\mathscr{S}_4^{1}$	$C_1(4)$, $3\,C_2(2)$, $\mathbf{4\,S_4}$
	9	$I\,\bar{4}$	$\mathscr{S}_4^{2}$	$C_1(8)$, $2\,C_2(4)$, $\mathbf{4\,S_4(2)}$
	8	$P\,4/m$	$\mathscr{C}_{4h}^{1}$	$C_1(8)$, $2\,C_s(4)$, $C_2(4)$, $2\,C_4(2)$, $\mathbf{2\,C_{2h}(2)}$, $\mathbf{4\,C_{4h}}$
	8	$P\,4_2/m$	$\mathscr{C}_{4h}^{2}$	$C_1(8)$, $C_s(4)$, $3\,C_2(4)$, $\mathbf{2\,S_4(2)}$, $\mathbf{4\,C_{2h}(2)}$
	8	$P\,4/n$	$\mathscr{C}_{4h}^{3}$	$C_1(8)$, $C_2(4)$, $\mathbf{2\,C_i(4)}$, $C_4(2)$, $\mathbf{2\,S_4(2)}$
	8	$P\,4_2/n$	$\mathscr{C}_{4h}^{4}$	$C_1(8)$, $2\,C_2(4)$, $\mathbf{2\,C_i(4)}$, $\mathbf{2\,S_4(2)}$
	9	$I\,4/m$	$\mathscr{C}_{4h}^{5}$	$C_1(16)$, $C_s(8)$, $C_2(8)$, $\mathbf{C_i(8)}$, $C_4(4)$, $\mathbf{S_4(4)}$, $\mathbf{C_{2h}(4)}$, $\mathbf{2\,C_{4h}(2)}$
	9	$I\,4_1/m$	$\mathscr{C}_{4h}^{6}$	$C_1(16)$, $C_2(8)$, $\mathbf{2\,C_i(8)}$, $\mathbf{2\,S_4(4)}$
	8	$P\,442$	$\mathscr{D}_4^{1}$	$C_1(8)$, $7\,C_2(4)$, $2\,C_4(2)$, $\mathbf{2\,D_2(2)}$, $\mathbf{4\,D_4}$
	8	$P\,42_12$	$\mathscr{D}_4^{2}$	$C_1(8)$, $3\,C_2(4)$, $C_4(2)$, $\mathbf{2\,D_2(2)}$
	8	$P\,4_122$	$\mathscr{D}_4^{3}$	$C_1(8)$, $3\,C_2(4)$
	8	$P\,4_12_12$	$\mathscr{D}_4^{4}$	$C_1(8)$, $C_2(4)$
	8	$P\,4_222$	$\mathscr{D}_4^{5}$	$C_1(8)$, $9\,C_2(4)$, $\mathbf{6\,D_2(2)}$
	8	$P\,4_22_12$	$\mathscr{D}_4^{6}$	$C_1(8)$, $4\,C_2(4)$, $\mathbf{2\,D_2(2)}$
	8	$P\,4_322$	$\mathscr{D}_4^{7}$	$C_1(8)$, $3\,C_2(4)$
	8	$P\,4_32_12$	$\mathscr{D}_4^{8}$	$C_1(8)$, $C_2(4)$
	9	$I\,422$	$\mathscr{D}_4^{9}$	$C_1(16)$, $5\,C_2(8)$, $C_4(4)$, $\mathbf{2\,D_2(4)}$, $\mathbf{2\,D_4(2)}$
	9	$I\,4_122$	$\mathscr{D}_4^{10}$	$C_1(16)$, $4\,C_2(8)$, $\mathbf{2\,D_2(4)}$
	8	$P\,4\,mm$	$\mathscr{C}_{4v}^{1}$	$C_1(8)$, $3\,C_s(4)$, $C_{2v}(2)$, $2\,C_{4v}$
	8	$P\,4\,bm$	$\mathscr{C}_{4v}^{2}$	$C_1(8)$, $C_s(4)$, $C_{2v}(2)$, $C_4(2)$
	8	$P\,4_2cm$	$\mathscr{C}_{4v}^{3}$	$C_1(8)$, $C_s(4)$, $C_2(4)$, $2\,C_{2v}(2)$

TABLE 1. (*Continued*)

Tetragonal (*Cont.*)	8	$P\,4_2nm$	$\mathscr{C}_{4v}{}^{4}$	$C_1(8)$, $C_s(4)$, $C_2(4)$, $C_{2v}(2)$
	8	$P\,4\,cc$	$\mathscr{C}_{4v}{}^{5}$	$C_1(8)$, $C_2(4)$, $2\,C_4(2)$
	8	$P\,4nc$	$\mathscr{C}_{4v}{}^{6}$	$C_1(8)$, $C_2(4)$, $C_4(2)$
	8	$P\,4_2mc$	$\mathscr{C}_{4v}{}^{7}$	$C_1(8)$, $2\,C_s(4)$, $3\,C_{2v}(2)$
	8	$P\,4_2bc$	$\mathscr{C}_{4v}{}^{8}$	$C_1(8)$, $2\,C_2(4)$
	9	$I\,4mm$	$\mathscr{C}_{4v}{}^{9}$	$C_1(16)$, $2\,C_s(8)$, $C_{2v}(4)$, $C_{4v}(2)$
	9	$I\,4cm$	$\mathscr{C}_{4v}{}^{10}$	$C_1(16)$, $C_s(8)$, $C_{2v}(4)$, $C_4(4)$
	9	$I\,4_1md$	$\mathscr{C}_{4v}{}^{11}$	$C_1(16)$, $C_s(8)$, $C_{2v}(4)$
	9	$I\,4_1cd$	$\mathscr{C}_{4v}{}^{12}$	$C_1(16)$, $C_2(8)$
	8	$P\,\bar{4}2m$	$\mathscr{D}_{2d}{}^{1} \equiv \mathscr{V}_d{}^{1}$	$C_1(8)$, $C_s(4)$, $5\,C_2(4)$, $2\,C_{2v}(2)$, $\mathbf{2\,D_2(2)}$, $\mathbf{4\,D_{2d}}$
	8	$P\,\bar{4}2c$	$\mathscr{D}_{2d}{}^{2}$	$C_1(8)$, $7\,C_2(4)$, $\mathbf{2\,S_4(2)}$, $\mathbf{4\,D_2(2)}$ $C_2(4)$
	8	$P\,\bar{4}2_1m$	$\mathscr{D}_{2d}{}^{3}$	$C_1(8)$, $C_s(4)$, $C_{2v}(2)$, $\mathbf{2\,S_4(2)}$
	8	$P\,\bar{4}2_1c$	$\mathscr{D}_{2d}{}^{4}$	$C_1(8)$, $2\,C_2(4)$, $\mathbf{2\,S_4(2)}$
	8	$P\,\bar{4}m2$	$\mathscr{D}_{2d}{}^{5}$	$C_1(8)$, $2\,C_s(4)$, $2\,C_2(4)$, $3\,C_{2v}(2)$, $\mathbf{4\,D_{2d}}$
	8	$P\,\bar{4}c2$	$\mathscr{D}_{2d}{}^{6}$	$C_1(8)$, $5\,C_2(4)$, $\mathbf{2\,S_4(2)}$, $\mathbf{2\,D_2(2)}$
	8	$P\,\bar{4}b2$	$\mathscr{D}_{2d}{}^{7}$	$C_1(8)$, $4\,C_2(4)$, $\mathbf{2\,D_2(2)}$, $\mathbf{2\,S_4(2)}$
	8	$P\,\bar{4}n2$	$\mathscr{D}_{2d}{}^{8}$	$C_1(8)$, $4\,C_2(4)$, $\mathbf{2\,D_2(2)}$, $\mathbf{2\,S_4(2)}$
	9	$I\,\bar{4}m2$	$\mathscr{D}_{2d}{}^{9}$	$C_1(16)$, $C_s(8)$, $2\,C_2(8)$, $2\,C_{2v}(4)$, $\mathbf{4\,D_{2d}(2)}$
	9	$I\,\bar{4}c2$	$\mathscr{D}_{2a}{}^{10}$	$C_1(16)$, $4\,C_2(8)$, $\mathbf{2\,S_4(4)}$, $\mathbf{2\,D_2(4)}$
	9	$I\,\bar{4}2m$	$\mathscr{D}_{2d}{}^{11}$	$C_1(16)$, $C_s(8)$, $3\,C_2(8)$, $C_{2v}(4)$, $\mathbf{S_4(4)}$, $\mathbf{D_2(4)}$, $\mathbf{2\,D_{2d}(2)}$
	9	$I\,\bar{4}2d$	$\mathscr{D}_{2d}{}^{12}$	$C_1(16)$, $2\,C_2(8)$, $\mathbf{2\,S_4(4)}$
	8	$P\,4/mmm$	$\mathscr{D}_{4h}{}^{1}$	$C_1(16)$, $5\,C_s(8)$, $7\,C_{2v}(4)$, $2\,C_{4v}(2)$, $\mathbf{2\,D_{2h}(2)}$, $\mathbf{4\,D_{4h}}$
	8	$P\,4/mcc$	$\mathscr{D}_{4h}{}^{2}$	$C_1(16)$, $C_s(8)$, $4\,C_2(8)$, $2\,C_4(4)$, $\mathbf{D_2(4)}$, $\mathbf{C_{2h}(4)}$, $\mathbf{2\,C_{4h}(2)}$, $\mathbf{2\,D_4(2)}$
	8	$P\,4/nbm$	$\mathscr{D}_{4h}{}^{3}$	$C_1(16)$, $C_s(8)$, $4\,C_2(8)$, $C_{2v}(4)$, $C_4(4)$, $\mathbf{2\,C_{2h}(4)}$, $\mathbf{2\,D_{2d}(2)}$, $\mathbf{2\,D_4(2)}$
	8	$P\,4/nnc$	$\mathscr{D}_{4h}{}^{4}$	$C_1(16)$, $4\,C_2\mathbf{(8)}$, $\mathbf{C_i(8)}$, $C_4(4)$, $\mathbf{S_4(4)}$, $\mathbf{D_2(4)}$, $\mathbf{2D_4(2)}$
	8	$P\,4/mbm$	$\mathscr{D}_{4h}{}^{5}$	$C_1(16)$, $3\,C_s(8)$, $3\,C_{2v}(4)$, $C_4(4)$, $\mathbf{2\,D_{2h}(2)}$, $\mathbf{2\,C_{4h}(2)}$

TABLE 1. (*Continued*)

Tetragonal (*Cont.*)	8	$P\,4/mnc$	$\mathscr{D}_{4h}^{6}$	$C_1(16)$, $C_s(8)$, $2\,C_2(8)$, $C_4(4)$, $\mathbf{D}_2\mathbf{(4)}$, $\mathbf{C}_{2h}\mathbf{(4)}$, $\mathbf{2\,C}_{4h}\mathbf{(2)}$
	8	$P\,4/nmm$	$\mathscr{D}_{4h}^{7}$	$C_1(16)$, $2\,C_s(8)$, $2\,C_2(8)$, $C_{2v}(4)$, $\mathbf{2\,C}_{2h}\mathbf{(4)}$, $C_{4v}(2)$, $\mathbf{2\,D}_{2d}\mathbf{(2)}$
	8	$P\,4/ncc$	$\mathscr{D}_{4h}^{8}$	$C_1(16)$, $2\,C_2(8)$, $\mathbf{C}_i\mathbf{(8)}$, $C_4(4)$, $\mathbf{S}_4\mathbf{(4)}$, $\mathbf{D}_2\mathbf{(4)}$
	8	$P\,4_2/mmc$	$\mathscr{D}_{4h}^{9}$	$C_1(16)$, $3\,C_s(8)$, $C_2(8)$, $7\,C_{2v}(4)$, $\mathbf{2\,D}_{2d}\mathbf{(2)}$, $\mathbf{4\,D}_{2h}\mathbf{(2)}$
	8	$P\,4_2/mcm$	$\mathscr{D}_{4h}^{10}$	$C_1(16)$, $2\,C_s(8)$, $3\,C_2(8)$, $4\,C_{2v}(4)$, $\mathbf{C}_{2h}\mathbf{(4)}$, $\mathbf{D}_2\mathbf{(4)}$, $\mathbf{2\,D}_{2d}\mathbf{(2)}$, $\mathbf{2\,D}_{2h}\mathbf{(2)}$
	8	$P\,4_2/nbc$	$\mathscr{D}_{4h}^{11}$	$C_1(16)$, $5\,C_2(8)$, $\mathbf{C}_i\mathbf{(8)}$, $\mathbf{S}_4\mathbf{(4)}$, $\mathbf{3\,D}_2\mathbf{(4)}$
	8	$P\,4_2/nnm$	$\mathscr{D}_{4h}^{12}$	$C_1(16)$, $C_s(8)$, $5\,C_2(8)$, $C_{2v}(4)$, $\mathbf{2\,C}_{2h}\mathbf{(4)}$, $\mathbf{2\,D}_2\mathbf{(4)}$, $\mathbf{2\,D}_{2d}\mathbf{(2)}$
	8	$P\,4_2/mbc$	$\mathscr{D}_{4h}^{13}$	$C_1(16)$, $C_s(8)$, $3\,C_2(8)$, $\mathbf{D}_2\mathbf{(4)}$, $\mathbf{S}_4\mathbf{(4)}$, $\mathbf{2\,C}_{2h}\mathbf{(4)}$
	8	$P\,4_2/mnm$	$\mathscr{D}_{4h}^{14}$	$C_1(16)$, $2\,C_s(8)$, $C_2(8)$, $3\,C_{2v}(4)$, $\mathbf{S}_4\mathbf{(4)}$, $\mathbf{C}_{2h}\mathbf{(4)}$, $\mathbf{2\,D}_{2h}\mathbf{(2)}$
	8	$P\,4_2/nmc$	$\mathscr{D}_{4h}^{15}$	$C_1(16)$, $C_s(8)$, $C_2(8)$, $\mathbf{C}_i\mathbf{(8)}$ $2\,C_{2v}(4)$, $\mathbf{2\,D}_{2d}\mathbf{(2)}$
	8	$P\,4_2/ncm$	$\mathscr{D}_{4h}^{16}$	$C_1(16)$, $C_s(8)$, $3\,C_2(8)$, $C_{2v}(4)$, $\mathbf{2\,C}_{2h}\mathbf{(4)}$, $\mathbf{S}_4\mathbf{(4)}$, $\mathbf{D}_2\mathbf{(4)}$
	9	$I\,4/mmm$	$\mathscr{D}_{4h}^{17}$	$C_1(32)$, $3\,C_s(16)$, $C_2(16)$, $4\,C_{2v}(8)$, $\mathbf{C}_{2h}\mathbf{(8)}$, $C_{4v}(4)$, $\mathbf{D}_{2d}\mathbf{(4)}$, $\mathbf{D}_{2h}\mathbf{(4)}$,$\mathbf{2\,D}_{4h}\mathbf{(2)}$
	9	$I\,4/mcm$	$\mathscr{D}_{4h}^{18}$	$C_1(32)$, $2\,C_s(16)$, $2\,C_2(16)$, $2\,C_{2v}(8)$, $C_4(8)$, $\mathbf{C}_{2h}\mathbf{(8)}$, $\mathbf{D}_{2h}\mathbf{(4)}$, $\mathbf{C}_{4h}\mathbf{(4)}$, $\mathbf{D}_{2d}\mathbf{(4)}$, $\mathbf{D}_4\mathbf{(4)}$
	9	$I\,4_1/amd$	$\mathscr{D}_{4h}^{19}$	$C_1(32)$, $C_s(16)$, $2\,C_2(16)$, $C_{2v}(8)$, $\mathbf{2\,C}_{2h}\mathbf{(8)}$, $\mathbf{2\,D}_{2d}\mathbf{(4)}$
	9	$I\,4_1/acd$	$\mathscr{D}_{4h}^{20}$	$C_1(32)$, $3\,C_2(16)$, $\mathbf{C}_i\mathbf{(16)}$, $\mathbf{D}_2\mathbf{(8)}$, $\mathbf{S}_4\mathbf{(8)}$
Trigonal	10	$P\,3$	$\mathscr{C}_{3}^{1}$	$C_1(3)$, $3\,C_3$
	10	$P\,3_1$	$\mathscr{C}_{3}^{2}$	$C_1(3)$
	10	$P\,3_2$	$\mathscr{C}_{3}^{3}$	$C_1(3)$
	10	$R\,3$	$\mathscr{C}_{3}^{4}$	$C_1(3)$, C_3

TABLE 1. (*Continued*)

Trigonal (*Cont.*)	10	$P\bar{3}$	$\mathcal{S}_6^1 \equiv \mathcal{C}_{3i}^1$	$C_1(6)$, $\mathbf{2\,C_i(3)}$, $2\,C_3(2)$, $\mathbf{2\,S_6}$
	10	$R\bar{3}$	$\mathcal{S}_6^2 \equiv \mathcal{C}_{3i}^2$	$C_1(6)$, $\mathbf{2\,C_i(3)}$, $C_3(2)$, $\mathbf{2\,S_6}$
	10	$P\,312$	$\mathcal{D}_3^1$	$C_1(6)$, $2\,C_2(3)$, $3\,C_3(2)$, $\mathbf{6\,D_3}$
	10	$P\,321$	$\mathcal{D}_3^2$	$C_1(6)$, $2\,C_2(3)$, $2\,C_3(2)$, $\mathbf{2\,D_3}$
	10	$P\,3_112$	$\mathcal{D}_3^3$	$C_1(6)$, $2\,C_2(3)$
	10	$P\,3_121$	$\mathcal{D}_3^4$	$C_1(6)$, $2\,C_2(3)$
	10	$P\,3_212$	$\mathcal{D}_3^5$	$C_1(6)$, $2\,C_2(3)$
	10	$P\,3_221$	$\mathcal{D}_3^6$	$C_1(6)$, $2\,C_2(3)$
	10	$R\,32$	$\mathcal{D}_3^7$	$C_1(6)$, $2\,C_2(3)$, $C_3(2)$, $\mathbf{2\,D_3}$
	10	$P\,3m1$	$\mathcal{C}_{3v}^1$	$C_1(6)$, $C_s(3)$, $3\,C_{3v}$
	10	$P\,31m$	$\mathcal{C}_{3v}^2$	$C_1(6)$, $C_s(3)$, $C_3(2)$, C_{3v}
	10	$P\,3c1$	$\mathcal{C}_{3v}^3$	$C_1(6)$, $3\,C_3(2)$
	10	$P\,31c$	$\mathcal{C}_{3v}^4$	$C_1(6)$, $2\,C_3(2)$
	10	$R\,3m$	$\mathcal{C}_{3v}^5$	$C_1(6)$, $C_s(3)$, C_{3v}
	10	$R\,3c$	$\mathcal{C}_{3v}^6$	$C_1(6)$, $C_3(2)$
	10	$P\,\bar{3}1m$	$\mathcal{D}_{3d}^1$	$C_1(12)$, $C_s(6)$, $2\,C_2(6)$, $C_3(4)$, $\mathbf{2\,C_{2h}(3)}$, $C_{3v}(2)$, $\mathbf{2\,D_3(2)}$ $\mathbf{2\,D_{3d}}$
	10	$P\,\bar{3}1c$	$\mathcal{D}_{3d}^2$	$C_1(12)$, $\mathbf{C_i(6)}$, $C_2(6)$, $2\,C_3(4)$, $\mathbf{S_6(2)}$, $\mathbf{3\,D_3(2)}$
	10	$P\,\bar{3}m1$	$\mathcal{D}_{3d}^3$	$C_1(12)$, $C_s(6)$, $2\,C_2(6)$, $\mathbf{2\,C_{2h}(3)}$ $2\,C_{3v}(2)$, $\mathbf{2\,D_{3d}}$
	10	$P\,\bar{3}c1$	$\mathcal{D}_{3d}^4$	$C_1(12)$, $C_2(6)$, $\mathbf{C_i(6)}$, $2\,C_3(4)$, $\mathbf{S_6(2)}$, $\mathbf{D_3(2)}$
	10	$R\,\bar{3}m$	$\mathcal{D}_{3d}^5$	$C_1(12)$, $C_s(6)$, $2\,C_2(6)$, $\mathbf{2\,C_{2h}(3)}$, $C_{3v}(2)$, $\mathbf{2\,D_{3d}}$
	10	$R\,\bar{3}c$	$\mathcal{D}_{3d}^6$	$C_1(12)$, $C_2(6)$, $\mathbf{C_i(6)}$, $C_3(4)$, $\mathbf{S_6(2)}$, $\mathbf{D_3(2)}$
Hexagonal	10	$P\,6$	$\mathcal{C}_6^1$	$C_1(6)$, $C_2(3)$, $C_3(2)$, C_6
	10	$P\,6_1$	$\mathcal{C}_6^2$	$C_1(6)$
	10	$P\,6_5$	$\mathcal{C}_6^3$	$C_1(6)$
	10	$P\,6_2$	$\mathcal{C}_6^4$	$C_1(6)$, $2\,C_2(3)$
	10	$P\,6_4$	$\mathcal{C}_6^5$	$C_1(6)$, $2\,C_2(3)$
	10	$P\,6_3$	$\mathcal{C}_6^6$	$C_1(6)$, $2\,C_3(2)$
	11	$P\,\bar{6}$	$\mathcal{C}_{3h}^1$	$C_1(6)$, $2\,C_s(3)$, $3\,C_3(2)$, $\mathbf{6\,C_{3h}}$
	11	$P\,6/m$	$\mathcal{C}_{6h}^1$	$C_1(12)$, $2\,C_s(6)$, $C_2(6)$, $C_3(4)$, $\mathbf{2\,C_{2h}(3)}$, $C_6(2)$, $\mathbf{2\,C_{3h}(2)}$, $\mathbf{2\,C_{6h}}$
	11	$P\,6_3/m$	$\mathcal{C}_{6h}^2$	$C_1(12)$, $C_s(6)$, $\mathbf{C_i(6)}$, $2\,C_3(4)$, $\mathbf{S_6(2)}$, $\mathbf{3\,C_{3h}(2)}$

TABLE 1. (*Continued*)

Hexagonal (*Cont.*)	11	$P\,622$	$\mathscr{D}_6^{\,1}$	$C_1(12)$, 5 $C_2(6)$, $C_3(4)$, **2 $\mathbf{D_2}$(3)**, $C_6(2)$, **2 $\mathbf{D_3}$(2)**, **2 $\mathbf{D_6}$**
	11	$P\,6_122$	$\mathscr{D}_6^{\,2}$	$C_1(12)$, 2 $C_2(6)$
	11	$P\,6_522$	$\mathscr{D}_6^{\,3}$	$C_1(12)$, 2 $C_2(6)$
	11	$P\,6_222$	$\mathscr{D}_6^{\,4}$	$C_1(12)$, 6 $C_2(6)$, **4 $\mathbf{D_2}$(3)**
	11	$P\,6_422$	$\mathscr{D}_6^{\,5}$	$C_1(12)$, 6 $C_2(6)$, **4 $\mathbf{D_2}$(3)**
	11	$P\,6_322$	$\mathscr{D}_6^{\,6}$	$C_1(12)$, 2 $C_2(6)$, 2 $C_3(4)$, **4 $\mathbf{D_3}$(2)**
	11	$P\,6mm$	$\mathscr{C}_{6v}^{\,1}$	$C_1(12)$, 2 $C_s(6)$, $C_{2v}(3)$, $C_{3v}(2)$, C_{6v}
	11	$P\,6cc$	$\mathscr{C}_{6v}^{\,2}$	$C_1(12)$, $C_2(6)$, $C_3(4)$, $C_6(2)$
	11	$P\,6_3cm$	$\mathscr{C}_{6v}^{\,3}$	$C_1(12)$, $C_s(6)$, $C_3(4)$, $C_{3v}(2)$
	11	$P\,6_3mc$	$\mathscr{C}_{6v}^{\,4}$	$C_1(12)$, $C_s(6)$, 2 $C_{3v}(2)$
	11	$P\,\bar{6}m2$	$\mathscr{D}_{3h}^{\,1}$	$C_1(12)$, 3 $C_s(6)$, 2 $C_{2v}(3)$, 3 $C_{3v}(2)$, **6 $\mathbf{D_{3h}}$**
	11	$P\,\bar{6}c2$	$\mathscr{D}_{3h}^{\,2}$	$C_1(12)$, $C_s(6)$, $C_2(6)$, 3 $C_3(4)$, **3 $\mathbf{C_{3h}}$(2)**, **3 $\mathbf{D_3}$(2)**
	11	$P\,\bar{6}2m$	$\mathscr{D}_{3h}^{\,3}$	$C_1(12)$, 3 $C_s(6)$, $C_3(4)$, 2 $C_{2v}(3)$, $C_{3v}(2)$, **2 $\mathbf{C_{3h}}$(2)**, **2 $\mathbf{D_{3h}}$**
	11	$P\,\bar{6}2c$	$\mathscr{D}_{3h}^{\,4}$	$C_1(12)$, $C_s(6)$, $C_2(6)$, 2 $C_3(4)$, **3 $\mathbf{C_{3h}}$(2)**, **$\mathbf{D_3}$(2)**
	11	$P\,6/mmm$	$\mathscr{D}_{6h}^{\,1}$	$C_1(24)$, 4 $C_s(12)$, 5 $C_{2v}(6)$, $C_{3v}(4)$, **2 $\mathbf{D_{2h}}$(3)**, $C_{6v}(2)$, **2 $\mathbf{D_{3h}}$(2)**, **2 $\mathbf{D_{6h}}$**
	11	$P\,6/mcc$	$\mathscr{D}_{6h}^{\,2}$	$C_1(24)$, $C_s(12)$, 3 $C_2(12)$, $C_3(8)$, **$\mathbf{C_{2h}}$(6)**, **$\mathbf{D_2}$(6)**, $C_6(4)$, **$\mathbf{C_{3h}}$(4)**, **$\mathbf{D_3}$(4)**, **$\mathbf{C_{6h}}$(2)**, **$\mathbf{D_6}$(2)**
	11	$P\,6_3/mcm$	$\mathscr{D}_{6h}^{\,3}$	$C_1(24)$, 2 $C_s(12)$, $C_2(12)$, $C_3(8)$, $C_{2v}(6)$, **$\mathbf{C_{2h}}$(6)**, $C_6(4)$, **$\mathbf{D_3}$(4)**, **$\mathbf{C_{3h}}$(4)**, **$\mathbf{D_{3d}}$(2)**, **$\mathbf{D_{3h}}$(2)**
	11	$P\,6_3/mmc$	$\mathscr{D}_{6h}^{\,4}$	$C_1(24)$, 2 $C_s(12)$, $C_2(12)$, $C_{2v}(6)$, **$\mathbf{C_{2h}}$(6)**, 2 $C_{3v}(4)$, **3 $\mathbf{D_{3h}}$(2)**, **$\mathbf{D_{3d}}$(2)**
Cubic	12	$P\,23$	$\mathscr{T}^1$	$C_1(12)$, 4 $C_2(6)$, $C_3(4)$, **2 $\mathbf{D_2}$(3)**, **2 T**
	14	$F\,23$	$\mathscr{T}^2$	$C_1(48)$, 2 $C_2(24)$, $C_3(16)$, **4 T(4)**
	13	$I\,23$	$\mathscr{T}^3$	$C_1(24)$, 2 $C_2(12)$, $C_3(8)$, **$\mathbf{D_2}$(6)**, **T(2)**

TABLE 1. (*Continued*)

Cubic (*Cont.*)	12	$P\,2_13$	$\mathscr{T}^4$	$C_1(12)$, $C_3(4)$
	13	$I\,2_13$	$\mathscr{T}^5$	$C_1(24)$, $C_2(12)$, $C_3(8)$
	12	$P\,m3$	$\mathscr{T}_h^{\,1}$	$C_1(24)$, $2\,C_s(12)$, $C_3(8)$, $4\,C_{2v}(6)$, $\mathbf{2\,D_{2h}(3)}$, $\mathbf{2\,T_h}$
	12	$P\,n3$	$\mathscr{T}_h^{\,2}$	$C_1(24)$, $2\,C_2(12)$, $C_3(8)$, $\mathbf{D_2(6)}$, $\mathbf{2\,S_6(4)}$, $\mathbf{T(2)}$
	14	$F\,m3$	$\mathscr{T}_h^{\,3}$	$C_1(96)$, $C_s(48)$, $C_2(48)$, $C_3(32)$, $C_{2v}(24)$, $\mathbf{C_{2h}(24)}$, $\mathbf{T(8)}$, $\mathbf{2\,T_h(4)}$
	14	$F\,d3$	$\mathscr{T}_h^{\,4}$	$C_1(96)$, $C_2(48)$, $C_3(32)$, $\mathbf{2\,S_6(16)}$, $\mathbf{2\,T(8)}$
	13	$I\,m3$	$\mathscr{T}_h^{\,5}$	$C_1(48)$, $C_s(24)$, $C_3(16)$, $2\,C_{2v}(12)$, $\mathbf{S_6(8)}$, $\mathbf{D_{2h}(6)}$, $\mathbf{T_h(2)}$
	12	$P\,a3$	$\mathscr{T}_h^{\,6}$	$C_1(24)$, $C_3(8)$, $\mathbf{2\,S_6(4)}$
	13	$I\,a3$	$\mathscr{T}_h^{\,7}$	$C_1(48)$, $C_2(24)$, $C_3(16)$, $\mathbf{2\,S_6(8)}$
	12	$P\,432$	$\mathscr{O}^1$	$C_1(24)$, $3\,C_2(12)$, $C_3(8)$, $2\,C_4(6)$, $\mathbf{2\,D_4(3)}$, $\mathbf{2\,O}$
	12	$P\,4_232$	$\mathscr{O}^2$	$C_1(24)$, $5\,C_2(12)$, $C_3(8)$, $\mathbf{3\,D_2(6)}$, $\mathbf{2\,D_3(4)}$, $\mathbf{T(2)}$
	14	$F\,432$	$\mathscr{O}^3$	$C_1(96)$, $3\,C_2(48)$, $C_3(32)$, $C_4(24)$, $\mathbf{D_2(24)}$, $\mathbf{T(8)}$, $\mathbf{2\,O(4)}$
	14	$F\,4_132$	$\mathscr{O}^4$	$C_1(96)$, $2\,C_2(48)$, $C_3(32)$ $\mathbf{2\,D_3(16)}$, $\mathbf{2\,T(8)}$
	13	$I\,432$	$\mathscr{O}^5$	$C_1(48)$, $3\,C_2(24)$, $C_3(16)$, $C_4(12)$, $\mathbf{D_2(12)}$, $\mathbf{D_3(8)}$, $\mathbf{D_4(6)}$, $\mathbf{O(2)}$
	12	$P\,4_332$	$\mathscr{O}^6$	$C_1(24)$, $C_2(12)$, $C_3(8)$, $\mathbf{2\,D_3(4)}$
	12	$P\,4_132$	$\mathscr{O}^7$	$C_1(24)$, $C_2(12)$, $C_3(8)$, $\mathbf{2\,D_3(4)}$
	13	$I\,4_132$	$\mathscr{O}^8$	$C_1(48)$, $3\,C_2(24)$, $C_3(16)$, $\mathbf{2\,D_2(12)}$, $\mathbf{2\,D_3(8)}$
	12	$P\,\bar{4}3m$	$\mathscr{T}_d^{\,1}$	$C_1(24)$, $C_s(12)$, $C_2(12)$, $2\,C_{2v}(6)$, $C_{3v}(4)$, $\mathbf{2\,D_{2d}(3)}$, $\mathbf{2\,T_d}$
	14	$F\,\bar{4}3m$	$\mathscr{T}_d^{\,2}$	$C_1(96)$, $C_s(48)$, $2\,C_{2v}(24)$, $C_{3v}(16)$, $\mathbf{4\,T_d(4)}$
	13	$I\,43m$	$\mathscr{T}_d^{\,3}$	$C_1(48)$, $C_s(24)$, $C_2(24)$, $C_{2v}(12)$, $\mathbf{S_4(12)}$, $C_{3v}(8)$, $\mathbf{D_{2d}(6)}$, $\mathbf{T_d(2)}$
	12	$P\,\bar{4}3n$	$\mathscr{T}_d^{\,4}$	$C_1(24)$, $3\,C_2(12)$, $C_3(8)$, $\mathbf{2\,S_4(6)}$, $\mathbf{D_2(6)}$, $\mathbf{T(2)}$

TABLE 1. (*Continued*)

Cubic (*Cont.*)	14	$F\bar{4}3c$	$\mathcal{T}_d^5$	$C_1(96)$, $2\,C_2(48)$, $C_3(32)$, $\mathbf{2\,S_4(24)}$, $\mathbf{2\,T(8)}$
	13	$I\bar{4}3d$	$\mathcal{T}_d^6$	$C_1(48)$, $C_2(24)$, $C_3(16)$, $\mathbf{2\,S_4(12)}$
	12	$P\,m3m$	$\mathcal{O}_h^1$	$C_1(48)$, $3\,C_s(24)$, $3\,C_{2v}(12)$, $C_{3v}(8)$, $2\,C_{4v}(6)$, $\mathbf{2\,D_{4h}(3)}$, $\mathbf{2\,O_h}$
	12	$P\,n3n$	$\mathcal{O}_h^2$	$C_1(48)$, $2\,C_2(24)$, $C_3(16)$, $C_4(12)$, $\mathbf{S_4(12)}$, $\mathbf{S_6(8)}$, $\mathbf{D_4(6)}$, $\mathbf{O(2)}$
	12	$P\,m3n$	$\mathcal{O}_h^3$	$C_1(48)$, $C_s(24)$, $C_2(24)$, $C_3(16)$, $3\,C_{2v}(12)$, $\mathbf{D_3(8)}$, $\mathbf{2\,D_{2d}(6)}$, $\mathbf{D_{2h}(6)}$, $\mathbf{T_h(2)}$
	12	$P\,n3m$	$\mathcal{O}_h^4$	$C_1(48)$, $C_s(24)$, $3\,C_2(24)$, $C_{2v}(12)$, $\mathbf{D_2(12)}$, $C_{3v}(8)$, $\mathbf{D_{2d}(6)}$, $\mathbf{2\,D_{3d}(4)}$, $\mathbf{T_d(2)}$
	14	$F\,m3m$	$\mathcal{O}_h^5$	$C_1(192)$, $2\,C_s(96)$, $3\,C_{2v}(48)$, $C_{3v}(32)$, $C_{4v}(24)$, $\mathbf{D_{2h}(24)}$, $\mathbf{T_d(8)}$, $\mathbf{2\,O_h(4)}$
	14	$F\,m3c$	$\mathcal{O}_h^6$	$C_1(192)$, $C_s(96)$, $C_2(96)$, $C_3(64)$, $C_4(48)$, $C_{2v}(48)$, $\mathbf{C_{4h}(24)}$, $\mathbf{D_{2d}(24)}$, $\mathbf{T_h(8)}$, $\mathbf{O(8)}$
	14	$F\,d3m$	$\mathcal{O}_h^7$	$C_1(192)$, $C_2(96)$, $C_s(96)$, $C_{2v}(48)$, $C_{3v}(32)$, $\mathbf{2\,D_{3d}(16)}$, $\mathbf{2\,T_d(8)}$
	14	$F\,d3c$	$\mathcal{O}_h^8$	$C_1(192)$, $2\,C_2(96)$, $C_3(64)$, $\mathbf{S_4(48)}$, $\mathbf{S_6(32)}$, $\mathbf{D_3(32)}$, $\mathbf{T(16)}$
	13	$I\,m3m$	$\mathcal{O}_h^9$	$C_1(96)$, $2\,C_s(48)$, $C_2(48)$, $2\,C_{2v}(24)$, $C_{3v}(16)$, $C_{4v}(12)$, $\mathbf{D_{2d}(12)}$, $\mathbf{D_{3d}(8)}$, $\mathbf{D_{4h}(6)}$, $\mathbf{O_h(2)}$
	13	$I\,a3d$	$\mathcal{O}_h^{10}$	$C_1(96)$, $2\,C_2(48)$, $C_3(32)$, $\mathbf{S_4(24)}$, $\mathbf{D_2(24)}$, $\mathbf{D_3(16)}$, $\mathbf{S_6(16)}$

Appendix F

Bravais Lattices and Primitive Cells

On the left of each of the following figures the primitive cell defined by the vectors $\mathbf{t}_1$, $\mathbf{t}_2$, and $\mathbf{t}_3$ [See Section VII, Chap. 3] is shown superimposed on the Bravais lattice. On the right-hand side the primitive-symmetric or Wigner–Seitz cell is shown, again superimposed on the Bravais lattice. In each case the volume of the primitive cell is equal to the volume of the Bravais cell divided by Z.

Each Bravais lattice is identified by a number (1 through 14) which refers to the entries in Table 1 of Appendix E. Hence, if the space group of a crystal has been determined by X-ray methods, the Bravais lattice can be found from Appendix E and the structure of the corresponding primitive cell from Appendix F. It is the primitive cell which must be used as the basis of a factor-group analysis, as indicated in Chapter 4.

Crystal System	Bravais Lattice		Z
Triclinic	1	Simple (P)	1
	Any lattice generated by three arbitrary primitive translation vectors		
Monoclinic	2	Simple (P)	1

$$\mathbf{t}_1 = t\mathbf{i} + s\mathbf{j}$$
$$\mathbf{t}_2 = v\mathbf{j}$$
$$\mathbf{t}_3 = r\mathbf{k}$$

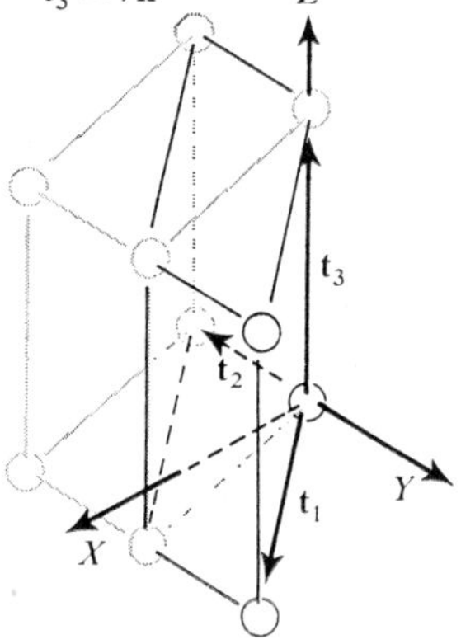

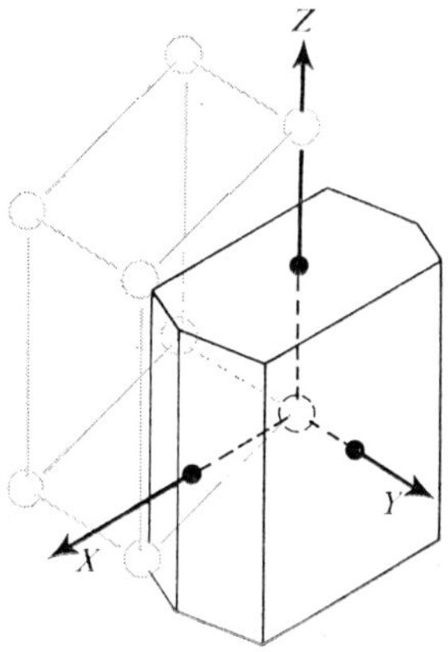

Monoclinic	3	Side-centered (C)	2

$$\mathbf{t}_1 = t\mathbf{i} + s\mathbf{j}$$
$$\mathbf{t}_2 = t\mathbf{i} - s\mathbf{j}$$
$$\mathbf{t}_3 = u\mathbf{i} + r\mathbf{k}$$

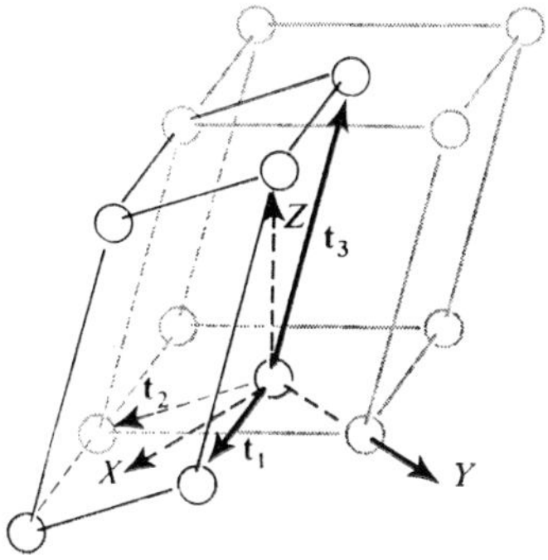

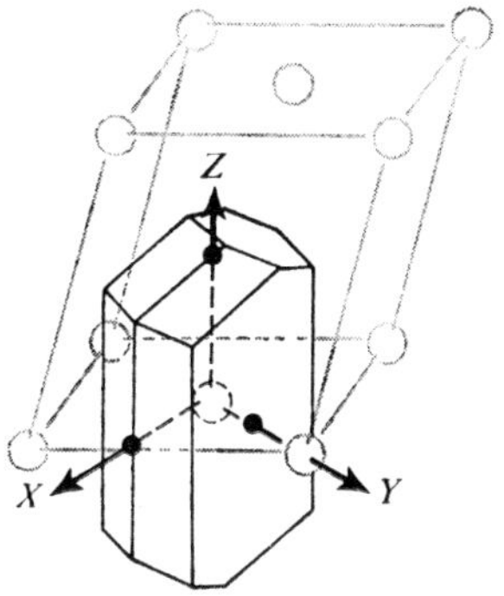

Crystal System	Bravais Lattice		Z
Orthorhombic	4	Simple (P)	1

$\mathbf{t}_1 = t\mathbf{i}$
$\mathbf{t}_2 = s\mathbf{j}$
$\mathbf{t}_3 = r\mathbf{k}$

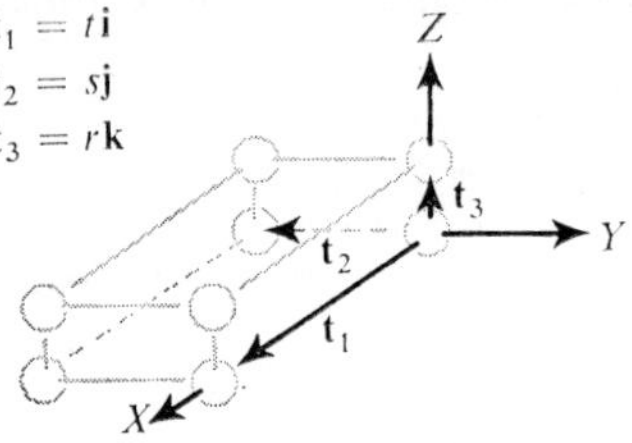

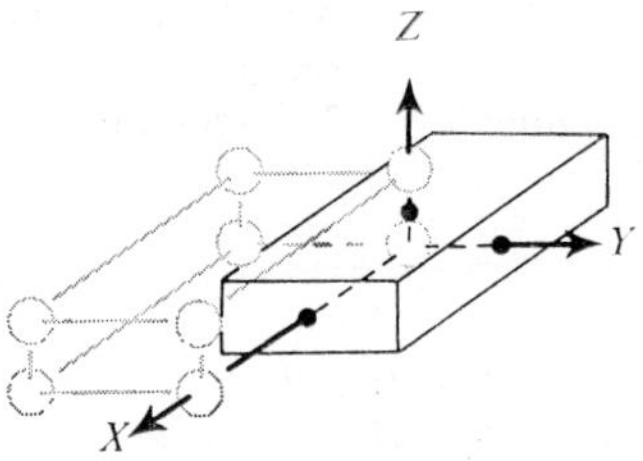

Crystal System	Bravais Lattice		Z
Orthorhombic	5	Side-centered (C)	2

$\mathbf{t}_1 = t\mathbf{i} + s\mathbf{j}$
$\mathbf{t}_2 = t\mathbf{i} - s\mathbf{j}$
$\mathbf{t}_3 = r\mathbf{k}$

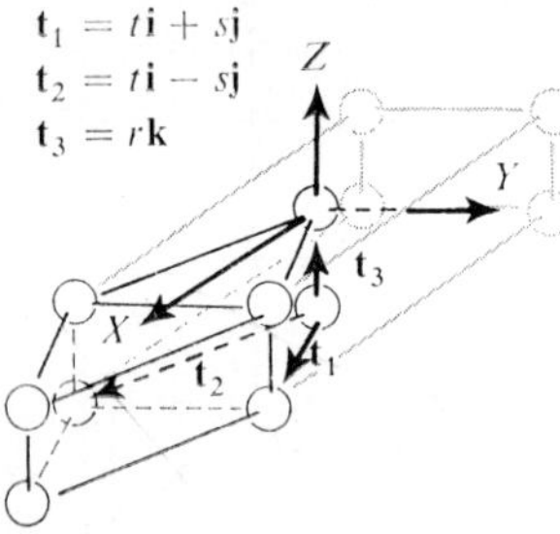

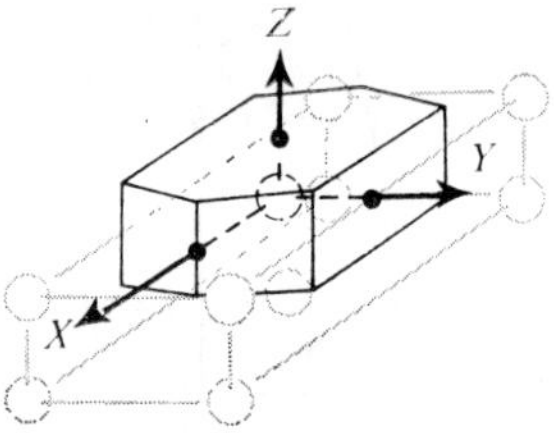

Crystal System	Bravais Lattice		Z
Orthorhombic	6	Body-centered (I)	2

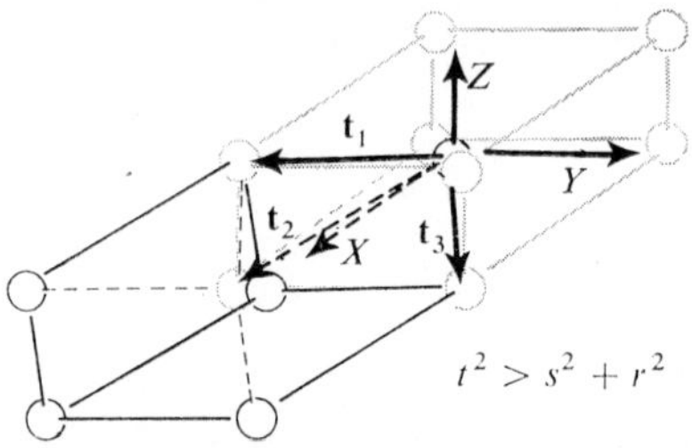

$t^2 > s^2 + r^2$

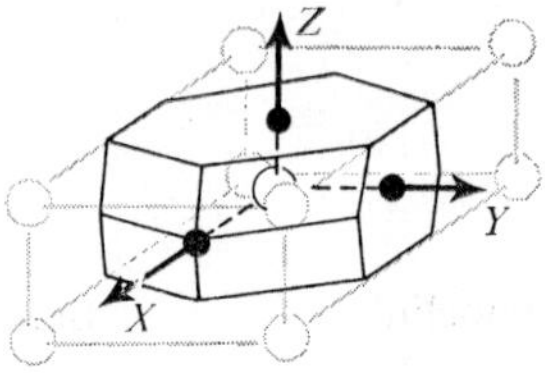

$\mathbf{t}_1 = t\mathbf{i} + s\mathbf{j} + r\mathbf{k}$
$\mathbf{t}_2 = t\mathbf{i} + s\mathbf{j} - r\mathbf{k}$
$\mathbf{t}_3 = t\mathbf{i} - s\mathbf{j} - r\mathbf{k}$

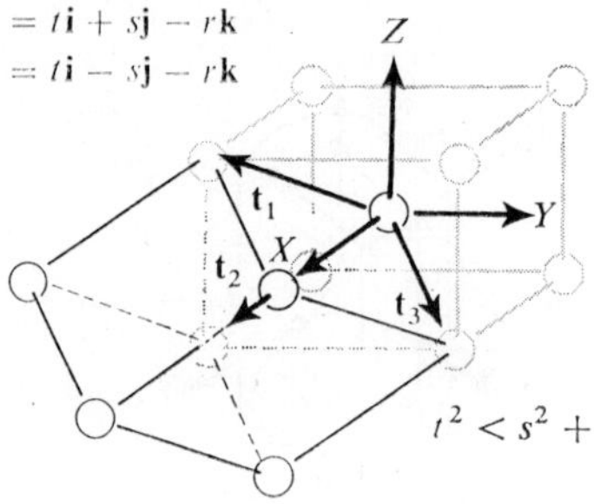

$t^2 < s^2 + r^2$

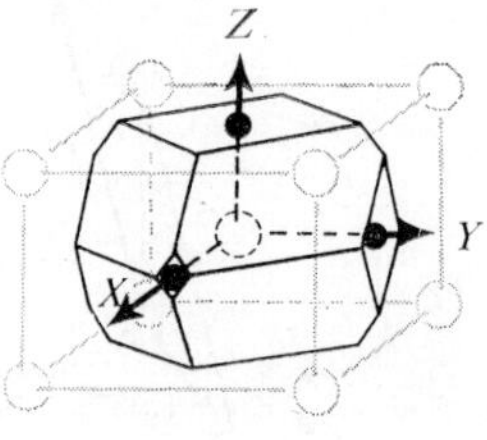

Crystal System	Bravais Lattice		Z
Orthorhombic	7	Face-centered (F)	4

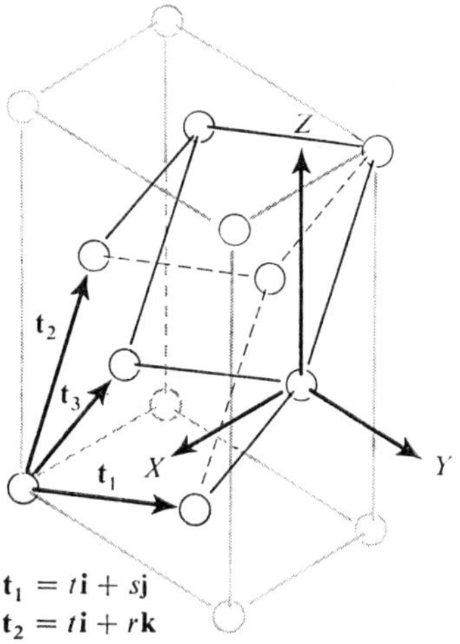

$\mathbf{t}_1 = t\mathbf{i} + s\mathbf{j}$
$\mathbf{t}_2 = t\mathbf{i} + r\mathbf{k}$
$\mathbf{t}_3 = s\mathbf{j} + r\mathbf{k}$

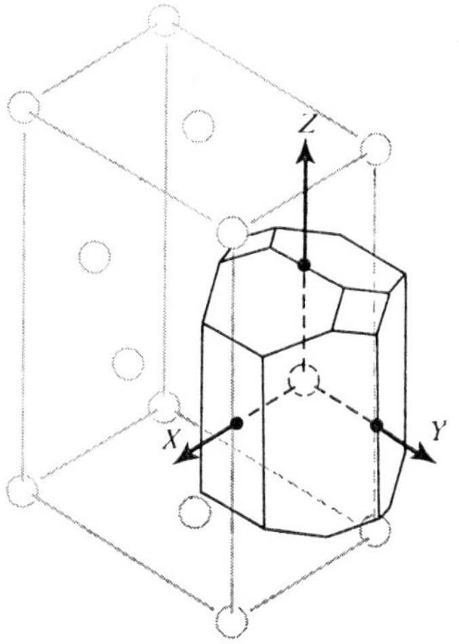

Crystal System	Bravais Lattice		Z
Tetragonal	8	Simple (P)	1

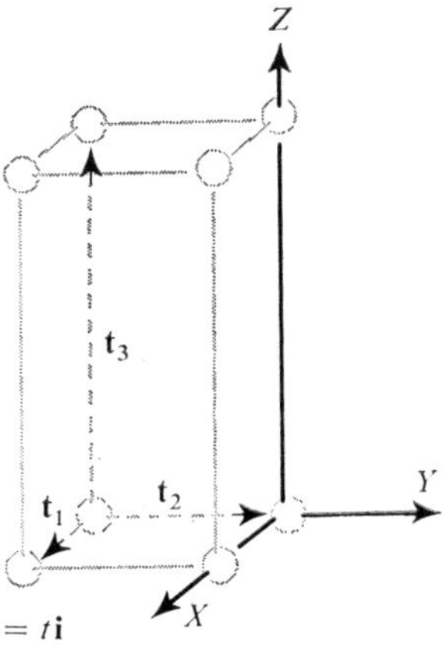

$\mathbf{t}_1 = t\mathbf{i}$
$\mathbf{t}_2 = t\mathbf{j}$
$\mathbf{t}_3 = s\mathbf{k}$

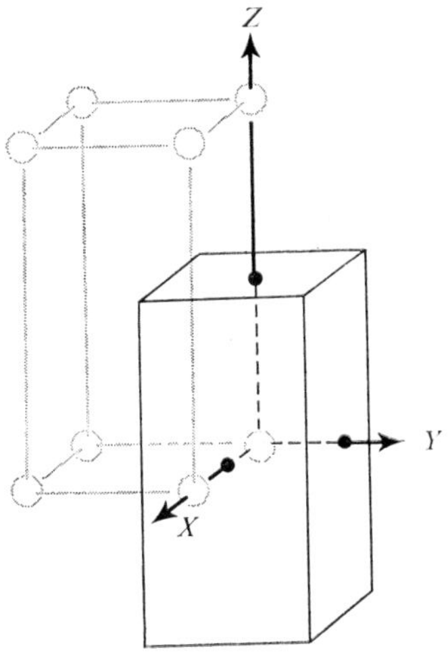

Crystal System		Bravais Lattice	Z
Tetragonal	9	Body-centered (I)	2

$\mathbf{t}_1 = t(\mathbf{i} + \mathbf{j}) + s\mathbf{k}$
$\mathbf{t}_2 = t(\mathbf{i} + \mathbf{j}) - s\mathbf{k}$
$\mathbf{t}_3 = t(\mathbf{i} - \mathbf{j}) - s\mathbf{k}$

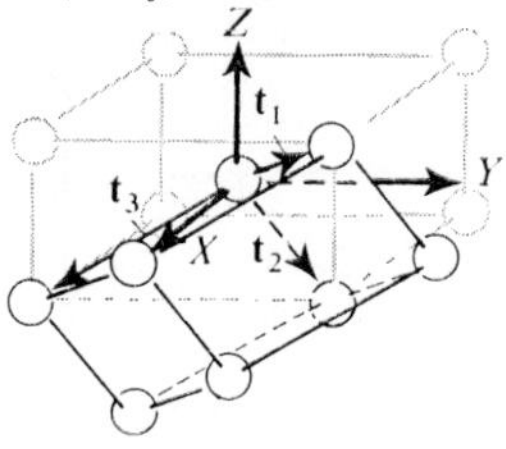

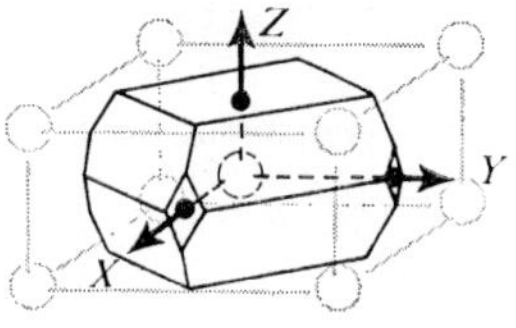

$s < \sqrt{2}t$

Crystal System		Bravais Lattice	Z
Tetragonal	9	Body-centered (I)	2

$\mathbf{t}_1 = t(\mathbf{i} + \mathbf{j}) + s\mathbf{k}$
$\mathbf{t}_2 = t(\mathbf{i} + \mathbf{j}) - s\mathbf{k}$
$\mathbf{t}_3 = t(\mathbf{i} - \mathbf{j}) - s\mathbf{k}$
$s > \sqrt{2}t$

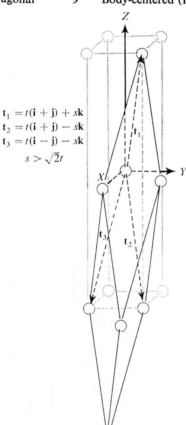

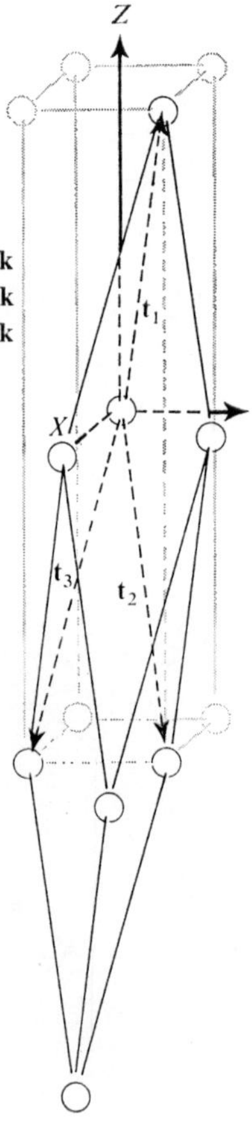

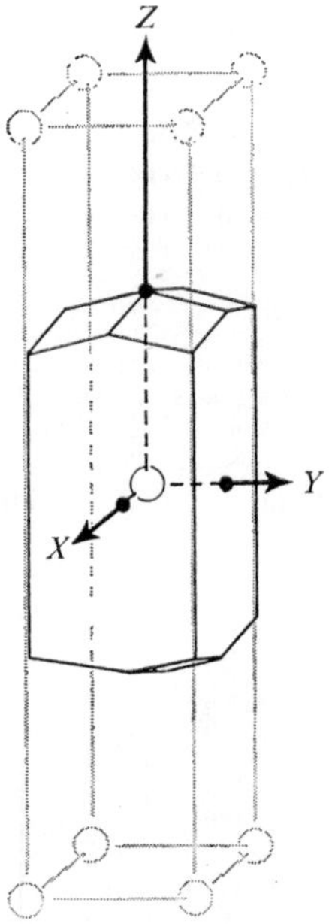

Crystal System	Bravais Lattice		Z
Trigonal	10	Simple (P or R)	

$\mathbf{t}_1 = s\mathbf{j} + r\mathbf{k}$
$\mathbf{t}_2 = \frac{1}{2}\sqrt{3}s\mathbf{i} - \frac{1}{2}s\mathbf{j} + r\mathbf{k}$
$\mathbf{t}_3 = -\frac{1}{2}\sqrt{3}s\mathbf{i} - \frac{1}{2}s\mathbf{j} + r\mathbf{k}$

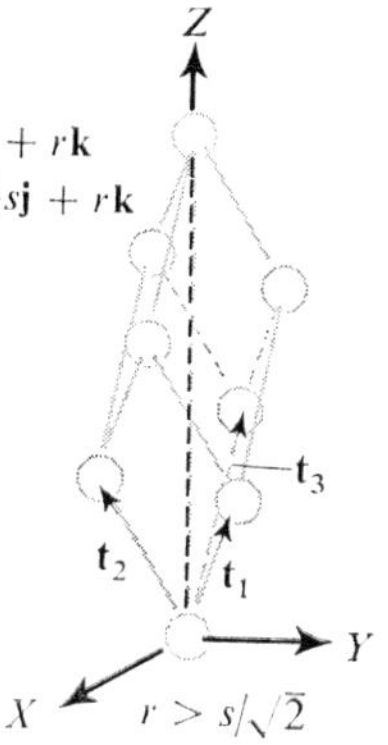

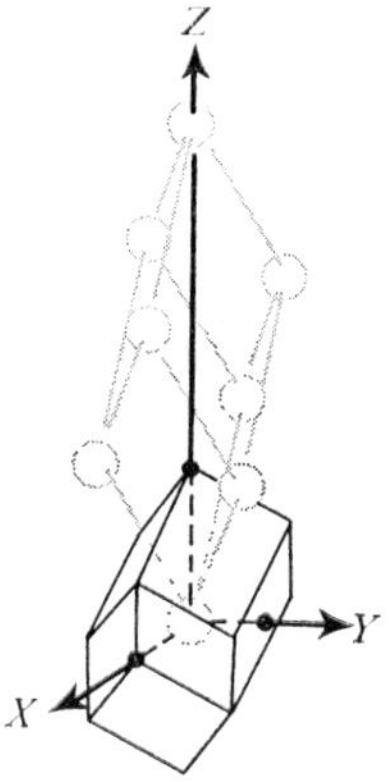

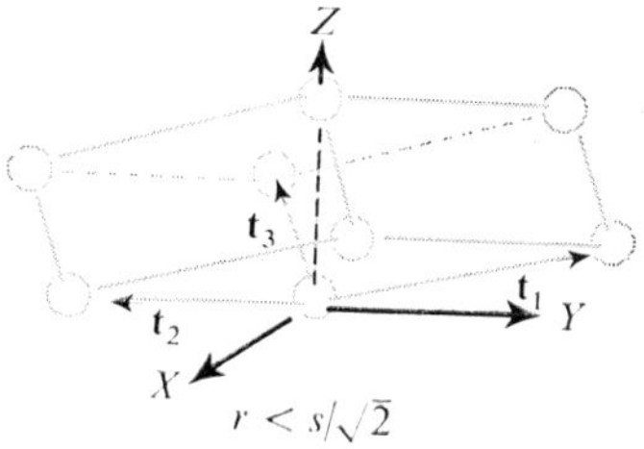

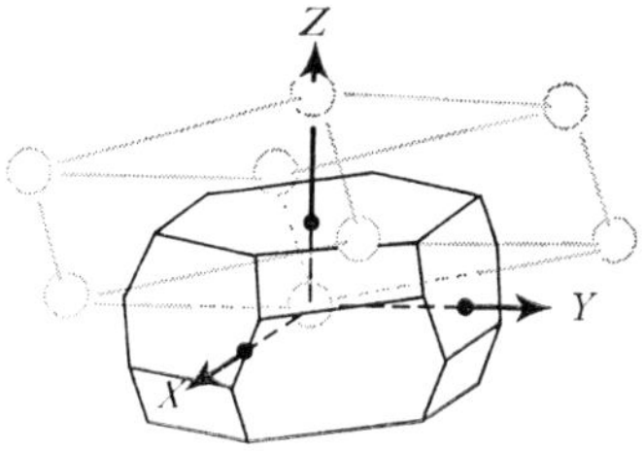

Crystal System	Bravais Lattice		Z
Hexagonal	11	Simple (P)	1

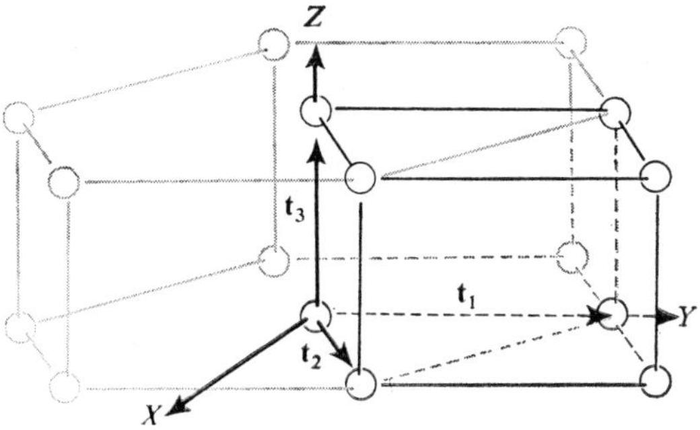

$\mathbf{t}_1 = s\mathbf{j}$
$\mathbf{t}_2 = \frac{1}{2}\sqrt{3}s\mathbf{i} - \frac{1}{2}s\mathbf{j}$
$\mathbf{t}_3 = r\mathbf{k}$

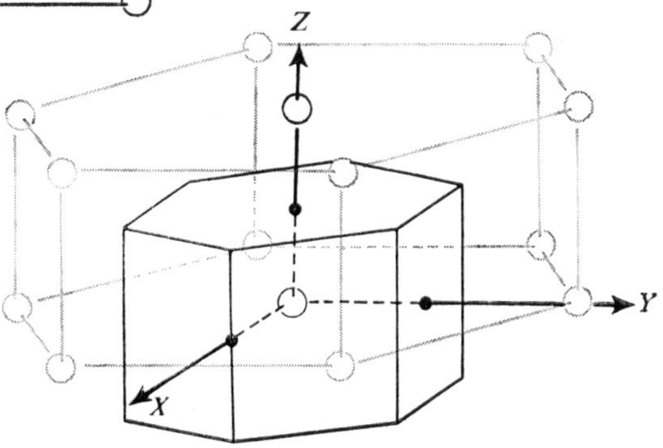

Crystal System	Bravais Lattice		Z
Cubic	12	Simple (P)	1

$\mathbf{t}_1 = t\mathbf{i}$
$\mathbf{t}_2 = t\mathbf{j}$
$\mathbf{t}_3 = t\mathbf{k}$

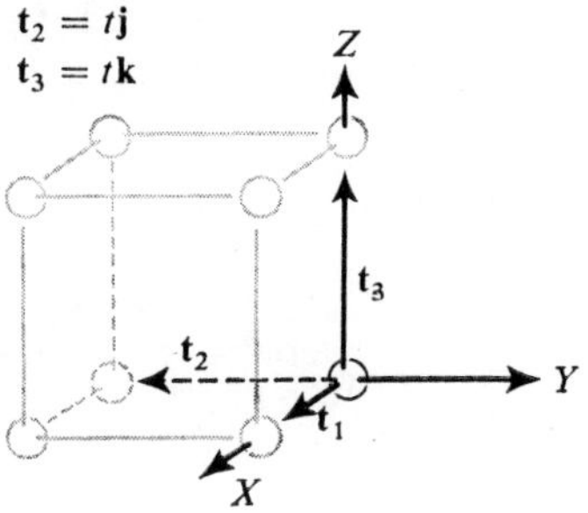

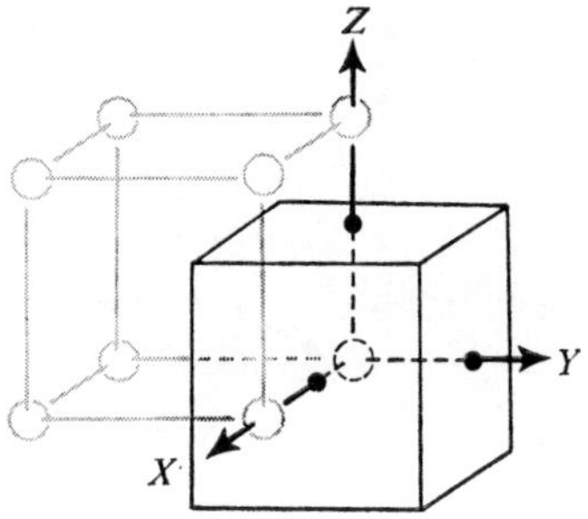

Cubic	13	Body-centered (I)	2

$\mathbf{t}_1 = t(\mathbf{i} + \mathbf{j} + \mathbf{k})/\sqrt{3}$
$\mathbf{t}_2 = t(\mathbf{i} + \mathbf{j} - \mathbf{k})/\sqrt{3}$
$\mathbf{t}_3 = t(\mathbf{i} - \mathbf{j} - \mathbf{k})/\sqrt{3}$

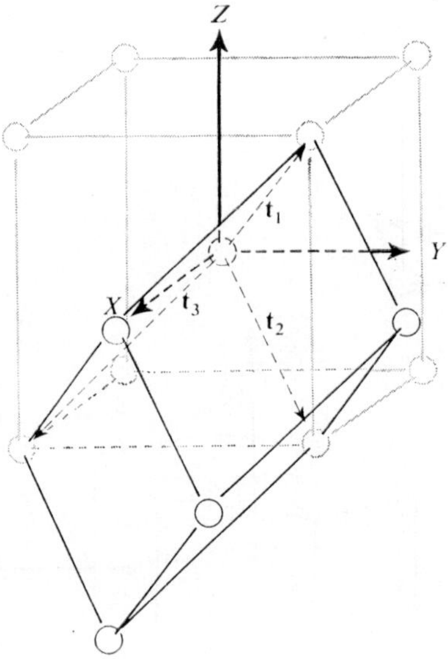

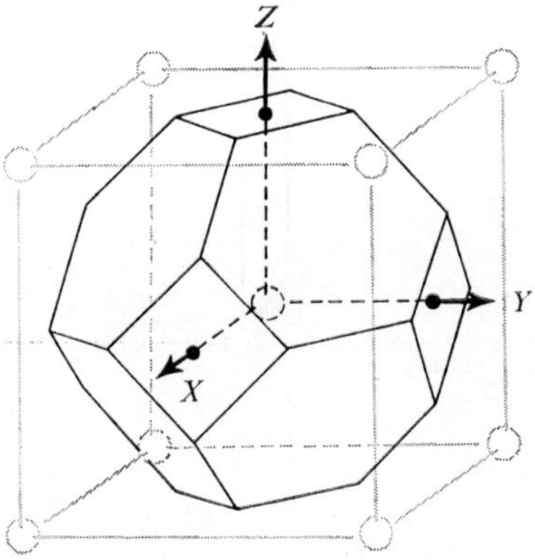

Crystal System	Bravais Lattice		Z
Cubic	14	Face-centered (F)	4

$\mathbf{t}_1 = t(\mathbf{i} + \mathbf{j})/\sqrt{2}$
$\mathbf{t}_2 = t(\mathbf{i} + \mathbf{k})/\sqrt{2}$
$\mathbf{t}_3 = t(\mathbf{j} + \mathbf{k})/\sqrt{2}$

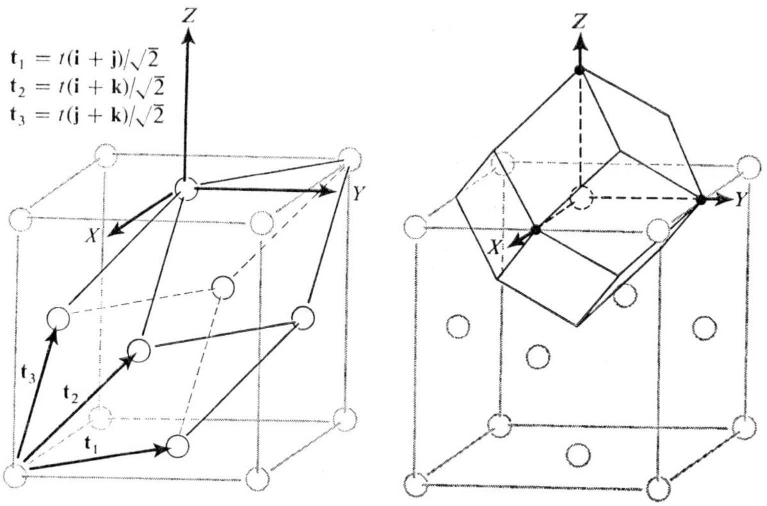

Appendix G

Polarizability Tensors for the 32 Crystal Classes

The irreducible representations of the Raman-active crystal vibrations are given in Table 1 opposite the symbols for each crystal class. The notation is consistent with that of Appendix C. If the symbols X, Y, or Z occur in parentheses after an irreducible representation, vibrations of this species are also infrared-active, with the indicated direction of polarization. Such vibrations are only possible in piezoelectric crystals, which lack a center of inversion (See Section X, Chap. 5). In centrosymmetric crystals only *gerade*-species vibrations can be Raman-active, while only *ungerade* vibrations have the possibility of being infrared-active. This principle is referred to by molecular spectroscopists as the *rule of mutual exclusion*.

Directly above the symbol for each irreducible representation is a matrix containing the nonvanishing components of the Raman-scattering tensor defined by Eqn (130), Chap. 5. The coordinates X, Y, and Z are the crystallographic axes chosen to be identical, respectively, to the principal axes x_1, x_2, and x_3 defined by Nye (**17**). In the case of triclinic symmetry the scattering tensor is a general symmetric tensor (**18**)–(**19**).

TABLE 1 [From reference (**18**)]

Raman-active vibrational symmetries and Raman tensors for the crystal symmetry classes

System	Class		Raman tensors			
Monoclinic			$\begin{pmatrix} a & & d \\ & b & \\ d & & c \end{pmatrix}$	$\begin{pmatrix} & e & \\ e & & f \\ & f & \end{pmatrix}$		
	2	$\mathscr{C}_2$	$A(Y)$	$B(X, Z)$		
	m	$\mathscr{C}_s$	$A'(X, Z)$	$A''(Y)$		
	$2/m$	$\mathscr{C}_{2h}$	A_g	B_g		
Orthorhombic			$\begin{pmatrix} a & & \\ & b & \\ & & c \end{pmatrix}$	$\begin{pmatrix} & d & \\ d & & \\ & & \end{pmatrix}$	$\begin{pmatrix} & & e \\ & & \\ e & & \end{pmatrix}$	$\begin{pmatrix} & & \\ & & f \\ & f & \end{pmatrix}$
	222	$\mathscr{D}_2$	A	$B_1(Z)$	$B_2(Y)$	$B_3(X)$
	$mm2$	$\mathscr{C}_{2v}$	$A_1(Z)$	A_2	$B_1(X)$	$B_2(Y)$
	mmm	$\mathscr{D}_{2h}$	A_g	B_{1g}	B_{2g}	B_{3g}
Trigonal			$\begin{pmatrix} a & & \\ & a & \\ & & b \end{pmatrix}$	$\begin{pmatrix} c & d & e \\ d & -c & f \\ e & f & \end{pmatrix}$	$\begin{pmatrix} d & -c & -f \\ -c & -d & e \\ -f & e & \end{pmatrix}$	
	3	$\mathscr{C}_3$	$A(Z)$	$E(X)$	$E(Y)$	
	$\bar{3}$	$\mathscr{S}_6$	A_g	E_g	E_g	

TABLE 1 (*Continued*)

System	Class		Raman tensors				
Trigonal (*cont.*)			$\begin{pmatrix} a & & \\ & a & \\ & & b \end{pmatrix}$	$\begin{pmatrix} c & & \\ & -c & d \\ & d & \end{pmatrix}$	$\begin{pmatrix} & -c & -d \\ -c & & \\ -d & & \end{pmatrix}$		
	32	$\mathcal{D}_3$	A_1	$E(X)$	$E(Y)$		
	3m	$\mathcal{C}_{3v}$	$A_1(Z)$	$E(Y)$	$E(-X)$		
	$\bar{3}m$	$\mathcal{D}_{3d}$	A_{1g}	E_g	E_g		
Tetragonal			$\begin{pmatrix} a & & \\ & a & \\ & & b \end{pmatrix}$	$\begin{pmatrix} c & d & \\ d & -c & \\ & & \end{pmatrix}$	$\begin{pmatrix} & & e \\ & & f \\ e & f & \end{pmatrix}$	$\begin{pmatrix} & & -f \\ & & e \\ -f & e & \end{pmatrix}$	
	4	$\mathcal{C}_4$	$A(Z)$	B	$E(X)$	$E(Y)$	
	$\bar{4}$	$\mathcal{S}_4$	A	$B(Z)$	$E(X)$	$E(-Y)$	
	4/m	$\mathcal{C}_{4h}$	A_g	B_g	E_g	E_g	
			$\begin{pmatrix} a & & \\ & a & \\ & & b \end{pmatrix}$	$\begin{pmatrix} c & & \\ & -c & \\ & & \end{pmatrix}$	$\begin{pmatrix} & d & \\ d & & \\ & & \end{pmatrix}$	$\begin{pmatrix} & & e \\ & & \\ e & & \end{pmatrix}$	$\begin{pmatrix} & & \\ & & e \\ & e & \end{pmatrix}$
	4mm	$\mathcal{C}_{4v}$	$A_1(Z)$	B_1	B_2	$E(X)$	$E(Y)$
	422	$\mathcal{D}_4$	A_1	B_1	B_2	$E(-Y)$	$E(X)$
	$\bar{4}2m$	$\mathcal{D}_{2d}$	A_1	B_1	$B_2(Z)$	$E(Y)$	$E(X)$
	4/mmm	$\mathcal{D}_{4h}$	A_{1g}	B_{1g}	B_{2g}	E_g	E_g

Hexagonal			$\begin{pmatrix} a & & \\ & a & \\ & & b \end{pmatrix}$	$\begin{pmatrix} & & c \\ & & d \\ c & d & \end{pmatrix}$	$\begin{pmatrix} & & -d \\ & & c \\ -d & c & \end{pmatrix}$	$\begin{pmatrix} e & f & \\ f & -e & \\ & & \end{pmatrix}$	$\begin{pmatrix} f & -e & \\ -e & -f & \\ & & \end{pmatrix}$	
	6	$\mathscr{C}_6$	$A(Z)$	$E_1(X)$	$E_1(Y)$	E_2	E_2	
	$\bar{6}$	$\mathscr{C}_{3h}$	A'	E''	E''	$E'(X)$	$E'(Y)$	
	$6/m$	$\mathscr{C}_{6h}$	A_g	E_{1g}	E_{1g}	E_{2g}	E_{2g}	
			$\begin{pmatrix} a & & \\ & a & \\ & & b \end{pmatrix}$	$\begin{pmatrix} & & \\ & & c \\ & c & \end{pmatrix}$	$\begin{pmatrix} & & -c \\ & & \\ -c & & \end{pmatrix}$	$\begin{pmatrix} & d & \\ d & & \\ & & \end{pmatrix}$	$\begin{pmatrix} d & & \\ & -d & \\ & & \end{pmatrix}$	
	622	$\mathscr{D}_6$	A_1	$E_1(X)$	$E_1(Y)$	E_2	E_2	
	$6mm$	$\mathscr{C}_{6v}$	$A_1(Z)$	$E_1(Y)$	$E_1(-X)$	E_2	E_2	
	$\bar{6}m2$	$\mathscr{D}_{3h}$	A_1'	E''	E''	$E'(X)$	$E'(Y)$	
	$6/mmm$	$\mathscr{D}_{6h}$	A_{1g}	E_{1g}	E_{1g}	E_{2g}	E_{2g}	
Cubic			$\begin{pmatrix} a & & \\ & a & \\ & & a \end{pmatrix}$	$\begin{pmatrix} b & & \\ & b & \\ & & b \end{pmatrix}$	$\begin{pmatrix} b & & \\ & b & \\ & & b \end{pmatrix}$	$\begin{pmatrix} & & \\ & & d \\ & d & \end{pmatrix}$	$\begin{pmatrix} & & d \\ & & \\ d & & \end{pmatrix}$	$\begin{pmatrix} & d & \\ d & & \\ & & \end{pmatrix}$
	23	$\mathscr{T}$	A	E	E	$F(X)$	$F(Y)$	$F(Z)$
	$m3$	$\mathscr{T}_h$	A_g	E_g	E_g	F_g	F_g	F_g
	432	$\mathscr{O}$	A_1	E	E	F_2	F_2	F_2
	$\bar{4}3m$	$\mathscr{T}_d$	A_1	E	E	$F_2(X)$	$F_2(Y)$	$F_2(Z)$
	$m3m$	$\mathscr{O}_h$	A_{1g}	E_g	E_g	F_{2g}	F_{2g}	F_{2g}

Appendix H

Linear-Response Theory and the Kramers–Kronig Relations†

As shown in Chapter 5, the effect of the application of an electric field to a dielectric material is expressed by the polarization, $\mathscr{P}$. In an isotropic medium the polarization is proportional to the applied field and the constant of proportionality is the complex electric susceptibility, χ. From Eqns (42) and (43), Chap. 5,

$$\mathscr{P} = \chi\mathscr{E} = \frac{\varepsilon(\nu) - 1}{4\pi}\,\mathscr{E} = (\chi_{\text{elec}} + \chi_{\text{ion}})\,\mathscr{E}. \tag{1}$$

The frequency-dependent part of the polarization, which is represented by the second term on the right-hand side of Eqn (1), becomes

$$\mathscr{P}_{\text{ion}} = \mathscr{P} - \mathscr{P}_{\text{elec}} = \left[\frac{\varepsilon(\nu) - 1}{4\pi} - \frac{\varepsilon(\infty) - 1}{4\pi}\right]\mathscr{E}$$

$$= \left[\frac{\varepsilon(\nu) - \varepsilon(\infty)}{4\pi}\right]\mathscr{E} = \chi_{\text{ion}}\,\mathscr{E}, \tag{2}$$

using Eqn (44), Chap. 5.

The behavior of a dielectric under the influence of an electric field is one example of various physical phenomena which can be treated using linear response theory **(24)**, **(25)**. In general, a system is said to be linear if its response to the sum of a number of independent applied signals is equal to the sum of its responses to each individual signal. Its behavior is in this case defined by a function $\Phi(t)$ which is its response at time t to a unit disturbance at $t = 0$. The general expression for the response $M(t)$ to a

† References **(20)**–**(23)**.

signal $S(\tau)$ is then given by

$$M(t) = \int_{-\infty}^{t} S(\tau)\,\Phi(t-\tau)\,\mathrm{d}\tau$$

$$= \int_{0}^{\infty} S(t-\tau)\,\Phi(\tau)\,\mathrm{d}\tau, \tag{3}$$

where $t - \tau$ has been replaced by τ in the second line of Eqn (3).

For a periodic signal of frequency ν, $S(t-\tau) = S_0\,\mathrm{e}^{2\pi i\nu(t-\tau)}$, and the response is given by

$$M(\nu, t) = S_0\,\mathrm{e}^{2\pi i\nu t}\int_{0}^{\infty} \mathrm{e}^{-2\pi i\nu\tau}\,\Phi(\tau)\,\mathrm{d}\tau$$

$$= S(\nu, t)\,G(\nu), \tag{4}$$

where

$$G(\nu) \equiv \int_{0}^{\infty} \mathrm{e}^{-2\pi i\nu\tau}\,\Phi(\tau)\,\mathrm{d}\tau \tag{5}$$

is the complex *admittance* of the system.

For the present example the admittance is simply the frequency-dependent part of the electric susceptibility of the medium. Thus, from Eqns (2) and (5)

$$\chi_{\mathrm{ion}} = \frac{\varepsilon(\nu) - \varepsilon(\infty)}{4\pi} = \int_{0}^{\infty} \mathrm{e}^{-2\pi i\nu\tau}\,\Phi(\tau)\,\mathrm{d}\tau \tag{6}$$

and, with $\varepsilon(\nu) = \varepsilon'(\nu) - i\,\varepsilon''(\nu)$, one finds

$$\frac{\varepsilon'(\nu) - \varepsilon(\infty)}{4\pi} = \int_{0}^{\infty} \Phi(\tau)\cos 2\pi\nu\tau\,\mathrm{d}\tau \tag{7}$$

and

$$\frac{\varepsilon''(\nu)}{4\pi} = \int_{0}^{\infty} \Phi(\tau)\sin 2\pi\nu\tau\,\mathrm{d}\tau, \tag{8}$$

by separating real and imaginary parts. The response function can be found from Eqns (7) or (8) by taking the Fourier transform. Thus,

$$\Phi(\tau) = \frac{1}{\pi}\int_{0}^{\infty} [\varepsilon'(\nu') - \varepsilon(\infty)]\cos 2\pi\nu'\tau\,\mathrm{d}\nu' \tag{9}$$

and

$$\Phi(\tau) = \frac{1}{\pi}\int_{0}^{\infty} \varepsilon''(\nu')\sin 2\pi\nu'\tau\,\mathrm{d}\nu'. \tag{10}$$

Substituting Eqn (10) into Eqn (7) one finds

$$\frac{\varepsilon'(\nu) - \varepsilon(\infty)}{4\pi} = \frac{1}{\pi}\int_0^\infty \left[\int_0^\infty \varepsilon''(\nu') \sin 2\pi\nu'\tau \, d\nu'\right] \cos 2\pi\nu\tau \, d\tau, \tag{11}$$

which can be written in the form

$$\frac{\varepsilon'(\nu) - \varepsilon(\infty)}{4\pi}$$

$$= \frac{1}{\pi}\lim_{s\to\infty}\int_0^\infty \varepsilon''(\nu')\left[\int_0^s \cos 2\pi\nu\tau \sin 2\pi\nu'\tau \, d\tau\right] d\nu'$$

$$= \frac{1}{\pi}\lim_{s\to\infty}\int_0^\infty \varepsilon''(\nu')\left[\frac{1}{2}\left(\frac{1-\cos[2\pi(\nu'+\nu)s]}{2\pi(\nu'+\nu)} + \frac{1-\cos[2\pi(\nu'-\nu)s]}{2\pi(\nu'-\nu)}\right)\right] d\nu'. \tag{12}$$

The cosine terms vanish as $s \to \infty$, yielding

$$\varepsilon'(\nu) - \varepsilon'(\infty) = \frac{2}{\pi}\int_0^\infty \frac{\nu' \, \varepsilon''(\nu')}{\nu'^2 - \nu^2} d\nu', \tag{13}$$

as given in Eqn (68), Chap. 5. The analogous expression for $\varepsilon''(\nu)$, which is given by Eqn (69), Chap. 5, can be easily found by substituting Eqn (9) into Eqn (8) and proceeding as above.

References

1. Wilson, E. B., Jr., Decius, J. C. and Cross, P. C. "Molecular Vibrations", McGraw-Hill, New York (1955).
2. Decius, J. C. *J. Chem. Phys.* **16,** 1025 (1948).
3. Miyazawa, T. *J. Chem. Soc. Japan* **76,** 1132 (1955).
4. King, W. T. Thesis, University of Minnesota (1956).
5. Mann, D. E., Shimanouchi, T., Meal, J. H. and Fano, L. *J. Chem. Phys.* **27,** 43 (1957).
6. Curtis, E. C. Thesis, University of Minnesota (1959).
7. Overend, J. and Scherer, J. R. *J. Chem. Phys.* **32,** 1289 (1960).
8. Schachtschneider, J. H. and Snyder, R. G. *Spectrochim. Acta* **19,** 117 (1963).
9. Long, D. A., Gravenor, R. B. and Woodger, M. *Spectrochim. Acta* **19,** 937 (1963).
10. Schachtschneider, J. H. "Vibrational Analysis of Polyatomic Molecules", Tech. Rep, No. 57–65, Shell Development Company, Emeryville, Calif. (1964).
11. Miyazawa, T. *J. Chem. Phys.* **29,** 246 (1958).
12. Koster, G. F., Dimmock, T. O., Wheeler, R. G. and Statz, H. "Properties of the Thirty-two Point Groups", M.I.T., Cambridge (1963).

13. Dimmock, J. O. and Wheeler, R. G. *Phys. Chem. Solids* **23,** 729 (1962).
14. Lazerev, A. N. "Kolebatel'n'ie Spektr'i i Strenie Silikatov", Izdatel'stov "Nauka", Lenningrad (1968).
15. Niggli, P. "Geometrische Kristallographie des Diskontinuums", Borntreger, Leipzig (1919).
16. Halford, R. S. *J. Chem. Phys.* **14,** 8 (1946).
17. Nye, J. F. "Physical Properties of Crystals". Oxford (1957).
18. Loudon, R. *Advan. Phys.* **13,** 423 (1964).
19. Ovander, L. N. *Opt. Spectry* (*U.S.S.R.*) **4,** 1078 (1962).
20. Kramers, H. A. *Att. Cong. Intern. Fis.* (*Como*) **2,** 545 (1927).
21. Kronig, R. de L. *J. Opt. Soc. Am.* **12,** 547 (1926).
22. Frolich, H. "Theory of Dielectrics", Oxford (1958).
23. Mathieu, J. P. "Optique, II. optique quantique", Société d'Edition d'Ensiegne ment Supérieur, Paris (1965).
24. Kubo, R. *J. Phys. Soc. Japan* **12,** 570 (1957).
25. Glarum, S. H. *J. Chem. Phys.* **33,** 1371 (1960).

Author Index

The numbers in italics refer to pages on which references are listed in full.

R

S

T

U

V

W

Subject Index

C

D

Q

R

T

U